Carbon Dioxide Capture, Utilization and Storage (CCUS)

Carbon Dioxide Capture, Utilization and Storage (CCUS)

Editors

Dongdong Feng
Jian Sun
Zijian Zhou

MDPI • Basel • Beijing • Wuhan • Barcelona • Belgrade • Manchester • Tokyo • Cluj • Tianjin

Editors
Dongdong Feng
College of Energy Science and Engineering
Harbin Institute of Technology
Harbin
China

Jian Sun
College of Energy and Mechanical Engineering
Nanjing Normal University
Nanjing
China

Zijian Zhou
State Key Laboratory of Coal Combustion
Huazhong University of Science and Technology Energy
Wuhan
China

Editorial Office
MDPI
St. Alban-Anlage 66
4052 Basel, Switzerland

This is a reprint of articles from the Special Issue published online in the open access journal *Energies* (ISSN 1996-1073) (available at: www.mdpi.com/journal/energies/special_issues/Carbon_Dioxide_Capture_Utilization_Storage_CCUS).

For citation purposes, cite each article independently as indicated on the article page online and as indicated below:

LastName, A.A.; LastName, B.B.; LastName, C.C. Article Title. *Journal Name* **Year**, *Volume Number*, Page Range.

ISBN 978-3-0365-7421-9 (Hbk)
ISBN 978-3-0365-7420-2 (PDF)

Contents

About the Editors

Dongdong Feng

Dongdong Feng, Deputy Director of Carbon Neutral Energy Technology Research Institute, Deputy Secretary of National Party Building Branch (Carbon Neutral Faculty Party Branch), China Postdoctoral Innovation Talent, Heilongjiang Province "Excellent Youth", Harbin Institute of Technology Young Top Talent, Excellent Civic Worker, University Excellent Expert, National Natural Science Foundation of China letter evaluation expert, member of "Eight Hundred Heroes" spirit preaching group, core member of Young Scientist Studio, "Gold Medal" master's student supervisor, 100 excellent undergraduate graduation project supervisor, national excellent undergraduate graduation design supervisor in energy and power category, with a comprehensive teaching evaluation: A+.

Jian Sun

Jian Sun, Lecturer, School of Energy and Mechanical Engineering, Nanjing Normal University; Editor, SCI Journal of Energy, "Carbon Dioxide Capture, Utilization and Storage"; Member, International Society of Combustion and Chinese Engineering Society. Carbon Dioxide Capture, Utilization and Storage (CUS); Reviewer of Applied Energy, Fuels, Cleaner Production, Chemical Engineering. Research interests: mainly engaged in the development of low carbon and clean energy technologies.

Zijian Zhou

Zijian Zhou, Lecturer, State Key Laboratory of Coal Combustion, School of Energy and Power Engineering, Huazhong University of Science and Technology, China. Main research interests: high-temperature thermochemical energy storage technology, the catalytic and efficient adsorption technology for coal combustion flue gas, and research on low-temperature SCR catalyst synergistic mercury removal technology.

Preface to "Carbon Dioxide Capture, Utilization and Storage (CCUS)"

With the rapid growth of the world economy, carbon dioxide emissions are increasing year by year. Currently, the increase in anthropogenic emissions of CO_2 is identified as a major contributor to global warming. In order to meet the emission reduction target and achieve near-zero CO_2 emissions by 2050, it is particularly necessary and urgent to accelerate CO_2 research.

Carbon dioxide capture, utilization and storage (CCUS) technology is considered to be one of the solutions in the near to medium term, and is gaining much attention and rapid development worldwide, while playing a key role in mitigating climate change. This Special Issue on "Carbon Dioxide Capture, Utilization and Storage (CCUS)" invites articles on the latest technologies and new developments in CCUS, including but not limited to pre-combustion carbon capture; post-combustion carbon capture; oxyfuel or chemical cycle combustion; conversion of CO_2 to synthetic fuels; biomass thermal conversion; storage of CO_2 BECCUS; and other abatement technologies.

We sincerely thank everyone who has helped.

Dongdong Feng, Jian Sun, and Zijian Zhou
Editors

Article

Techno-Economic Analysis of Carbon Dioxide Separation for an Innovative Energy Concept Towards Low-Emission Glass Melting

Sebastian Gärtner [1,2,*], Thomas Marx-Schubach [3,4], Matthias Gaderer [2], Gerhard Schmitz [4] and Michael Sterner [1]

1 Research Center on Energy Transmission and Energy Storage (FENES), Technical University of Applied Sciences (OTH) Regensburg, Seybothstrasse 2, 93053 Regensburg, Germany
2 Chair of Regenerative Energy Systems (RES), Campus Straubing for Biotechnology and Sustainability, Technical University Munich, Schulgasse 16, 94315 Straubing, Germany
3 XRG Simulation GmbH, Harburger Schlossstrasse 6-12, 21079 Hamburg, Germany
4 Institute of Engineering Thermodynamics, Hamburg University of Technology, Denickestrasse 15, 21073 Hamburg, Germany
* Correspondence: s.gaertner@tum.de

Abstract: The currently still high fossil energy demand is forcing the glass industry to search for innovative approaches for the reduction in CO_2 emissions and the integration of renewable energy sources. In this paper, a novel power-to-methane concept is presented and discussed for this purpose. A special focus is on methods for the required CO_2 capture from typical flue gases in the glass industry, which have hardly been explored to date. To close this research gap, process simulation models are developed to investigate post-combustion CO_2 capture by absorption processes, followed by a techno-economic evaluation. Due to reduced flue gas volume, the designed CO_2 capture plant is found to be much smaller (40 m^3 absorber column volume) than absorption-based CO_2 separation processes for power plants (12,560 m^3 absorber column volume). As there are many options for waste heat utilization in the glass industry, the waste heat required for CO_2 desorption can be generated in a particularly efficient and cost-effective way. The resulting CO_2 separation costs range between 41 and 42 EUR/t CO_2, depending on waste heat utilization for desorption. These costs are below the values of 50–65 EUR/t CO_2 for comparable industrial applications. Despite these promising economic results, there are still some technical restrictions in terms of solvent degradation due to the high oxygen content in flue gas compositions. The results of this study point towards parametric studies for approaching these issues, such as the use of secondary and tertiary amines as solvents, or the optimization of operating conditions such as stripper pressure for further cost reductions potential.

Keywords: power-to-gas; methanation; oxyfuel; glass industry; CO_2-separation; economic evaluation

Citation: Gärtner, S.; Marx-Schubach, T.; Gaderer, M.; Schmitz, G.; Sterner, M. Techno-Economic Analysis of Carbon Dioxide Separation for an Innovative Energy Concept Towards Low-Emission Glass Melting. *Energies* **2023**, *16*, 2140. https://doi.org/10.3390/en16052140

Academic Editors: Dongdong Feng, Zijian Zhou and Jian Sun

Received: 13 January 2023
Revised: 8 February 2023
Accepted: 10 February 2023
Published: 22 February 2023

1. Introduction

Commercial glass production is a very energy-intensive industrial process. Converting raw materials, such as silica sand, sodium carbonate, lime, dolomite, etc., into molten glass requires high process temperatures of up to 1600 °C. This melting process is the central phase of glass production and accounts for 50–80% of total energy demand in overall glass production. The dominating source for achieving the required process temperatures has been the combustion of fossil fuels such as natural gas (NG) and crude oil for a long time [1].

Due to the dominance of fossil fuels, glass production is currently still associated with high carbon dioxide (CO_2) emissions. The container and flat glass industry can be considered the dominating glass-producing sector and currently emit over 60 million tons of CO_2-emissions a year. This is more than the annual emissions of Portugal. The glass industry in the EU, which is the world's largest producer [2], emits more than 20 million tons of CO_2-emissions a year, with an energy demand of more than 350 PJ. Approximately 75% of these emissions can be located in combustion processes, while the remaining 25% are contributed by the dissociation of carbonate raw materials [3]. The level of glass manufacturing-related CO_2 emissions in Germany has stagnated at a constant level of approximately 4 mio. t CO_2-eq. per year since 2007 [4]. The CO_2 emission budget from 2020 to 2050 of the entire German glass industry is 17.7 Mt CO_2-eq. for a strict 1.5 °C climate target and 112.0 Mt CO_2-eq. for a 2 °C climate target [5]. Due to the current high CO_2 emissions of the global glass production, as well as these severely limited remaining emission budgets, fast and effective measures to reduce emissions in the glass industry are of vital importance.

Of course, there are efforts towards greater sustainability and climate protection in the glass industry. Accordingly, there are many research approaches and developments for the decarbonization and the integration of renewable energies into the glass-melting process. The most important concepts are all-electric melting and the switch to hydrogen (H_2) as combustion fuel instead of NG [6]. However, all of these concepts, despite their promising CO_2 emissions reduction potential, have certain restrictions:

All-electric melting is well established for small-scale glass melting systems, but large-scale applications are still controversial [7]. Recent all-electric melting projects have made some progress and could reach melting capacities of up to 250 t/d . However, the complex melting tank design and extensive heat control strategies of such large scale melting tanks are challenging and result in high investment and operating costs [8]. In addition, a low glass production rate (so called pull rate) flexibility, short melting tank lifetime, high electricity costs, and low operating experience are disadvantages. Besides that, not all glass types are feasible for all electric melting, for example, non-ionic glasses [6].

Recent projects, such as HyGlass [9], HyNet [10], and Kopernikus P2X [11], have investigated hydrogen as a fuel substitution option. Since the combustion of H_2 does not cause any CO_2 emissions, this would significantly reduce the CO_2 footprint of glass melting. However, the total CO_2 emissions of the glass industry are not only caused by fossil fuel combustion but also by carbonate reactions during batch to melt conversion. As a switch to H_2-combustion will not influence the batch composition of glass manufacturing, further CO_2 emissions will remain. The CO_2-emission reduction of H_2-combustion-based melting systems will significantly depend on the means of hydrogen production (green, blue, grey,... H_2). To date, only H_2 production based on renewable energies can guarantee a large reduction in CO_2 emissions (green H_2). Current scientific debates mainly focus on the climate impact of blue hydrogen. Ref. [12] states that blue hydrogen comes with only 9–12% less CO_2 emissions compared to conventional (i.e., grey) hydrogen. However, according to [13], the climate impact of blue hydrogen significantly depends on the key parameters such as methane emissions during the natural gas supply chain, CO_2 capture rate during production, and the applied global warming metrics. Thus, the authors concluded that blue hydrogen can be competitive to green hydrogen production in terms of climate impacts, if state-of-the-art process technologies and metrics are applied [13].

However, both green and blue H_2 are still very cost-intensive and are therefore not fully economically competitive with established fossil fuels. In addition, H_2 shows a very different combustion behavior, including higher flame temperatures, different flame velocity, faster ignition behavior, and changed heat radiation properties [8]. Some of these effects are compensated by oxyfuel combustion, for example, changes in adiabatic flame temperature, as shown in [8]. Nevertheless, there are still some uncertainties, such as the effects of higher water content in the flue gases in the firing chamber on glass quality, and the melting tank life time.

Scope of This Work

Due the limitations of currently discussed decarbonization options, there is an urgent need for innovative energy concepts which allow the integration of renewable energies, a reduction in CO_2 emissions, and which ideally have no impact on the established melting processes. This work is the first to demonstrate the integration of a power-to-methane (PtM) system into the glass-melting process that meets all of these requirements. Since the PtM concept has been known for a long time as a sector-coupling option in energy systems [14,15], there are a lot of established technical options for most of the process steps. The state of the art of these established options is briefly described, before the remaining open process of CO_2 capture from exhaust gases is discussed in detail. Since CO_2 capture from exhaust gases in the glass industry has not yet been considered in this particular application, this process shows the most extensive research and development demand. Therefore, a technology concept for the integration of amine-based CO_2 capture processes into the already existing flue gas treatment systems of the glass industry is presented, and various possibilities of waste heat utilization are investigated. The main focus of this work was the techno-economic analysis of the CO_2 separation concept, in order to provide a basis for the economic analyses of the entire PtM system in future work.

2. Integration of Power-to-Methane into Oxyfuel Glass Melting

Figure 1 shows a simplified process flow sheet for the integration of PtM into oxyfuel glass melting processes.

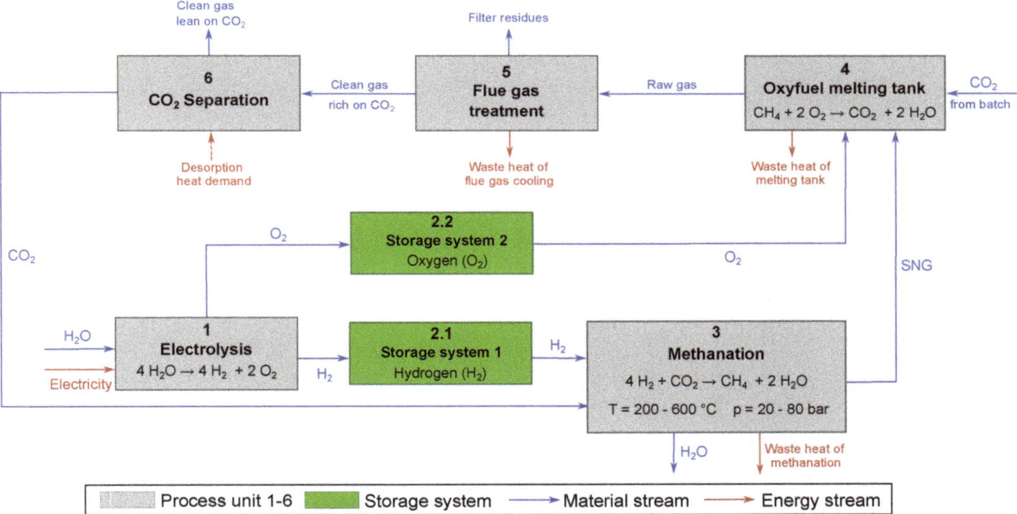

Figure 1. Simplified flow sheet of the integration of power-to-methane (PtM) into oxyfuel glass melting processes.

In the first process unit (Figure 1, 1), water (H_2O) is separated into (H_2) and oxygen (O_2) by electrolysis. Both H_2 and O_2 are subsequently stored in compressed gas storage systems (Figure 1, 2.1 and 2.2, details in Section 2.1.2). Besides H_2, O_2 is also an important product for downstream PtM processes, as it can be used in oxyfuel combustion (see Section 2.1.4).

The intermediate storage of H_2 is necessary due to the technological restrictions on the flexibility of the downstream catalytic methanation process (Figure 1, 3, details on which can be found in Section 2.1.3). In methanation, H_2 reacts with CO_2 in a pressure range of 20–80 bar and a temperature range of 200–600 °C in a catalytic process to form CH_4 and steam. The exact thermodynamic characteristics of methanation reactors may vary for each specific reactor design. As CH_4 is the main component of fossil natural gas compositions, the methanation product stream is also referred to as synthetic natural gas (SNG).

SNG is subsequently used in the oxyfuel glass melting furnace (Figure 1, 4) with O_2 from electrolysis. Thus, the technique of O_2 production in glass manufacturing processes via energy intensive air separation units established to date can be avoided, by utilizing combined effects in the overall process. In oxyfuel glass melting, CO_2 is mainly emitted from combustion (approximately 90%), but also from carbonate reactions during the batch-to-melt conversion (approximately 10%, depending on batch composition) [16]. CO_2 is subsequently emitted from the melting tank by the exhaust gases.

In the flue gas treatment process (Figure 1, 5), pollutants such as NO_x and SO_x are removed due to regulatory air pollution control laws. Such flue gas cleaning processes were introduced in the 1990s and aim to improve air quality, protect the health of residents, and reduce environmental pollution [6] . The used technologies are scrubbers, electrostatic, and cloth filters, as well absorption-based processes [17]. Since the separation of pollutants such as NO_x and SO_x is important for air pollution control, but not for the described integration concept for PtM processes, the separation technology and its costs will not be discussed in detail.

However, these processes do not focus on CO_2-separation from flue gases. Therefore, an additional separation process is included in the concept to remove CO_2 from cleaned flue gases Figure 1, 6. The separated and purified CO_2 is subsequently used as a product stream for the methanation process. Holistically, this concept for integrating PtM into glass melting processes creates an almost closed carbon cycle (depending on CO_2 capture rate), which enables a significant reduction in greenhouse gas emissions. Additionally, the established oxyfuel glass melting processes, associated with highly stable glass quality and output, can still be used. Moreover, a switch to a renewable electricity supply from fluctuating sources such as wind and solar power can be achieved, without losing proven know-how and associated process stability.

However, the multiple process steps such as electrolysis, storage, and methanation lead to losses in the overall efficiency of the system. Thus, access to low-cost energy from renewable sources, as well as high costs for natural gas or CO_2 emission certificates, are crucial for the economic viability of the overall concept. In addition, only a renewable energy supply for electrolysis will lead to a reduction in total CO_2 emissions (see [8]).

There are adequate technical options with a significant degree of technological readiness for the most important PtM process steps electrolysis, H_2 and O_2 storage, and methanation (see Section 2.1). However, CO_2 capture from the flue gases of the glass industry has not been investigated in detail yet. To close this research gap, this work discusses the technical background and options for CO_2 separation processes in the glass industry in detail (see Section 2.2, provides a simulation-based design approach for such plants, (Section 4), and investigates the techno-economic evaluation of CO_2 separation from the flue gases of the glass industry (Section 5.2).This techno-economic analysis focuses explicitly on CO_2 capture, not on the overall concept. The economic evaluation of the overall system that includes electrolysis, storage concepts, and methanation will be investigated in subsequent work.

2.1. State-of-the-Art of Process Components

2.1.1. Electrolysis

Alkaline (AEL), solid oxide (SOEC), and polymer electrolyte membrane electrolysis (PEMEL) can currently be considered the most technically established processes for this purpose [14,18]. AEL has the highest technology readiness level (TRL) 9, but only offers limited part-load capability (30–100%) due to thermal restrictions [18]. SOEC would offer the highest efficiency of all options (up to 95% [19]), but requires an operating temperature level of 600–1000 °C [18]. This temperatures can be provided in an energy-efficient way by integrating the waste heat from catalytic methanation. However, the heating and cooling creates a thermal inertia of SOEC, which impedes rapid adaptation to fluctuations of renewable energy sources. In addition, the TRL of 5–6 is the lowest of all technical options described here. The most promising technology for the application proposed in this work is PEMEL, as it has the widest part-load capability (0–100%) and can thus adapt to the fluctuations of renewable energy sources such as wind power and photovoltaic. While the TRL of AEL is considered to be mature, the TRL of PEM is slightly behind [20]. Nevertheless, PEM electrolyzers are already available in the power categories of multiple MW (e.g., Siemens Silyzer 300 [21]) and the technology can therefore be considered to be of commercial technology status [20]. Development potential is primarily focused on the industrial series production of this electrolysis technology, which is expected to result in a significant reduction in costs [14].

2.1.2. Gas Storage

In general, there are various options for H_2 storage:

- Physical storage technologies such as compressed gas, liquid, or cryo-compressed H_2 storage [22].
- Adsorption technologies, relying on carbon-based materials such as multi-walled carbon nanotubes (MWCNTs), metal–organic frameworks (MOFs), or zeolites [22].
- Adsorption technologies, based on metal hydrates, like iron-oxide pellets [23].
- Chemical H_2 storage including liquid organic hydrogen carriers (LOHCs) [24],

For the process described in this work, H_2 storage in pressurized gas tanks is currently the most suitable option. It is described as technologically mature and therefore a comparatively cost-effective option, especially in combination with battery storage systems [22]. Nevertheless, the application of innovative H_2 storage technologies within the described PtM process should be focused on in further research. The especially high waste heat potential in glass melting processes offers attractive options for an energy-efficient use of LOHC or iron-oxide based H_2 storage technologies.

Storage system 2 in Figure 1 is necessary to ensure a constant flow of O_2 for a stable combustion process in the melting tank. Various options for O_2 storage have existed at the industrial scale for a long time and can therefore be considered robust [25]. To date, the O_2 required for oxyfuel combustion in the glass industry has primarily been produced by air separation using vacuum pressure swing adsorption (VPSA) processes [25]. This is performed either directly at the industrial site, or by delivery and storage in compressed gas tanks (liquid O_2). In both cases, oxygen storage facilities are usually available on site.

2.1.3. Methanation

Currently, thermocatalytic and biological methanation are the main technologies being discussed for use in power-to-gas systems [18,26]. Due to its higher TRL and technical performance, catalytic methanation is mainly used in large-scale PtM plants [14].

However, this methanation process requires constant operating conditions to achieve a high CH_4 content in the synthetic natural gas (SNG). For this purpose, the feed gases CO_2 and H_2 must be supplied in the ideal stoichiometric ratio of 1:4. In addition, pressure and temperature conditions must be constant to avoid damages on the catalysts through hot-spot formation. From a thermodynamic point of view, low temperatures, and high pressure would be ideal for a high conversion rate. However, the exothermic nature of the dominant methanation reactions challenges the operation at the thermodynamic optimum. For many commercially available fixed-bed reactors, an operating temperature of 300 °C and a pressure of 20 bar have proven to be suitable operating conditions. Nevertheless, the operating temperatures of established reactor concepts vary between 200 and 600 °C and a pressures range of 20–80 bar [18,26]. In any case, the effective cooling of the reactor is required. In the PtM system described above, this waste heat could be used for SOEC or CO_2-desorption (as can be seen in Section 2.2).

In addition to the technical properties mentioned above, the purity of CO_2 is crucial for a stable methanation process. In order to investigate the effects of CO_2 impurities on methanation, experiments investigating the direct methanation of flue gases for different types of power plants were performed. The flue gases investigated were obtained both from lignite-fired power plants with conventional combustion and from pilot plants with oxyfuel combustion. In both cases, the direct methanation of the flue gases led to heavy damage on the catalyst, so that the experiments had to be stopped. However, oxyfuel flue gases could ensure stable methanation after purification and treatment. Within these investigations, mainly SO_2 and halogen compounds (F, Cl, Br, At, TS) have been described as strong-acting catalyst toxicants [27,28].

2.1.4. Oxyfuel Glass Melting Tank

In oxyfuel combustion processes, a fuel such as natural gas is burned in a pure oxygen atmosphere instead of ambient air. Oxyfuel combustion is more energy-efficient because the nitrogen content in the furnace atmosphere is significantly reduced, thus lowering the thermal capacity of the exhaust gases. This also enables improved mass transport in the combustion chamber. In addition, a higher adiabatic flame temperature is achieved [29]. The higher overall efficiency of this combustion technology also enables a reduction in CO_2 emissions compared to the regenerative processes that are primarily used at present. A major limitation of this technology has been the energy-intensive production of oxygen and the associated higher investment costs. Nevertheless, oxyfuel furnaces are established state of the art, especially in the special glass production sector. Due to the much more energy-intensive glasses that are molten for this purpose, the higher efficiency of oxyfuel melting offers an economic advantage [8].

2.1.5. Flue Gas Purification Systems

Currently, mainly fabric filters and electrostatic precipitators are established for flue gas purification in the glass industry. These cleaning systems are primarily aimed for compliance with air pollutant limits set by country-specific regulations. The air pollutants relevant for the glass industry are mostly dust resulting from turbulence in the furnace, carbon monoxide (CO), sulfur and nitrogen oxides (SO_x, NO_x), as well as hydrochloride and fluoric acid (HCl, HF). While NO_x and CO emissions are mainly caused by combustion, SO_x-, HCl-, and HF-emissions occur due to glass batch impurities. To remove these air pollutants from the flue gas mixture, sorbents such as lime (calcium hydroxide, $Ca(OH)_2$) are used, which then deposit as so-called filter dust. In order to prevent the filter systems from being damaged by excessively high gas temperatures, a heat exchanger is usually installed upstream to cool the uncleaned flue gases. Fabric filters in particular are much more sensitive to high gas temperatures than electrostatic filters.

2.2. *CO_2 Separation*

Technologies for CO_2 capture from combustion processes can be categorized as pre-combustion, post-combustion processes, and processes that are directly connected to combustion (in this work, further referred to as "in-combustion" processes). Pre-combustion processes primarily involve the separation of CO_2 from fuels such as natural gas or pulverized coal, in order to prevent CO_2 emissions prior to combustion. These are mainly absorption processes with solvents such as rectisol, selexol, or purisol. Such processes are widely established in the petrochemical industry, for example, to avoid the acidification of natural gas and to be able to offer a high-purity product [30].

In-combustion processes such as chemical looping or so-called oxyfuel processes are directly involved in the combustion process. In the chemical looping process, metal oxides are added to the combustion chamber in fluidized bed reactors in order to effect a redox reaction, binding the CO_2 [31]. Oxyfuel processes are closely related to the combustion process described above. When natural gas is burned in pure oxygen, the exhaust gas theoretically consists only of CO_2 and water vapor. The H_2O can then be condensed by cooling and a highly pure CO_2 stream would remain [30]. However, both processes are unsuitable for the glass industry. Metal oxides in the combustion chamber would also react with the glass batch, thus negatively affecting the glass quality. The separation of CO_2 by cooling is unsuitable, since the exhaust gases in the glass industry also contain evaporation from the molten glass (Section 3). Accordingly, the CO_2 product stream would be contaminated after the condensation of H_2O.

Post-combustion processes concern the capture of CO_2 from combustion flue gases. Membrane processes as well as the adsorption and absorption processes are discussed for this purpose. Membrane processes provide for the separation of CO_2 from the flue gas stream by means of molecular lattices. However, these membranes are currently not yet sufficiently temperature-stable and scalable for rapid implementation. Adsorption processes are currently mainly used for CO_2 capture from ambient air, known as direct air capture [32]. However, de- and adsorption cycles are still too slow for application in exhaust gas streams and are not available in sufficient capacity [30]. For CO_2 capture from exhaust gases, absorption processes are currently being discussed as the main option [30]. In particular, the use of monoethanolamine (MEA) as an absorbent is widely investigated and can be considered a standard. However, MEA can suffer from degradation, caused by carbamate polymerization, as well as oxidative and thermal degradation. Therefore, a switch to more stable and efficient solvents is widely discussed [33,34].

Nevertheless, CO_2 absorption processes can be considered the most promising technology for the glass industry, since they have been investigated for a long time and have already been tested in several worldwide projects for the separation of CO_2 from flue gases. They were also proposed for applications in other industry sectors, such as the cement industry [35].

The integration of such a post-combustion CO_2 absorption plant is preferably realized after the already existing flue gas purification systems. At this point, the exhaust gases are already cleaned of harmful residues such as dust and a large proportion of the SO_2 emissions. A flowsheet for this integration is shown in Figure 2a).

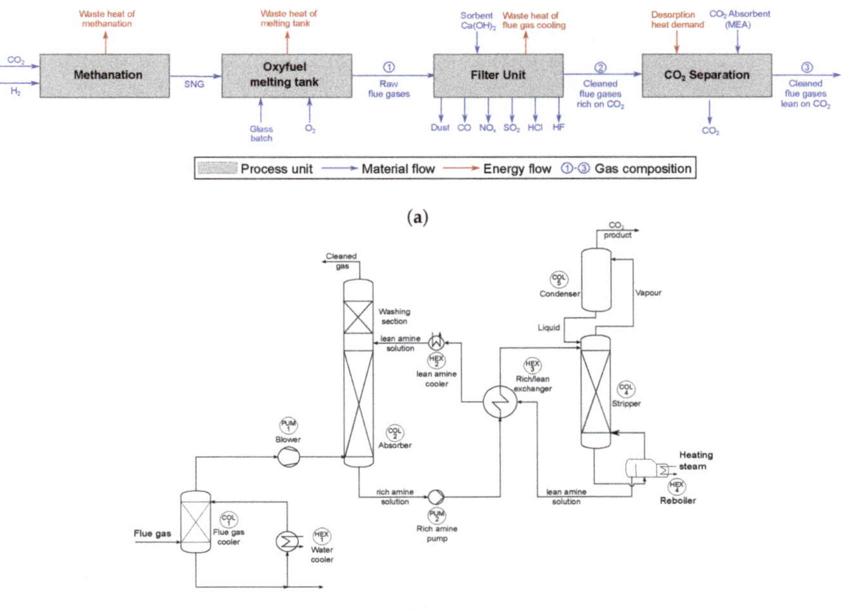

Figure 2. **(a)** Flowsheet for the conceptual integration of a CO_2 separation processes in glass melting systems. **(b)** Principal flow sheet of the components of a CO_2 absorption plant with an amine-based solvent.

CO_2 separation from flue gases using absorption process can be divided into two main process steps. First, CO_2 reacts with the lean aqueous amine solution (30 wt.-% MEA in this case), in the absorber column. The rich solvent, containing chemically bounded CO_2, enters the stripper column by passing a lean/rich heat exchanger. This preheats the rich solution close to the stripper operating temperature of approximately 120 °C and consequently cools the lean solution. In the stripper column, the MEA solution is regenerated by an endothermic process. Therefore, low-pressure steam is supplied to the reboiler to maintain stable regeneration conditions. A temperature level of at least 150 °C and a pressure of 2 bar are required to ensure an efficient desorption process.

This thermal energy demand for the desorption of CO_2 is considered to be the main cost factor during the long-term operation of amine-based CO_2 separation plants. In power plant technology, usually steam from intermediate stages of turbines is used for this purpose. This reduces the electrical power output of the turbines and thus has a significant impact on the increased electricity production cost of carbon capture power plants. However, the PtM concept introduced in this work offers several options for waste heat utilization to meet the desorption heat demand in an energy-efficient way.

2.3. Heat Supply for CO_2 Desorption

The ability to provide the necessary heat for desorption process is crucial for the economic viability of a post-combustion CO_2 absorption plant. Especially for heat-intensive processes at common temperature levels of the glass industry, the use of waste heat from these processes for steam generation is evident. Figure 2a shows the most promising waste heat sources of the PtM process for the glass industry, namely (i) methanation, (ii) melting process, and (iii) the flue gas filter unit. These options will be described in greater detail below.

It should be mentioned that waste heat is also generated during PEM electrolysis. Depending on the performance and design of PEM electrolysis systems, either water or air cooling is used. The temperature level of this waste heat is at approximately 80 °C [19]. Therefore, the utilization of this waste heat potential for steam generation is limited. Further processing, for example, by means of high-temperature heat pumps would be required to generate steam on the required temperature and pressure levels of CO_2 desorption. Due to these limitations, the waste heat option of PEM electrolysis is not considered in detail in this work.

2.3.1. Waste Heat of Thermochemical Methanation

For thermochemical methanation processes, the continuous cooling of reactors is required to prevent the degeneration of established catalyst materials. Otherwise, the exothermic nature of the dominant reaction processes would lead to catalyst particle sintering. Commercially established tubular bulb reactors for methanation processes are usually operated at a temperature level of 300 °C and a pressure of 20 bar. Usually water cooling is a common way to maintain constant operating conditions in these reactors, generating steam at a temperature of approximately 260 °C and a pressure of 45 bar [36]. A Sankey diagram of tubular bulb reactors for methanation and their cooling process is shown in Figure 3.

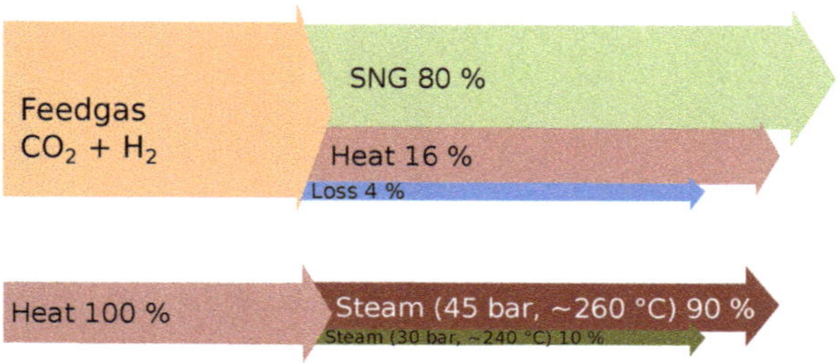

Figure 3. Sankey diagram of a tubular bundle methanation reactor [36].

2.3.2. Waste Heat of Melting Tank

Due to the melting temperatures of up to 1600 °C, as well as an efficiency of approximately 42% of oxyfuel furnaces [8], a high waste heat potential occurs in the direct furnace environment. However, the use of this heat reservoir is limited by a number of restrictions. For example, the installation of heat exchangers close to the furnace wall is restricted, since permanent access for industrial maintenance and servicing must be ensured. The remaining option is to use the heated ambient air close to the tank environment. However, the low-temperature level of the ambient air (below 100 °C) is an obstacle for the generation of process steam at the temperature and pressure level required for CO_2 desorption [17].

2.3.3. Waste Heat of Flue Gas Treatment

In many cases, the waste gases from the furnaces are still at too high a temperature level to be directly passed on to a filter system. In particular, the textiles in the cloth filter systems are sensitive to excessively high temperatures. Therefore, heat exchangers are installed in many flue gas cleaning systems to cool flue gases to a suitable level for filter systems. The heat recovered in this way is used, for example, to supply heat to office and administration buildings at glass industry sites [37]. In some cases, however, the reheating of the waste gases is also necessary to ensure sufficient flow conditions for the removal of waste gases through pipes and stacks [17,38].

3. Flue Gas Properties and Composition

Detailed data on the flue gas composition and properties must be available to enable an evaluation of the CO_2 separation process. However, the composition of flue gases in the glass industry is highly variable. The main influencing factors are: batch composition, cullet fraction, glass type, melting technology, and fuel, as well as flue gas treatment and cooling. Therefore, a literature screening was conducted for data from the flue gas analyses of melting systems, suitable for the application of the PtM process presented in this paper. The exhaust gas study by Roger et al. [39] was identified as the most promising source.

In this study, flue gas analyses were performed on several special glass melting tanks. The aim was to investigate the potential for the removal of boron compounds (namely boric acid (H_3BO_3) and meta boric acid (HBO_2)) from the flue gases generated during the melting of the typical borosilicate glasses. Although these investigations do not fully meet the scope of this work, several oxyfuel melting systems were investigated that would be well suited for the use of a PtM system. The most promising system is the one with the data given in Table 1. For the studies in this work, a conventional soda-lime glass, was assumed to be molten in this furnace. Compared to borosilicate glass, soda-lime glass changes the chemical composition of the glass batch while maintaining the technical design of the melting tank. Melting a soda-lime glass further results in the absence of boron compounds in the flue gas. The effects of boron compounds, such as H_3BO_3) and HBO_2, on currently established absorbents such as MEA have not been investigated to date and are therefore unpredictable.

Table 1. Key figures of the melting system investigated in [39].

Parameter	Value
Melting technology	Oxyfuel
Operating temperature	1600 °C
Thermal power	6.08 MW
Fuel	Natural gas H
Glass type	borosilicate
Nominal pull rate	40 t/d
Operating hours	8760 h/year

The flue gas properties and composition shown in Table 2 were used for the examination of the CO_2 absorption plant in this work. Additional thermodynamic properties for the flue gases were calculated using the open source software Cantera with GriMech 3.0 reaction mechanism [40,41].

Table 2. Flue gas composition and properties for the examination of the CO_2 absorption plant. Based on the experimental results of [39], additional properties are calculated using [40,41]. STP = standard temperature and pressure conditions (0 °C; 1.0135 bar). Numbers of gas compositions refer to Figure 2. ① raw gas before flue gas treatment. ② cleaned gas after flue gas treatment unit, entering the CO_2 separation process. ③ clean gas composition after CO_2 separation.

Properties	Unit	① Raw Gas	② Clean Gas Rich on CO_2	③ Clean Gas Lean on CO_2
Temperature	°C	470	226	26
Volume flow rate:				
- Dry	m^3/h	4780	6910	4290
- Wet	m^3/h	5910	8020	4437
Density at STP	kg/m^3	1.266	1.268	1.161
Spec. heat capacity (c_p)	J/(kg·K)	1219.2	1138.4	1022.5
Composition				
H_2O	Vol%	19.1	13.8	2.3
O_2	Vol%	19.8	20.5	24.8
CO_2	Vol%	11.3	7.0	0.8
N_2	Vol%	49.8	58.7	71.1
Trace Substances				
Dust	mg/m^3	1482	0.3	-
HCl	mg/m^3	2.6	0.2	-
HF	mg/m^3	72.7	0.5	-
SO_2	mg/m^3	19.0	8.1	-

It should be noted that the volume fractions of CO_2 in the flue gas are very low, while N2 and O_2 are very high for an oxyfuel combustion system. It can be concluded that in this particular case, high amounts of ambient air are entering the flue gas system. The rise of O_2 fractions from raw gas to clean gas may also be explained due to ambient air leaks in the system. It should be noted, that such flue gas compositions are based on experimental results from existing glass melting plants, with no further explanations for the presumed high ambient air content provided by the study [39]. However, for the sake of the most energy-efficient overall process, this high ambient air content should be avoided, as it can result in a higher CO_2 concentration in the flue gas. This would have a positive effect on the specific heat requirement (MJ/kg CO_2) of the CO_2 separation.

4. Materials and Methods

4.1. Simulation Approach

The technical and economic analysis of CO_2 capture from the flue gases of the glass industry requires the design and sizing of the plant components. For this work, the process simulation tool Aspen Plus was used for component design. Numerous other studies found this approach to be useful and validated [42,43]. The most important equipment of the CO_2 capture plant, namely the absorber and stripper column (as can be seen in Figure 2), were designed in detail, using the simulation results of the Aspen Plus model (v12).

The flowsheet of the Aspen Plus model is depicted in Figure 4. The columns are modeled in detail using a rate-based approach. The other components, namely the compressor, the pump, or the heat exchanger were modeled in a simplified way. The ELECNRTL property method in Aspen is used to calculate the media data using the eNRTL model for the liquid phase and the Redlich–Kwong equation of state for the vapor phase.

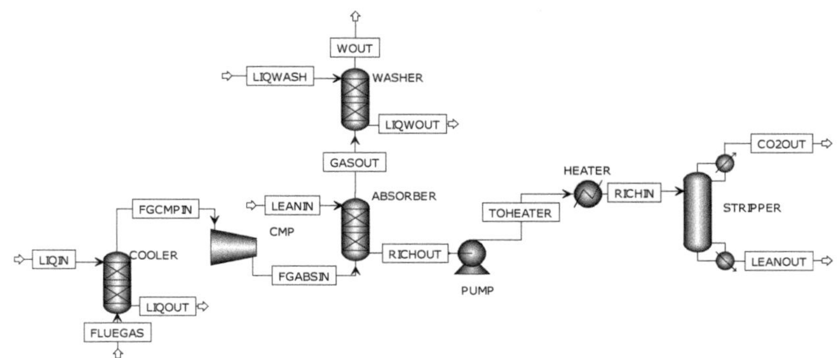

Figure 4. Main flow sheet overview of Aspen Plus Model used for the simulation.

The dissociation reactions are considered by defining the equilibrium reactions, whereas the kinetics are considered for all reactions including CO_2 as a reactant using built-in power laws in Aspen Plus. Because of the fast reaction of the CO_2, the liquid film is discretized in five parts. As structured packings, the Melapak 250Y from Sulzer [44] are used in all columns. The mass transfer coefficients, interfacial areas, and column hold-up are calculated using the correlation of Bravo et al. [45,46]. The heat transfer coefficients were determined according to the theory of Chilton and Coburn [47].

In order to ensure the validity of the mass balance, it would be an obvious step to close the loop in the Aspen model. However, this significantly decreases the numerical stability of the model. Thus, the mass balance is controlled using the built-in optimizer in Aspen Plus. The CO_2 mass balance is controlled by the heat flow rate to the stripper. All other minor components such as oxygen or nitrogen are varied in such a way that the mass balance is fulfilled. The carbon capture rate is controlled by using the solvent flow rate. To ensure that the plant is operated at the optimal operation point, the lean CO_2 loading (CO_2 loading at absorber entry) is varied by the optimization function.

In order to calculate the column diameters, the fractional approach to maximum capacity is used in Aspen Plus [48]. A factor of 0.65 is chosen. A constant capture rate of 90% is set using the lean solvent mass flow. The heat flow into the reboiler in the stripper column is varied in such a way that the CO_2 mass balance is met. The lean solvent loading is varied to ensure an optimal operation point of the plant and minimum heat demand in the reboiler. The packed column height was set according to [42], and the relevant parameters of the plant are listed in Table 3.

Table 3. Assumed design parameters of amine scrubbing plant.

Parameter	Value
Solvent temperature at absorber inlet	35 °C
MEA-concentration	30 w. %
Desorber pressure	2 bar
Absorber flue gas outlet pressure	1 bar
CO_2 product temperature	25 °C
Washing fluid temperature	25 °C
Desorber solvent inlet temperature	105 °C
Carbon capture rate (set point)	90%

In order to calculate the equipment costs, a comparison with an identical system is required (as can be seen in Section 4.3). Therefore, a simulation model was developed for the CO_2 separation plant with the exhaust conditions ② shown in Table 2 and another was developed for the reference CO_2 absorption plant described in [43,49]. In addition, the operating parameters shown in Table 4 are assumed for the CO_2 separation plant. Due to the continuously operated glass melting tank, operating hours of 8688 h/year are assumed for the CO_2 separation plant. These high operating hours are necessary to guarantee a sufficient separation rate of 90% CO_2. This assumption for operating time is higher than the reported 7500 operating hours from [49]. However, operating hours of more than 8500 h/year are a common assumption in other studies, investigating CO_2 separation from industrial processes [50].

Table 4. Operation parameter of the designed CO_2 absorption plant. All parameters are based on [49], unless indicated otherwise. * Own simulations.

Parameter	Value
Specific desorption heat demand	928 kWh/t CO_2
Reboiler heat power	915 kW
Life time	20 years
Operating hours	8688 h/year
MEA degradation rate	1.5 kg/t CO_2
Cooling water make up	1.0 m^3/ GJ thermal
Water cooler duty	640 kW
CO_2 product temperature	26 °C
CO_2 product pressure	2 bar
CO_2 emissions before [8]	9601 t CO_2/year
CO_2 separation rate	90%
CO_2 avoided *	8641 t CO_2/year
CO_2 emissions after *	960 t CO_2/year

4.2. CO_2 Compression

After the CO_2 separation unit (Figure 1, 6), the separated CO_2 is available at a pressure of approximately 2 bar and a temperature of approximately 26 °C. Thus, to reach the operating pressure of the methanation reactor of approximately 20 bar, CO_2 compression is required. The required pressure of the methanation reactor is much lower than the usual pressures for the underground storage of CO_2 in established carbon capture and storage concepts of up to 120 bar [51]. To date, to achieve this high compression from 2 to 120 bar, four-stage compression processes have been suggested for CCS concepts [51], which might no longer be required for the reduced compression ratio from 2 to 20 bar for the PtM concept. Due to the significant deviations in the compression processes of the PtM concept described in Figure 1 and the described process in [43], a detailed evaluation of the compression energy demand and investment cost is conducted. In [51], the required power demand for CO_2 compression is calculated as:

$$P_{el} = \alpha_{el} \cdot \ln\left(\frac{P_{out}}{P_{in}}\right) \cdot \dot{m}_{CO_2} \tag{1}$$

where P_{el} is the electrical power demand in kW, α_{el} is the compressor power constant of 87.85 kW/kg CO_2, P_{in} and P_{out} are the inlet and outlet pressure in Pa, respectively, and $\dot{m}_{CO_2}$ is the inlet CO_2 mass flow in kg/s. Moreover, the calculation of investment costs in a CO_2 compressor is suggested as [51]:

$$I = \left(\alpha_1 \cdot \dot{m}_{CO_2}{}^{\alpha_2} + \alpha_3 \cdot \ln\left(\frac{P_{out}}{P_{in}}\right) \cdot \dot{m}_{CO_2}{}^{\alpha_4}\right) \cdot \dot{m}_{CO_2} \tag{2}$$

where I is the investment cost in EUR and $\dot{m}_{CO_2}$ is the CO_2 mass flow in kg/s. α_n are empiric constants, determined as $\alpha_1 = 0.1 \cdot 10^6$ in EUR/(kg/s), $\alpha_2 = -0.71$, $\alpha_3 = 1.1 \cdot 10^6$ in EUR/(kg/s) and $\alpha_4 = -0.60$ [51]. The results of the calculations obtained from Equations (1) and (2) were validated against vendor information [52].

4.3. Economic Analysis

The costs of the plant components were calculated using the "estimating equipment cost by scaling" method, described in [53]. For this purpose, the costs of closely related reference plant equipment k_r is compared to the equipment cost of the proposed plant k_e by an exponential factor m:

$$k_e = k_r \left(\frac{G}{G_r} \right)^m \tag{3}$$

The basis of the characteristic component parameters of the reference equipment G_r include, for example, the column volume, pump mass flow rate, or heat transfer area. These parameters are in relation to that of the proposed plant G. Various degression exponents $m < 1$ for the respective components allow the cost comparison for each equipment type.

Other components of the investment costs, such as installation, instrumentation and control, piping, or electricity supply can be subsequently calculated as a proportion of total equipment costs, such as that proposed by [53] and already used for similar approaches for cost calculations of CO_2 absorption plants [49]. The well-established equivalent annual cost (EAC) method was used to calculate the costs of owning and operating assets over their entire life time. Further assumptions for cost parameters are shown in Table 5.

Table 5. Assumptions for cost calculations of CO_2 absorption plant. All parameters are based on [49], unless indicated otherwise.

Parameter	Value
Interest rate	5.0% p.a.
Maintenance cost	4.0% of FCI
MEA price	1000 EUR/t
Cooling water cost	0.20 EUR/m^3
Operating labor cost, based on [54,55]	45.00 EUR/h
Electricity costs [8]	0.12 EUR/kWh

According to [54], hourly gross salaries in Germany for operating labor in maintenance and repair are between 15.89 and 30.96 EUR/h, depending on the company, industry sector, and location. Ref. [55] stated a gross annual salary of EUR 47,756 for maintenance and repair operating labor. Given the standard 40 h work week in Germany, this results in a gross hourly wage of 22.26 EUR/h. In Germany, however, companies have to pay additional insurance, taxes, and levies (ITL) on top of the gross wages of employees. To account for these costs, an additional ITL-factor of 1.7 can be considered on the gross salary. The 45 EUR/h, named in Table 5 thus results in a gross hourly wage of 26.47 EUR/h, which meets both the published data of [54,55].

The "ProcessNet—Chemical Plant Price Index for Germany" was used to take inflation-related price development since 2007 as well as increased material, distribution, and logistics costs into account. According to this index, prices have risen by 21.5% since 2008. Due to the impact of the global COVID-19 pandemic, the effects of production cutbacks, and limited logistics capacities are particularly evident from 2020 onward. By 2019, the increase would have been only 16% (see Figure A1) [56].

5. Results and Discussion

5.1. Simulation Results

For CO_2 absorption plants, the absorber and stripper column, as well as the gas blower and the cooling water pump for the pre-scrubber are the main cost factors with a share of more than 80% [49]. Therefore, relevant comparative parameters for Equation (3) were considered with particular detail during simulations. The most important results are shown in Table 6. In addition, columns 4 and 6 in Table 7 show the simulation results for the CO_2 absorption plant considered in this work, as well as the reevaluation of the reference plant, considered in [43,49].

Table 6. Major simulation results for the design of the amine scrubbing plant

Parameter	Value
Absorber diameter	0.95 m
Desorber diameter	0.5 m
Scrubber diameter	0.95 m
Solvent flow rate	2.28 kg/s
Heat demand in stripper	556.5 kW
Specific heat demand in stripper	3.354 MJ/kg CO_2
CO_2 purity	99.3%
CO_2 product mass flow	0.167 kg/s
Washing fluid temperature	25 °C
Desorber solvent inlet temperature	105 °C
Carbon capture rate	90%

The sizing of the effective column diameters of the absorber and desorber is crucial for the adequate operation of CO_2 capture processes from flue gases. These were determined in the developed ASPEN Plus model by linear optimization with a 90% CO_2 capture rate as a boundary condition. For the desorber, this results in a diameter of 0.95 m for the flue gas volume flow rate and composition specified in Table 2. The effective desorber column diameter is determined as 0.50 m. The required heat output of the desorber of 556.5 kW, or the specific heat requirement of 3354 MJ/kg CO_2 must be provided by the reboiler. The CO_2 capture system can avoid 8641 t CO_2/year.

5.2. *Cost Calculation Results*

Cost calculations for the CO_2 absorption plant were determined considering the following steps:(i) calculation of equipment cost (Section 5.2.1); (ii) calculation of capital expenditures (CAPEX, Section 5.2.2)); (iii) calculation of operational expenditures (OPEX, Section 5.2.2); and (iv) the calculation of resulting CO_2 separation cost (Section 5.4).

5.2.1. Equipment Cost

Table 7 gives an overview of the equipment cost for both the modeled CO_2 absorption plants. Details concerning the reference plant are given in [49]. The plant for this work was designed for the same CO_2 capture rate of 90% and stripper operating conditions of 2 bar and 130 °C.

The absorber column is the most cost-intensive plant component, both in the original CO_2 separation plant and in the newly designed plant for the glass industry. The costs for the absorber are approximately EUR 340,000. The second most important contribution to equipment costs is related to the stripper column, at approximately EUR 110,000. Accordingly, another 20% of the total costs are contributed by the stripper. To date, 85% of the total equipment costs can be attributed to the most important columns in the process. Other important and cost-relevant items are the cold water pump, blower and DC water cooler with a combined investment cost of approximately EUR 51,000. The previously described parts cover 96% of the total component costs. The remaining plant components such as rich/lean heat exchanger, solvent pumps, reboiler, or the chiller for the exhaust gas cooler cooling supply can be attributed a total cost of EUR 20, 000, based on the given percentage share of [49]. Since these components account for only 4% of the total cost, this summarized cost estimate is sufficient for these components. This cost allocation pattern is consistent with the data in [49] and can thus be considered valid.

Table 7. Equipment cost calculation. Indicators (Ind.) refer to Figure 2b for the easier identification of equipment. * Remaining heat exchanger and pumps with less significant influence on the total component costs.

Ind.	Equipment	Type	Exponent	Reference Plant		Own Plant	
			m	G_r	k_r in millions of EUR	G	k_e in EUR
COL 2	Absorber	Column	0.60 [57]	12,560 m^3	10.94	39.25 m^3	340,000
COL 4	Stripper	Column	0.60 [57]	3391 m^3	3.43	8.75 m^3	110,000
–	Cold water pump	Pump, reciprocating	0.34 [53]	1.21 m^3/s	2.04	0.0015 m^3/s	21,000
PUM 1	Blower	Blower, centrifugal	0.59 [53]	485.44 m^3/s	3.10	2.236 m^3/s	13,000
COL 1	Flue gas cooler	Column	0.60 [53]	1570 m^3	0.54	4.75 m^3	17,000
–	Others *	mixed	-	-	3.89	-	20,000
			Total equipment cost:		23.94		521,000
	CO_2 compressor	Compressor, rotary, two stage [52]		31.49 m^3/s	31.73	0.07761 m^3/s 0.167 kg/s	120,000

The total costs for CO_2 compression after separation, calculated as described in Section 4.2, were approximately EUR 153,000. Alternatively, using the second approach described in Section 4.3 would yield investment costs of EUR 69,000, but only for a single-stage compression process. A single-stage compressor, however, would have to achieve a compression ratio of 1:10. Such a high compression ratio is unfavorable for thermodynamic reasons in single-stage design and would lead to high heat generation, high compressor operating power, and thus high energy demand. Thus, the single-stage compressor is extended to a two-stage compression process, yielding a compression ratio of 1:5 in the first compressor stage. Therefore, the costs calculation of method 2 would double, yielding costs of EUR 138,000. Finally, the costs were validated against vendor information, giving estimated costs of EUR 120,000 for a two-stage reciprocating compressor [52]. For further calculations, the cost estimated by the vendor was used, as this can be considered the most practical approach.

5.2.2. Capital Expenditures

In addition to the equipment cost, additional investments for installation, control engineering, pipes, and electrical installation have to be considered. These cost are calculated according to the methodology shown in Section 4.3. The cost elements can be divided into direct costs for materials and further equipment, and indirect costs for services such as engineering and construction expenses of the plant. The direct cost elements can be further structured according to inside battery limit (ISBL) cost, and outside battery limit (OSBL) cost. ISBL includes material costs, which are directly associated to the plant, while ISBL are secondary expenses for building, yard improvement, or service facilities.

Ranges for the assumed percentages of ISBL costs are given in [53]. Table 8 shows the parameters taken into account and their influence on the total CAPEX. Due to the comparably small size of the CO_2 capture plant, the lower limits of the ranges proposed in [53] were selected. Thus, the additional costs for construction and installation, instrumentation, and control, as well as piping and electrical equipment are approximately EUR 354,000, or approximately 68% of the component costs. The total material and equipment cost of the CO_2 separation plant for the glass industry are approximately EUR 875,000.

The OSBL costs for buildings, yard improvements, and service facilities were assumed to be lower than recommended by [53]. Due to the low column volumes and the resulting compactness of the plant, a container solution for plant housing is attractive. Considering this option, OSBL costs of over EUR 100,000 building infrastructure seems justified, or even on the safe side. The excavation and concrete construction work will be limited with this simplified type of building design. Service facilities such as changing rooms and showers for personnel or workshops can be provided by the existing infrastructure at a glass manufacturer site. The total direct costs thus amount to approximately EUR 980,000.

For indirect cost, ranges suggested by [53,58] are used. This results in total indirect costs of approximately EUR 107,000. For the calculation of the total plant investment costs (FCI), the investment costs of the CO_2compressor must also be considered. In total, the FCI add to approximately EUR 1.15 million.

In addition to the FCI, costs for labor arising in the company, start-up, and initial material costs must be taken into account. The costs for start-up and initial material costs amount to approximately EUR 92.000. The material cost is mainly for the initial loading of the solvent circuit. This consists in a 30 wt. % solution of monoethanolamine (MEA) in water.

The component costs were determined by cost comparison with a CO_2 separation plant for coal-fired power plants, which was designed as early as 2007. Since then, due to inflation, as well as global geopolitical events such as the COVID-19 pandemic, non-negligible price increases have occurred in all areas of public life, which also affected chemical engineering. In order to take these developments into account, a price index correction, based on [56] was applied. This price index correction since 2007 has a significant impact of 21% on total CAPEX. The extent to which prices developed according to current influences such as recent conflicts, should be closely examined. These effects have not yet been taken into account in the current values of [56] and may require reassessment for future works.

The CAPEX of plants for CO_2 capture from flue gases of power plants are far above the values calculated in this study. Abu-Zahra and Singh proposed EUR 147 million and EUR 179 million, respectively [49,59]. With respect to CAPEX, there is a factor of 90 between both plants, and with respect to the absorber column dimensions, there is even a factor of 300. This is mainly due to the significantly lower flue gas mass flow of the investigated glass melting tank. In the investigated case study, 0.569 kg/s are present at the glass industrial plant, while 616.0 kg/s are present at the reference plant of the coal-fired power plant. To allow a more specific comparison, the required absorber column volume per ton of CO_2 captured is calculated. The glass industry plant needs $4.54 \cdot 10^{-3}$ m^3 absorber/t CO_2 captured, while the reference plant needs only $3.44 \cdot 10^{-3}$ m^3 absorber/t CO_2 captured. Accordingly, for the specific amount of flue gas occurring in the glass industry, a larger absorber column is necessary. This is most likely due to the high levels of ambient air flowing into the investigated flue gas filter unit. The ambient air increases the exhaust gas volume while reducing the CO_2 partial pressure in the cleaned flue gases rich on CO_2.

Nevertheless, the large scaling factor between the reference plant and the designed plant is outside the limit proposed by [53] for the applicability of the cost comparison method. However, the determined cost frameworks appear plausible by comparing specific CAPEX to other literature sources. Ref. [35] investigated the CO_2 capture for carbon-intensive industrial processes and reported a specific CAPEX of 160 EUR/t CO_2 and year for amine-based post-combustion CO_2 capture in cement industry processes. The resulting specific CAPEX of this work is 188 EUR/t CO_2 per year. Considering the price increases since 2012 (approximately 16% according to [56], as can also be seen in Figure A1), the deviation of 15% calculated in this work seems justified.

It should be noted that the applied boundary conditions and assumptions of this work may have a significant impact on the implementation costs of a CO_2 capture plant in glass melting processes. However, the figures proposed herein should provide a sufficiently detailed basis for an initial cost estimation that could serve as a valuable basis for further investigations.

Table 8. Total capital investment (CAPEX) for MEA CO_2 separation plant.

	Range [53] %	**Used** %	**Cost** **EUR**
Direct cost			
Inside battery limit (ISBL) cost			
Total equipment cost (EC)	-	-	521,000
Installation	25–55	25	130,250
Instrumentation and control	8–50	8	41,680
Pipes	20–80	20	104,200
Electrical equipment	15–30	15	78,150
		Total ISBL:	875,250
Outside battery limit (OSBL) cost			
Building and building services	10–80	5	26,050
Yard improvements	10–20	5	26,050
Service facilities	30–80	10	52,100
		Total OSBL:	104,200
		Total direct cost:	979,480
Indirect cost			
Engineering	10	10	52,100
Construction expenses	10	10	52,100
Contractor's fee	0.5	0.5	2605
Contingency	17	17	88,570
		Total indirect cost:	106,805
CO_2 compressor equipment [52]			120,000
		Fixed capital investment (FCI):	1,155,285
Working investment	12–28	12	136,634
Start-up cost and MEA cost	8–10	8	92,423
Price index correction (2007–2022);% of total direct and indirect cost		21.5	248,386
		Total capital investment (CAPEX):	1,634,278

5.2.3. Operational Expenditures

The operating expenses (OPEX) for the CO_2 capture plant can be further divided into fixed charge, direct production costs, plant overhead costs, and general expenses. Table 9 shows an overview of the calculated OPEX. According to the methodology described in [53,58] , ranges or fixed percentages for the respective cost categories are given.

The fixed costs of the CO_2 capture plant include local taxes, fees, and charges, as well as insurance rates. Lower percentages of the range suggested in references [53,58] were specified for this purpose, as there are currently no specific tax rates or fees for such plants in Germany. The insurance coverage will primarily include fire insurance, the costs of which are based on the investment volume, fire risk, and the business activity. Based on the investment costs of EUR 1.63 million determined in Section 5.2.2 , the low fire risk of the plant (essentially short circuit and cable fire, hardly flammable materials) and the low fire-prone glass industry, the resulting fixed costs of approximately 18,000 EUR/year (1.1% of CAPEX) appear sufficient.

Table 9. Total operating expenditures (OPEX) for MEA CO_2 separation plant. Ranges are from [53], others are named.

	Range %	Used %	Cost EUR/Year
Fixed charge			
Local taxes	1.0–4.0% of FCI	1.0%	12,063
Insurance	0.5–1.0 % of FCI	0.5%	6031
			18,094
Direct production cost			
Cooling water			4003
MEA makeup [49]		1.5 kg/t CO_2	14,401
Maintenance (M)	1.0–10% of FCI	1.0%	12,063
Operating labor (OL) [54]	0.1 job/shift	45 EUR/h	39,096
Supervision and support labor (S)	30% of total labor cost	30%	11,729
Operating supplies	15% of maintenance	15%	1809
Laboratory charges	10–20% of operating labor	10%	3910
			87,012
Plant overhead cost	50–70% of M + OL + S	50%	31,444
General expenses			
Administrative cost	15–20% of OL	15%	5864
Distribution and marketing	2–20% of OPEX	2%	2848
Research and development cost	2–20% of OPEX	2%	2848
	Total operating expenditures (OPEX):		148,111

The direct production cost included the material requirements for plant operation, such as the cooling water, MEA makeup caused by degradation, and maintenance materials. The cooling water demand is based on simulation results. The MEA degradation was calculated based on the data given in Tables 3 and 5. The cost of the MEA makeup is approximately 14,000 EUR/year or 18% of the total OPEX. Other sources have reported an OPEX cost share of up to 32% [49,59]. Due to the size of the plant, the reduced cost share for the CO_2 separation plant in the glass industry seems plausible. However, the high O_2 content in the flue gas (see Table 2) may cause increased MEA degradation due to oxidation. For this reason, switching to more stable absorbents such as secondary or tertiary amines appears promising. These exhibit increased thermal and chemical stability against oxidative degeneration, as well as increased energy efficiency (as can be seen in Section 5.5). Besides these beneficial effects on operating parameters, the reduced energy demand and increased stability of advanced amines may result in a significant cost reduction.

The operating labor cost for the maintenance, operation, and monitoring of the plant, amounting to approximately EUR 39,000 per year, is the largest cost component (approximately 27%) in the annual OPEX. For this work, we assumed a 0.1 job per shift for the operation of the CO_2 separation plant. In the glass industry, a three-shift operation of 8 h each for 365 days per year is common to maintain labor supply. The assumption that 10% of the working time of a maintenance employee per shift is spent on the maintenance and servicing of the CO_2 separation plant thus corresponds to a total time expenditure of approximately 2.5 h/day or 912.5 h/year. Other studies such as [49,59] assumed a two job per shift staff expenditure for plant operation. However, as the separation plant discussed in this work is much smaller, such a reduction in OL expenses seems justified. In addition, OL costs influence supervision and support labor costs, laboratory charges, and plant overhead costs. Taking all of this influence into account, OL costs contribute to approximately 60% of total OPEX.

Additionally, general expenses for distribution and marketing, as well as ongoing research and development, add up to approximately EUR 11,000 per year. As a result, the annual operating expenses (OPEX) amount to approximately EUR 148,000 in total. This represents approximately 9% of the total CAPEX. [35] assumed an OPEX range of 7–12% of total CAPEX for a post-combustion MEA separation plant in the cement industry. Therefore, the OPEX calculated in this work can be considered consistent with other literature sources and even indicate potential for further optimization.

5.3. *Cost for Desorption Heat Generation*

The supply of the necessary heat for desorption is crucial for the economic viability of a post-combustion CO_2 absorption plant (Section 2.3). As shown in Table 4, the heat demand for the designed CO_2 separation process is 928 kW, at a steam temperature and pressure of 150 °C and 2 bar.

In particular, the methanation process as well as the flue gas filtration systems are the most promising options regarding the use of waste heat for steam generation.

5.3.1. Methanation Waste Heat Utilization

Considering the thermal power demand of the melting tank described in Table 1, a 7.60 MW methanation reactor is required to ensure a constant heat supply. Such a reactor can provide high pressure and temperature steam (HP-steam) and low pressure and temperature steam (LP-steam, see Figure 3). Table 10 shows the power distribution of a tubular bulb reactor based on vendor information [36]. It is evident that the HP steam of the methanation reactor can already cover approximately 120% of the heat demand for CO_2 desorption. According to vendor information, a specific CAPEX of 40 EUR/kW thermal can be expected for the cooling equipment of the methanation reactor. Based on the total steam power of 1.22 MW, a total CAPEX of approximately EUR 49,000 can be considered for the cooling equipment of the methanation reactor in the considered PtM system of this work. Thus, assuming a lifetime of 20 years and an interest rate of 5%, an exact thermal cost of 0.0004 EUR/kWh was achieved.

Table 10. Waste heat potential of the methanation reactor equipment. HP = high pressure and temperature; LP = low pressure and temperature.

	Reference Point	**Power MW**
Thermal power	Heating value of SNG	6.08
HP-steam	45 bar, 260 °C	1.09
LP-steam	30 bar, 240 °C	0.12
Power loss	Heat losses	0.30
Methanation reactor	Combined	7.60

5.3.2. Flue Gas Treatment Waste Heat Utilization

In addition to methanation, waste heat recovery from flue gas cleaning for desorption heat demand is a promising option. However, considering the properties shown in Table 2, the enthalpy content in the investigated flue gas is approximately 850 kW for a ΔT of 247 °C. This is not sufficient to provide the required thermal power of the reboiler of 915 kW.

The flue gas needs to be cooled to approximately 130 °C to ensure sufficient heat output for adequate steam generation. At this ΔT of 340 °C, an enthalpy of 1171 kW can be used. Based on vendor information and calculations, a five bundle plain tube heat exchanger is sufficient to generate steam at the required desorption conditions. Depending on the exhaust gas properties, conventional steel or, in the case of highly corrosive exhaust gases, stainless steel would be suitable as a construction material for this heat exchanger. A cost estimate for these material options is approximately EUR 40,000 for conventional steel, and EUR 90,000 for stainless steel, respectively, [37]. Assuming a lifetime of 20 years and an interest rate of 5%, specific costs of 0.0007 EUR/kWh thermal for common steel and 0.001 EUR/kWh thermal for stainless steel are achieved.

The lower outlet temperature of the flue gases from the filter system would also affect the CO_2 separation system. As a result, the capacity of the flue gas cooler could be reduced, which in turn would result in lower investment costs. However, this would also influence flow characteristics in the flue gas channel. Consequently, existing flue gas systems may no longer be able to achieve the required flow capacities and may require re-planning and design.

5.4. Cost of CO_2 Separation

With a lifetime of 20 years and an interest rate of 5%, the annual depreciation or EAC costs of the CO_2 separation plant designed in this work are approximately 135,000 EUR/year. As shown in Table 9, there are additional OPEX of approximately 146,000 EUR/year. The power required for CO_2 compression can be calculated to 73 kW using the method described in Section 4.2. However, vendor information named a compressor peak power of 55 kW and an operating power of 41 kW. Using the vendor information, CO_2 compression has an energy demand of approximately 357 MWh/year, or approximately 41 kWh/t of CO_2 captured, for the compressor operating hours of Table 4. Specific electricity costs are taken from [8] where 0.12 EUR/kWh were assumed for glass industry companies. The three cost factors EAC of CAPEX, OPEX, and electricity demand for the CO_2 compression add up to an annual cost of EUR 353,000 for the CO_2 separation plant designed in this work.

The heat demand for desorption in year-round operation is 8017 MWh/year, or 928 kWh/t CO_2. The described CO_2 capture plant can avoid 8641 t CO_2/year (Table 4). The three options for waste heat utilization, described in Section 5.3, were used to investigate their influence on CO_2 separation costs. The results are shown in Table 11. The CO_2 separation costs are 38.07 EUR/t CO_2 for the waste heat utilization of the methanation process. For the flue gas heat exchanger option, slightly higher costs of 38.42 EUR/t CO_2 or 38.73 EUR/t CO_2 were achieved. Above all, the lower heat output that these heat exchangers can generate, but also the higher material costs for the stainless steel version have an impact on additional costs. Nevertheless, these cost increases are very small and illustrate the potential for cost-effective waste heat utilization for absorption-based CO_2 capture in the glass industry.

Table 11. Calculation of CO_2 separation costs for different waste heat utilization options. EAC = equivalent annual costs of CAPEX.

	Specific Cost		Total Cost
			EUR/Year
EAC of CAPEX		189 EUR/(t CO_2 ·year)	134,895
OPEX			146,440
Electricity for CO_2 compression	0.12 EUR/kWh	4,95 EUR/t CO_2	42,745
		Fixed costs:	325,751
Heat supply cost options			
1. Methanation reactor	0.0004 EUR/kWh	0.37 EUR/t CO_2	3207
		Total annual costs:	328,958
		CO_2 separation cost:	38.07 EUR/t CO_2

Table 11. *Cont.*

2. Flue gas heat exchanger conventional steel	0.0007 EUR/kWh	0.65 EUR/t CO_2	6235
		Total annual costs:	331,986
		CO_2 separation cost:	38.42 EUR/t CO_2
3. Flue gas heat exchanger stainless steel	0.0010 EUR/kWh	0.93 EUR/t CO_2	8908
		Total annual costs:	334,659
		CO_2 separation cost:	38.73 EUR/t CO_2

Other studies reported CO_2 separation costs from industrial waste gases to range from 25 to 135 EUR/t CO_2. However, these reported values depend very much on the used CO_2 capture technology. Comparable process designs in the cement industry with post-combustion CO_2 capture using MEA, assume short- to mid-term CO_2 separation costs to be approximately 65 EUR/t CO_2 at a low cost for external steam import [35]. In studies investigating CO_2 capture from industrial point sources for methanation, 50 EUR/t CO_2 was found [60]. However, they assumed a lower CO_2 compression level of 10 bar and no detailed specification of industrial point sources was given.

5.5. Options for CO_2 Separation Process Improvement

The CO_2 separation costs of 38–39 EUR/t CO_2 found indicate the promising potential for low-cost and energy-efficient integration options for the post combustion CO_2 separation within the introduced PtM process for the glass industry. In addition, there are further options for optimizing the CO_2 capture process that have not yet been investigated:

In this work, a 30 wt. % aqueous solution of MEA was assumed. Other studies have shown that increasing the MEA content in the aqueous solution up to 40 wt.l% can further reduce the CO_2 separation costs. This can be achieved by the reduction in the energy requirement for desorption and lower investment cost for the plant equipment because of reduced liquid flow rates [49].

MEA is considered to be the standard solvent for post-combustion CO_2 absorption. However, as mentioned in Section 2.2, MEA shows disadvantages in terms of thermal and chemical stability. Chemical degradation caused by oxidation can especially be an issue for the application of MEA-based CO_2 separation processes in the glass industry. The high O_2-content of 20.5 vol-% in the investigated flue gas composition might increase oxidative degeneration (see Table 2). These effects can be avoided by changing the absorbent. The substitution of the primary amine MEA with secondary or tertiary amines such as methyl-diethanolamine (MDEA) or mixtures such as methyl-diethanolamine with piperazine (MDEA-PZ) are discussed in the literature for more stable CO_2 absorption processes. These solvents are not improved with regard to higher thermal or chemical resistance, but also enable a higher CO_2 loading capacities, while at the same time reducing the desorption heat demand. For example, the desorption heat demand of MDEA-PZ is reported as 2.6 GJ/t CO_2, while the heat demand of MEA is 3.4 GJ/t CO_2 [61,62]. Thus, a change of absorbents could not only improve the chemical stability but also further reduce the energy demand and CO_2 capture costs, respectively.

In addition to solvent replacement and changing solution properties, an advanced process design could also positively influence the costs of CO_2 separation. Improved designs for the rich/lean heat exchanger [63], vapor recompression, and split-stream processes [64], have been investigated and shown promising potential for enhancing the CO_2 removal capacity for reduced energy demand.

Combinations of these improved process designs and careful solvent selection could even further optimize the costs and performance of absorption-based CO_2 separation from flue gases of glass melting processes.

6. Conclusions

In this work, an innovative concept for the integration of a PtM process into oxyfuel glass melting processes is presented. The state of the art of the main process components is described and recommendations for suitable technology options are given. The required capture of CO_2 from the glass industry flue gases were identified as the process that requires more detailed research, as it has not been discussed before. Post-combustion CO_2 absorption capture processes, which have also been applied in power plant technology, were identified as the most promising option. A process simulation of this technology allowed a detailed design of the most important plant components and their cost evaluation. The following conclusions can be drawn:

- Amine-based CO_2 capture processes allow a reduction of −90% of CO_2 emissions of combustion-based glass melting processes. The designed CO_2 absorption plant for the glass industry is approximately 400 times smaller than the comparable concepts for power plants, due to the much lower volume of produced flue gas.
- This enables cost reductions in the required building infrastructure, as well as in staff costs for maintenance and monitoring. At the same time, cost-reducing scaling effects for large plants have a negative impact on the small plant described here.
- Various options for waste heat utilization are available within the introduced PtM process for oxyfuel glass melting. For CO_2 desorption, the methanation process can be seen as the most suitable option as it provides steam at sufficient temperature and pressure, with no impact on the existing glass industry infrastructure.
- The resulting low operating costs result in low CO_2 separation costs of 38–39 EUR/t CO_2. These could be further reduced by described optimization approaches, such as changes in solvent loading, the use of advanced solvents, and/or advanced CO_2 absorption processes.

Further studies should start with these optimization approaches and can thus provide a basis for later cost-efficient experiments on real glass melting plants.

Author Contributions: Conceptualization, S.G.; methodology, S.G. and T.M.-S.; software, S.G., T.M.-S., G.S and M.S.; validation, S.G. and T.M.-S.; formal analysis, S.G. and M.S.; investigation, S.G.; resources, M.G., G.S. and M.S.; data curation, S.G. and T.M.-S.; writing—original draft preparation, S.G. and T.M.-S.; writing—review and editing, T.M.-S., M.G., G.S. and M.S. ; visualization, S.G.; supervision, M.G., G.S. and M.S.; project administration, M.G., G.S. and M.S.; funding acquisition, M.G., G.S. and M.S. All authors have read and agreed to the published version of the manuscript.

Funding: This research was funded by a scholarship from the Nagelschneider Foundation, Munich. The authors would like to acknowledge the financial support received from the Nagelschneider Foundation.

Acknowledgments: Christian Schuhbauer of MAN Energy Solutions SE, Germany, should be thanked for their valuable advice on methanation reactor designs, energetic analysis, and cost evaluations. The authors would also like to thank Bernhard Lattner and Rudolf Steiner of Kelvion Ltd., Austria, for their contribution on flue gas heat exchanger design and cost evaluation. We would also like to thank Jörg Mehrer and the Mehrer Compression GmbH for the valuable advice on CO_2 compressor design and evaluation.

Conflicts of Interest: The authors declare no conflict of interest.

Appendix A

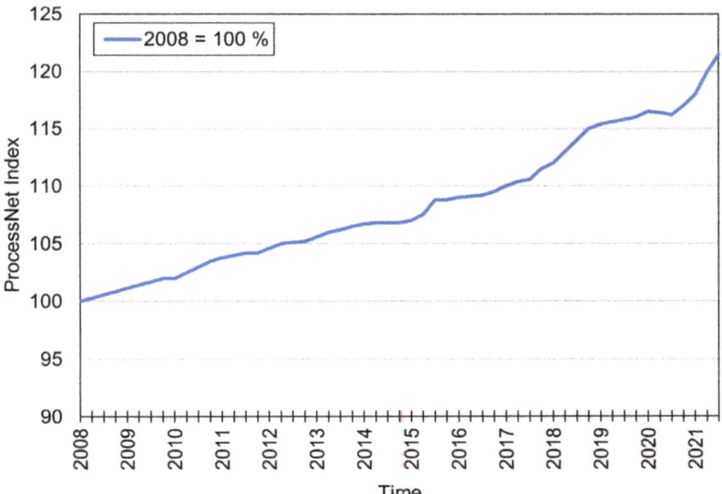

Figure A1. Quarterly development of the ProcessNet—Chemical Plant Price Index Germany since 2008.

References

1. Scalet, B.M.; Garcia Muñoz, M.; Sissa Aivi, Q.; Roudier, S.; Luis, D.S. *Best Available Techniques (BAT) Reference Document for the Manufacture of Glass*; Publication Office of the Eurpoean Union: Luxembourg, 2013; p. 485. [CrossRef]
2. Schmitz, A.; Kamiński, J.; Maria Scalet, B.; Soria, A. Energy consumption and CO_2 emissions of the European glass industry. *Energy Policy* **2011**, *39*, 142–155. [CrossRef]
3. Rio, D.D.F.D.; Sovacool, B.K.; Foley, A.M.; Griffiths, S.; Bazilian, M.; Kim, J.; Rooney, D. Decarbonizing the glass industry: A critical and systematic review of developments, sociotechnical systems and policy options. *Renew. Sustain. Energy Rev.* **2022**, *155*, 111885. [CrossRef]
4. Deutsche Emissionshandelsstelle (DEHSt). *Treibhausgasemissionen 2021. Emissionshandelspflichtige Stationäre Anlagen und Luftverkehr in Deutschland (VET-Bericht 2021)*; DEHst im Umweltbundesamt: Berlin, Germany, 2022.
5. Zier, M.; Pflugradt, N.; Stenzel, P.; Kotzur, L.; Stolten, D. Industrial decarbonization pathways: The example of the German glass industry. *Energy Convers. Manag. X* **2023**, *17*, 100336. [CrossRef]
6. Zier, M.; Stenzel, P.; Kotzur, L.; Stolten, D. A review of decarbonization options for the glass industry. *Energy Convers. Manag. X* **2021**, *10*, 100083. [CrossRef]
7. Wesseling, J.H.; Lechtenböhmer, S.; Åhman, M.; Nilsson, L.J.; Worrell, E.; Coenen, L. The transition of energy intensive processing industries towards deep decarbonization: Characteristics and implications for future research. *Renew. Sustain. Energy Rev.* **2017**, *79*, 1303–1313. [CrossRef]
8. Gärtner, S.; Rank, D.; Heberl, M.; Gaderer, M.; Dawoud, B.; Haumer, A.; Sterner, M. Simulation and techno-economic analysis of a power-to-hydrogen process for oxyfuel glass melting. *Energies* **2021**, *14*, 8603. . [CrossRef]
9. NRW Energy 4 Climate. HyGlas Project. 2022. Available online: https://www.energy4climate.nrw/themen/best-practice/hyglass (accessed on 17 February 2022).
10. Glass International. NSG to Test Hydrogen Fuel for Glassmaking at St Helens Site. 2020. Available online: https://www.glass-international.com/news/nsg-to-use-hydrogen-fuel-for-glassmaking-at-st-helens-site (accessed on 17 February 2022).
11. Bundesminsterium für Bildung und Forschung (BMBF); Projekte: Kopernikus-Projekt: P2X; 2022. Available online: https://www.kopernikus-projekte.de/projekte/p2x (accessed on 17 February 2022).
12. Howarth, R.W.; Jacobson, M.Z. How green is blue hydrogen? *Energy Sci. Eng.* **2021**, *9*, 1676–1687. . [CrossRef]
13. Bauer, C.; Treyer, K.; Antonini, C.; Bergerson, J.; Gazzani, M.; Gencer, E.; Gibbins, J.; Mazzotti, M.; McCoy, S.T.; McKenna, R.; et al. On the climate impacts of blue hydrogen production. *Sustain. Energy Fuels* **2021**, *6*, 66–75. [CrossRef]
14. Thema, M.; Bauer, F.; Sterner, M. Power-to-Gas: Electrolysis and methanation status review. *Renew. Sustain. Energy Rev.* **2019**, *112*, 775–787. [CrossRef]

15. Sterner, M.; Specht, M. Power-to-Gas and Power-to-X—The History and Results of Developing a New Storage Concept. *Energies* **2021**, *14*, 6594. [CrossRef]
16. Conradt, R. The industrial glass melting process. In *The SGTE Casebook: Thermodynamics at Work*; Hack, K., Ed.; Elsevier: Amsterdam, The Netherlands, 2008; Chapter II.24, pp. 282–303.
17. Brenner, W.; Weig, M. Schott AG, Business Unit Tubing, Mitterteich, Germany. Personal communication, 2022.
18. Ghaib, K.; Ben-Fares, F.Z. Power-to-Methane: A state-of-the-art review. *Renew. Sustain. Energy Rev.* **2018**, *81*, 433–446. [CrossRef]
19. Sterner, M.; Stadler, I. *Handbook of Energy Storage: Demand, Technologies, Integration*; Springer: Berlin/Heidelberg, Germany, 2018.
20. Miller, H.A.; Bouzek, K.; Hnat, J.; Loos, S.; Bernäcker, C.I.; Weißgärber, T.; Röntzsch, L.; Meier-Haack, J. Green hydrogen from anion exchange membrane water electrolysis: A review of recent developments in critical materials and operating conditions. *Sustain. Energy Fuels* **2020**, *4*, 2114–2133. [CrossRef]
21. Siemens Energy. Hydrogen Solutions-Renewable Energy-Siemens Energy Global. 2022. Available online: https://www.siemens-energy.com/global/en/offerings/renewable-energy/hydrogen-solutions.html (accessed on 17 February 2022).
22. Hassan, I.A.; Ramadan, H.S.; Saleh, M.A.; Hissel, D. Hydrogen storage technologies for stationary and mobile applications: Review, analysis and perspectives. *Renew. Sustain. Energy Rev.* **2021**, *149*, 111311. [CrossRef]
23. Gamisch, B.; Gaderer, M.; Dawoud, B. On the Development of Thermochemical Hydrogen Storage: An Experimental Study of the Kinetics of the Redox Reactions under Different Operating Conditions. *Appl. Sci.* **2021**, *11*, 1623. [CrossRef]
24. Teichmann, D.; Arlt, W.; Wasserscheid, P.; Freymann, R. A future energy supply based on Liquid Organic Hydrogen Carriers (LOHC). *Energy Environ. Sci.* **2011**, *4*, 2767–2773. [CrossRef]
25. Linde AG. Liquid Air Oxygen Storage. 2022. Available online: https://www.linde-gas.de/shop/de/de-ig/fluessigluftenergiespeicherung (accessed on 2 February 2022)
26. Götz, M.; Lefebvre, J.; Mörs, F.; Koch, A.M.; Graf, F.; Bajohr, S.; Reimert, R.; Kolb, T. Renewable Power-to-Gas: A technological and economic review. *Renew. Energy* **2016**, *85*, 1371–1390. [CrossRef]
27. Rachow, F. Prozessoptimierung für die Methanisierung von CO_2. Ph.D. Dissertation, Brandenburgischen Technischen Universität Cottbus–Senftenberg, Cottbus, Germany, 2017.
28. Fleige, M. *Direkte Methanisierung von CO_2 aus dem Rauchgas Konventioneller Kraftwerke*; Springer: Berlin/Heidelberg, Germany, 2015. [CrossRef]
29. Conradt, R. Prospects and physical limits of processes and technologies in glass melting. *J. Asian Ceram. Soc.* **2019**, *7*, 377–396. [CrossRef]
30. Bui, M.; Adjiman, C.S.; Bardow, A.; Anthony, E.J.; Boston, A.; Brown, S.; Fennell, P.S.; Fuss, S.; Galindo, A.; Hackett, L.A.; et al. Carbon capture and storage (CCS): The way forward. *Energy Environ. Sci.* **2018**, *11*, 1062–1176. [CrossRef]
31. Zerobin, F.; Bertsch, O.; Penthor, S.; Pröll, T. Concept Study for Competitive Power Generation from Chemical Looping Combustion of Natural Gas. *Energy Technol.* **2016**, *4*, 1299–1304. [CrossRef]
32. ClimeWorks AG. Direct Air Capture Technology and Carbon Removal. 2022. Available online: https://climeworks.com/direct-aircapture (accessed on 2 February 2022).
33. Davidson, R.M. *Post-Combustion Carbon Capture from Coal Fired Plants—Solvent Scrubbing*; IEA Clean Coal Centre: London, UK, 2007.
34. Shan, Z.; Shujuan, W.; Chenchen, S.; Changhe, C. SO2 effect on degradation of MEA and some other amines. *Energy Procedia* **2013**, *37*, 896–904. [CrossRef]
35. Kuramochi, T.; Ramírez, A.; Turkenburg, W.; Faaij, A. Comparative assessment of CO_2 capture technologies for carbon-intensive industrial processes. *Prog. Energy Combust. Sci.* **2012**, *38*, 87–112. [CrossRef]
36. Schuhbauer, C. MAN Energy Solutions SE, Deggendorf, Germany. Personal communication, 2022.
37. Lattner, B.; Steiner, R. Kelvion GmbH, Heat Exchanger, Gaspoltshofen, Austria. Personal communication, 2022.
38. Weißert, T. Luehr Filter GmbH, Technology Management, Stadthagen, Germany. Personal communication, 2022.
39. Roger, U.; Conradt, R. *Schlussbericht IGF Vorhaben: Effiziente Abgasreinigung von Boro-Silicatglas-Schmelzanlagen*; Technical Report; Hüttentechnische Vereinigung der Glasindustrie (HVG) e.V.: Offenbach, Germany, 2014.
40. Goodwin, D.G.; Speth, R.L.; Moffat, H.K.; Weber, B.W. Cantera: An Object-Oriented Software Toolkit for Chemical Kinetics, Thermodynamics, and Transport Processes. Version 2.5.1. 2021. Available online: https://www.cantera.org (accessed on 22 February 2022).
41. Smith, G.P.; Golden, D.M.; Frenklach, M.; Moriarty, N.W.; Eiteneer, B.; Goldenberg, M.; Bowman, C.T.; Hanson, R.K.; Song, S.; Gardiner, W.C.; et al. GRI-MECH 3.0 Available online: http://combustion.berkeley.edu/gri-mech/version30/text30.html#cite (accessed on 11 February 2022).
42. Marx-Schubach, T.; Schmitz, G. Modeling and simulation of the start-up process of coal fired power plants with post-combustion CO_2 capture. *Int. J. Greenh. Gas Control.* **2019**, *87*, 44–57. [CrossRef]
43. Abu-Zahra, M.R.; Schneiders, L.H.; Niederer, J.P.; Feron, P.H.; Versteeg, G.F. CO_2 capture from power plants. Part I. A parametric study of the technical performance based on monoethanolamine. *Int. J. Greenh. Gas Control.* **2007**, *1*, 37–46. [CrossRef]
44. *Mellapak and MellapakPlus*; Sulzer Ltd.: Winterthur, Switzerland, 2022. Available online: https://www.sulzer.com/en/shared/products/mellapak-and-mellapakplus (accessed on 20 February 2023).
45. Bravo, J.L.; Rocha, J.A.; Fair, J.R. Mass Transfer in Gauze Packings. *Hydrocarb. Process.* **1985**, *64*, 91–95.

46. Bravo, J.L.; Rocha, J.A.; Fair, J.R. A comprehensive model for the performance of columns containing structured packings. *Inst. Chem. Eng. Symp. Ser.* **1992**, *129*, A439–A657.
47. Chilton, T.H.; Colburn, A.P. Mass Transfer (Absorption) Coefficients: Prediction from Data on Heat Transfer and Fluid Friction. *Ind. Eng. Chem.* **1934**, *26*, 1183–1187. [CrossRef]
48. Agbonghae, E.O.; Hughes, K.J.; Ingham, D.B.; Ma, L.; Pourkashanian, M. Optimal Process Design of Commercial-Scale Amine-Based CO_2 Capture Plants. *Ind. Eng. Chem. Res.* **2014**, *53*, 14815–14829. [CrossRef]
49. Abu-Zahra, M.R.; Niederer, J.P.; Feron, P.H.; Versteeg, G.F. CO_2 capture from power plants: Part II. A parametric study of the economical performance based on mono-ethanolamine. *Int. J. Greenh. Gas Control.* **2007**, *1*, 135–142. [CrossRef]
50. Kuramochi, T.; Faaij, A.; Ramírez, A.; Turkenburg, W. Prospects for cost-effective post-combustion CO_2 capture from industrial CHPs. *Int. J. Greenh. Gas Control.* **2010**, *4*, 511–524. [CrossRef]
51. Hendriks, C.; Crijns-Graus, W.H.J.; van Bergen, F. *Global Carbon Dioxide Storage Potential and Costs*. Ecofys, Rijksinstituut voor Volksgezondheid en Milieu, 2004. Available online: https://www.researchgate.net/publication/260095614 (accessed on 3 October 2022).
52. Mehrer, J. (Mehrer Compression GmbH, Rosenfelder Strasse 35, 72336 Balingen, Germany). Personal communication, 2022.
53. Peters, M.; Timmerhaus, K.; West, R. *Plant Design and Economics for Chemical Engineers*; McGraw-Hill Chemical Engineering Series; McGraw-Hill Education: New York, NY, USA, 2003.
54. Indeed.com. Maintainance Labour Vages in Germany Based on 2600 Average Salerys in Relevant Industries. 2022. Available online: https://https://de.indeed.com/career/mitarbeiter-instandhaltung/salaries (accessed on 22 February 2022).
55. Stepstone. Stepstone Gehaltsreport 2021-Deutschland. 2022. Available online: https://www.stepstone.de/Ueber-StepStone/wp-content/uploads/2021/02/StepStone-Gehaltsreport-2021.pdf (accessed on 20 December 2022).
56. DECHEMA; VDI-ECG. ProcessNet-ProcessNet Chemieanlagenindex Deutschland PCD. 2022. Available online: https://processnet.org/pcd.html (accessed on 22 February 2022).
57. Lühe, C. Modulare Kostenschätzung als Unterstützung der Anlagenplanung für die Angebots- und frühe Basic Engineering Phase. Ph.D. Dissertation, Faculty for Process Engineering, Technical University of Berlin, Berlin, Germany, 2012.
58. Ripperger, S.; Nikolaus, K. *Entwicklung und Planung Verfahrenstechnischer Anlagen*; VDI-Buch; Springer: Berlin/Heidelberg, Germany, 2020.
59. Singh, D.; Croiset, E.; Douglas, P.L.; Douglas, M.A. Techno-economic study of CO_2 capture from an existing coal-fired power plant: MEA scrubbing vs. O2/CO_2 recycle combustion. *Energy Convers. Manag.* **2003**, *44*, 3073–3091. [CrossRef]
60. Gorre, J.; Ruoss, F.; Karjunen, H.; Schaffert, J.; Tynjälä, T. Cost benefits of optimizing hydrogen storage and methanation capacities for Power-to-Gas plants in dynamic operation. *Appl. Energy* **2020**, *257*, 113967. . [CrossRef]
61. Closmann, F.; Nguyen, T.; Rochelle, G.T. MDEA/Piperazine as a solvent for CO_2 capture. *Energy Procedia* **2009**, *1*, 1351–1357. [CrossRef]
62. Leung, D.Y.; Caramanna, G.; Maroto-Valer, M.M. An overview of current status of carbon dioxide capture and storage technologies. *Renew. Sustain. Energy Rev.* **2014**, *39*, 426–443. [CrossRef]
63. Aromada, S.A.; Eldrup, N.H.; Normann, F.; Øi, L.E. Techno-Economic Assessment of Different Heat Exchangers for CO_2 Capture. *Energies* **2020**, *13*, 6315. [CrossRef]
64. Øi, L.E.; Brathen, T.; Berg, C.; Brekne, S.K.; Flatin, M.; Johnsen, R.; Moen, I.G.; Thomassen, E. Optimization of Configurations for Amine based CO_2 Absorption Using Aspen HYSYS. *Energy Procedia* **2014**, *51*, 224–233. [CrossRef]

Article

Effect of External Mineral Addition on PM Generated from Zhundong Coal Combustion

Shizhang Wang, Junjie Wang, Yu Zhang *, Linhan Dong, Heming Dong, Qian Du and Jianmin Gao

School of Energy Science and Engineering, Harbin Institute of Technology, Harbin 150001, China
* Correspondence: zhang.y@hit.edu.cn

Abstract: The effect of intrinsic metal mineral elements in the combustion process of pulverized coal on the formation and transformation mechanism of PM was investigated in a drop-tube furnace in air atmospheres at 1200 °C, which laid a solid foundation for the control of particulate pollutants. The results show that reducing the evaporation of mineral elements or the generated PM_1 aggregating to form PM_{1-10} or particles bigger than 10μm can reduce the emission of PM_1 in the coal combustion process. The amount of $PM_{0.2}$, $PM_{0.2-1}$, $PM_{1-2.5}$ and $PM_{2.5}$ produced by the raw coal-carrying Mg are reduced by 36.7%, 17.4%, 24.6% and 21.6%, respectively. The amount of PM_{10} is almost unchanged. The addition of Mg increases the viscosity of submicron particles effectively, making it easier to aggregate and bond together to form ultra-micron particles. The amount of $PM_{0.2}$, $PM_{0.2-1}$, $PM_{1-2.5}$, $PM_{2.5}$ and PM_{10} produced by the raw coal-carrying Ca are reduced by 36.3%, 33.0%, 42.8%, 38% and 17.7%, respectively. The effect of adding Ca compounds on the particles is better than that of Mg. The amount of $PM_{0.2}$, $PM_{0.2-1}$, $PM_{1-2.5}$, $PM_{2.5}$ and PM_{10} produced by the raw coal-carrying Fe are reduced by 15.6%, 16.2%, 31.1%, 22.4% and 5%, respectively. While the production of $PM_{2.5-10}$ increased from 0.17 mg/g to 0.34 mg/g, it is clear that a significant fraction of the submicron particles produced during the combustion of the raw coal-carrying Fe are transformed into ultra-micron particles. After comparing the particulate matter produced by raw coal-carrying Mg, Ca and Fe, it shows that the addition of these three elements can effectively reduce the ash melting point, so that during the process of coal combustion, part of the sub-micron are transformed into ultra-micron particles, which are easy to remove.

Keywords: external mineral addition; PM; coal; combustion; Mg; Ca; Fe

Citation: Wang, S.; Wang, J.; Zhang, Y.; Dong, L.; Dong, H.; Du, Q.; Gao, J. Effect of External Mineral Addition on PM Generated from Zhundong Coal Combustion. *Energies* **2023**, *16*, 730. https://doi.org/10.3390/en16020730

Academic Editor: Islam Md Rizwanul Fattah

Received: 27 October 2022
Revised: 24 December 2022
Accepted: 27 December 2022
Published: 8 January 2023

1. Introduction

Currently, China's air quality situation is extremely severe, particulate matter (PM) pollution is the main reason. Particulate matter (PM) has become a major air pollutant in urban China in recent years and is a leading factor contributing to decreased atmospheric visibility, global climate change and photochemical smog [1]. Particulate matter emitted from coal combustion is one of the sources of atmospheric particulates [2]. Fine particles generated during coal combustion are easy to enrich toxic heavy metals such as Pb, Cd, As, Se and organic pollutants such as PAHs [3–5]. In addition, the electrostatic precipitator has relatively low collection efficiency for these fine particles [6]. Coal burning particulates easily accumulate heavy metal trace elements, which are harmful to the environment and human health [7,8]. Reducing particulate matter generated by coal combustion is one of the key points of pollutant emission. The premise of a number of pollutant control means playing a role is fully understanding the PM formation and transformation mechanism. In the process of PM formation, the presence of mineral elements largely controls the formation of PM, which occupies a large proportion of PM particulates after combustion [9,10]. The particulate matter formation process is more complex; it is generally believed that the volatile part of atomic inorganic elements in coal and the external minerals on the surface of the coal will gasify due to high temperature and reducing atmosphere, mainly in the form of atomic or in the form of sub-oxide [11]. Submicron particles (<1 μm) are composed of the following three components: metal oxides (Na_2O and K_2O), refractory oxides (SiO_2, Fe_2O_3, CaO), sulfides and so on [12,13]. The submicron particles of high-rank bituminous coal are mainly composed of refractory oxide SiO_2, followed by Na_2O, Fe_2O_3 or SO_3. The content of refractory oxides in low-grade lignite submicron particles is low, and the main components are alkali metal and the oxidation of alkaline earth metal and sulfides. The results show that the content of Si in Zhundong coal ultrafine particles is very small [14], mainly S and Na (more than 15%) and volatile elements such as K, Ca, Mg, Cl and F. The main component in submicron particles is sulfate and aluminosilicate based on Na and Ca; the spherical particles of micron particles are mostly aluminosilicate based on Ca, Mg and Na [15]. In the process of pulverized coal combustion, volatiles are internally oxidized by the outward diffusion of oxygen into the corresponding oxide. If the steam in the flue gas reaches the supersaturation state, the steam will form a large number of fine particles (<0.01 μm) by homogeneous nucleation. Small particles grow mainly in two ways, one way is the formation of inorganic vapors on the surface of fine particles in the condensation, so that the particulate matter volume increases [16], another way is the collision of fine particles with each other, forming larger particles [11]. The presence of Na, K and the easy gasified elements enriched on the submicron particles and other difficult gasified elements like Si, Al, Fe, Ca and Mg existed on the submicron particles indicates that the gasification-condensation mechanism of inorganic mineral elements is the most important form of submicron particulate formation [17–19]. The direct conversion of intrinsic minerals in raw coal and the polymerization of volatile mineral vapors are the two kinds of main ways to form ultrafine particles [20]. The gasification of Ca and Mg has an important influence on the formation of submicron particles during the combustion of low-order sub-bituminous coal and lignite. The occurrence of Ca and Mg in low-rank coal is the main reason for its gasification behavior. Ca and Mg in subbituminous coal mainly exist in the form of carbonate, and the amount of gasification is very low. Ca and Mg in the lignite mainly exist in the atomic state, which is prone to gasification in the combustion process; therefore, they occupy a higher content in the fine particles [21].

The above studies have proved the effect of intrinsic metal mineral elements in the combustion process of pulverized coal on the formation and transformation mechanism of PM, which laid a solid foundation for the control of particulate pollutants [22]. However, there is little research on the mechanism of PM formation and transformation and its subsequent control mechanism under the condition of externally added mineral elements [23]. In this paper, we use the method of external loading mineral elements was adopted to study the physical and chemical characteristics of the particles formed during the burning process of the Chinese Zhundong coal with external mineral load in the one-dimensional settling furnace reactor [24].

2. Experiment

2.1. Sample Preparation

Chinese Zhundong coal, which was first dried at 80 °C for 12 h, was used in the experiment. The proximate and ultimate analyses of the coal sample are shown in Table 1. In order to explore the influence of different mineral elements on the particles generated in the combustion process of Zhundong coal, the compound loading was carried out. The sample was loaded with compounds of different metal elements by the immersion method. The loading steps are as follows. The Zhundong coal particles were mixed thoroughly with 1 mol/L of mineral salts ($Fe(NO_3)_3{\cdot}9H_2O$, $MgAc_2$, $CaAc_2$) at a liquid to solid ratio of 30:1 [25]. The mixture was stirred for 24 h in an inert atmosphere, and then the mixed slurry was filtered and the precipitation was dried in a vacuum oven at 80 °C for 24 h, crushed and collected, then stored it at 4 °C. After carrying out an analytical test on the Zhundong coal sample carrying the compound, the content of Mg in the coal was increased from 1.26 mg/g to 2.61 mg/g in the raw coal after carrying $MgAc_2$. The content of Ca was increased from 2.61 mg/g to 4.12 mg/g after loading $CaAc_2$, the content of Fe increased from 2.83 mg/g to 4.21 mg/g after carrying $Fe(NO_3)_3{\cdot}9H_2O$.

Table 1. Proximate and ultimate analyses of raw Zhundong coal.

Proximate Analysis (wt%)				Ultimate Analysis (wt%)				
M_{ad}	A_{ad}	V_{ad}	FC_{ad}	$C_{daf.}$	$H_{daf.}$	$O_{daf.(diff.)}$	$N_{daf.}$	$S_{t,daf.}$
11.31	5.31	24.37	61.19	73.20	4.10	21.86	0.74	0.10

Note: *diff.*: by difference; *ad*: air-dried basis; *daf.*: dry ash-free basis.

2.2. Experimental System

A drop-tube furnace (shown in Figure 1) was used for experiments with Chinese Zhundong coal. The experiment was carried out in air atmospheres at 1200 °C. The furnace was composed of an 80 mm diameter corundum tube of length 2000 mm. It can withstand high temperatures above 1500 °C. The B indexing thermocouple is arranged on the outer surface for temperature monitoring and control. The experimental system is provided with two air supplies. The primary air is mainly used for conveying pulverized coal and the oxygen demand at the initial stage of pulverized coal combustion. The secondary air is fully mixed and preheated inside the burner to supplement the oxygen required for complete combustion of pulverized coal to enable the pulverized coal to burn out. The pulverized coal sample was pneumatically conveyed into the furnace chamber at a constant rate of 1.0 g/min by a primary air stream. A secondary air stream was then mixed in at the burner to ensure complete combustion occurred during the coal's residence in the furnace. Fly ash was collected at the bottom of the reactor to investigate the PM [26].

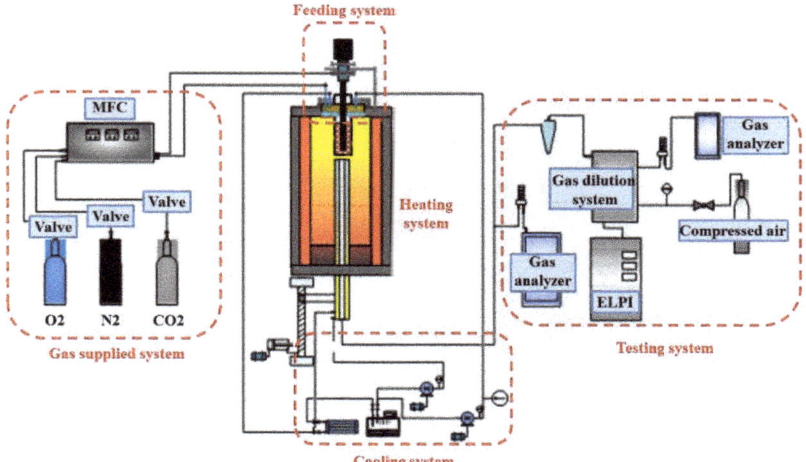

Figure 1. Schematic diagram of the experimental furnace.

2.3. Sampling System and Characteristics Analysis

The experimental set up employed the fixed combustion particle dilution sampling system developed by Tsinghua University. PM samples were simultaneously sampled from the collected ash and classified for analysis [27]. To study the morphology and composition, a poly-ammonia carbonate membrane was used for imaging and electron microscope analysis [28]. An electrical low-pressure impactor (ELPI) (97 2E, NO.24423, Dekati Ltd., Kangasala, Finland) was used to measure the ash particle size distribution. The particle morphology and composition were analyzed using a combined scanning electron microscopy–energy dispersive X-ray spectroscopy approach (SEM–EDX; EVO18).

3. Results and Discussion

3.1. Effect of Magnesium on the Formation of Fine Particles

3.1.1. Effect of Magnesium on the Particle Morphology and Particle Size Distribution

Morphologies of the particulate matter generated by the combustion of raw coal and raw coal-carrying Mg are shown in Figures 2 and 3, respectively. Compared with raw coal combustion, the particulates generated by the combustion of raw coal-carrying Mg shows accumulation state [29], and it is evident that a part of the submicron particles adhered to the surface of the coarse-mode particles, which makes the particle size generally increase. Yoshihiko [30] observed the coal-carrying magnesium-based compounds by scanning electron microscopy, and it showed that magnesium-based compounds uniformly dispersed in the coal surface, the average particle size was among the sub-micron level, the magnesium-based additive is an intrinsic mineral element in coal and not an external mineral in physically mixed form (this extraneous mineral and coal particles have relatively few connections). Therefore, it can be considered that the loaded magnesium element exists mainly in the form of an intrinsic mineral [31–33].

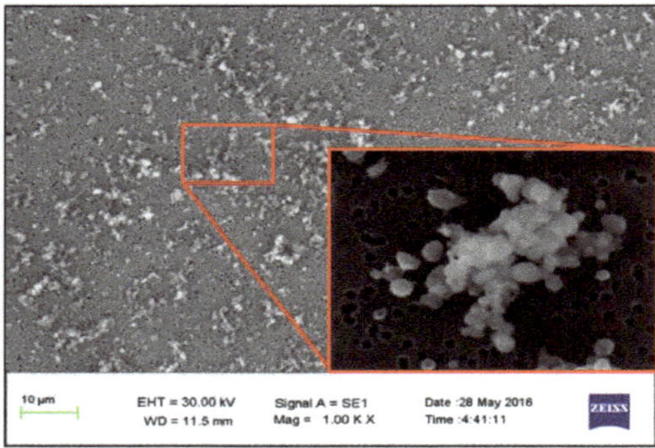

Figure 2. Particle morphology of raw coal.

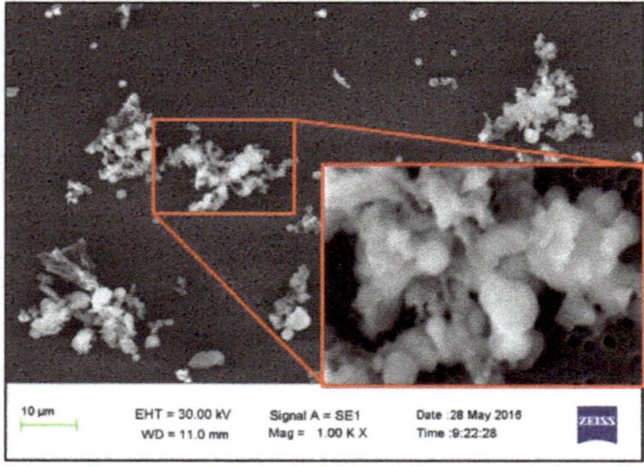

Figure 3. Morphology of particles after carrying magnesium.

Figure 4 is the generating particulate matter mass size distribution by the original coal and the original coal with the addition of Mg. It can be seen that the addition of Mg element shows the changes in particle size. Primitive minerals (including particle size and chemical composition) to some extent affect the effect of adding Mg minerals. Figure 5 shows the amounts of $PM_{0.2}$, $PM_{0.2-1}$, $PM_{1-2.5}$ generated from raw coal and Mg-carrying raw coal. The production of $PM_{0.2}$, $PM_{0.2-1}$, $PM_{1-2.5}$ were reduced from 0.32 mg/g, 0.31 mg/g, 0.45 mg/g to 0.25 mg/g, 0.26 mg/g and 0.34 mg/g, respectively. After adding Mg, the amount of

$PM_{0.2}$, $PM_{0.2–1}$, $PM_{1–2.5}$ decreased by 36.7%, 17.4% and 24.6%, respectively, and the amount of $PM_{2.5}$ was reduced by 21.6%. While the $PM_{2.5–10}$ production increased from 0.17 mg/g to 0.41 mg/g. The total mass of PM_{10} was 1.24 mg/g and 1.26 mg/g, almost no changes at all. It is clear that the chemical composition of PM_{10} has changed.

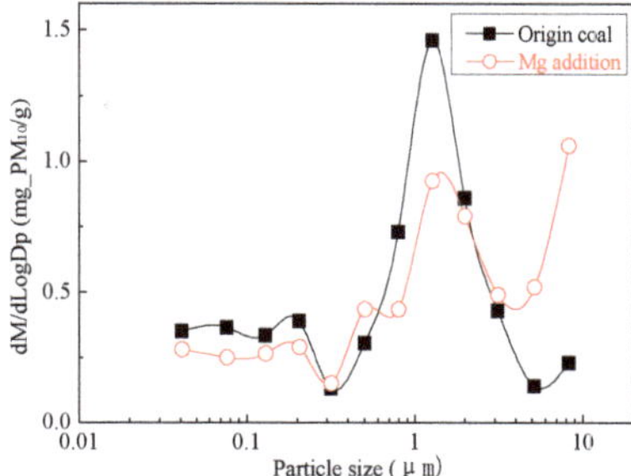

Figure 4. Particle size distribution (Mg addition).

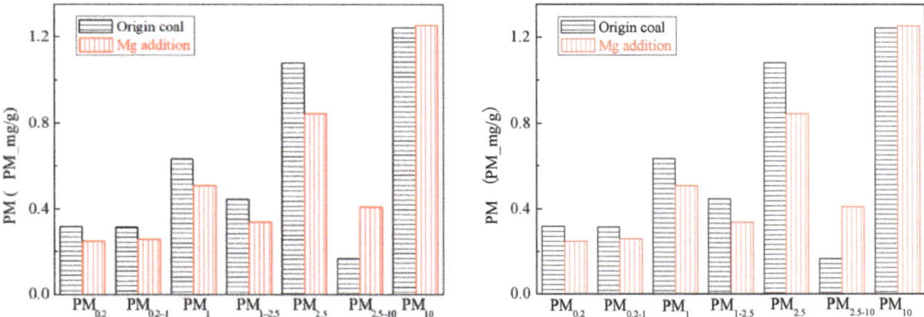

Figure 5. Particulate matter production (Mg addition).

3.1.2. Effect of Magnesium on the Chemical Composition of Particles

The main elemental compositions of the particulates generated by the combustion of pulverized coal-carrying Mg are Fe, Mg, Ca, Al and Si, and the main volatile elements is Na, which are shown in Figures 6–11. The decrease in PM_1 is mainly due to the decrease in Fe, Ca, Al and Si, and the reduction of volatile elements also play a certain role. The reduction of $PM_{1–2.5}$ is mainly due to the reduction of the insoluble Al, Si and Fe elements, and the reduction of volatile elements is little, which is almost negligible [34]. The Al, Si and Fe elements in part of the particles below 1 μm transformed coarse mode particles above 2.5 μm, which is the main reason for the decrease in PM 2.5 generation [35,36]. According to the analysis of particle composition under different particle sizes, the results show that the main reason for the decrease in PM_1 is the transfer of Ca, Mg, Al and Si from submicron particles to ultra-micrometer particles.

The addition of Mg-based compounds can significantly affect the mineral conversion during pulverized coal combustion. The morphology of the particle and energy spectrum are shown in Figure 12. The SEM results show that the spherical particles indicate that they are formed by droplets in the combustion process; semi-molten irregular particles show that they are from the transition of the solid or incomplete liquid particles. The forming liquid material may provide a viscosity that allows the submicron particles or fine particles to adhere to the coarse mode particles, which allows the conversion of these elements to particles of different particle sizes. The fine particle morphology indicates that they are formed by the accumulation of small particles, the increase in the mass percentage of the liquid material is the primary reason for the formation of suitable viscosities [37], the submicron particles adhere to the coarse mode particles and they can capture submicron particles which can reduce the amount of sub-micron particulate matter. From the three points of the first figure, the Mg content is the highest at point 1 which can reach 87%, the Mg content at point 2 is 30%, and the Mg content at point 3 is 33%. Mg-Ca-Fe-Al-Si, Mg-Ca-Al-Si and Mg-Al-Si are the main components of the magnesium-containing particles [38].

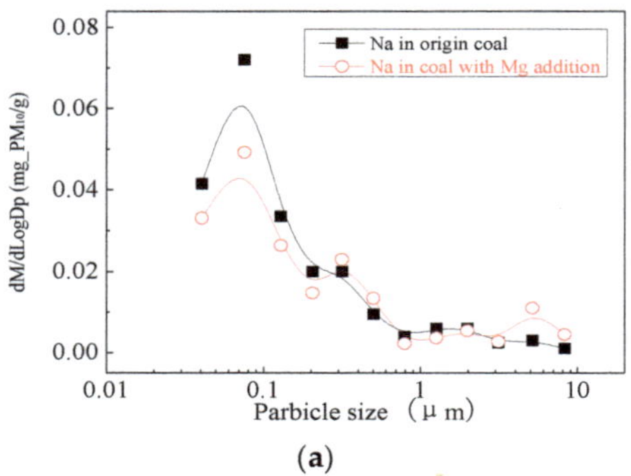

(a)

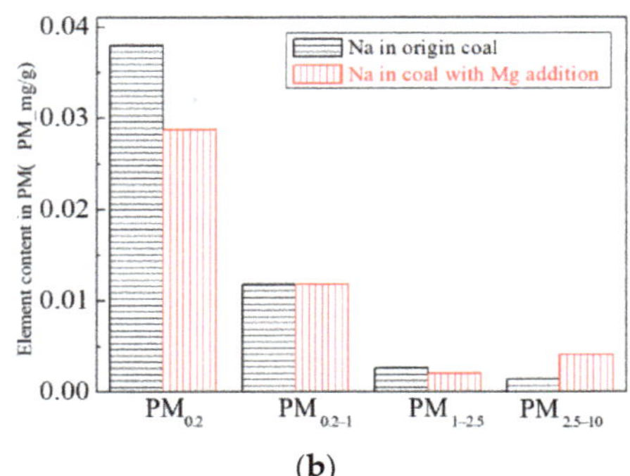

(b)

Figure 6. Mass particle size distribution of Na (Mg addition). (**a**) The curves of mass particle size distribution. (**b**) Sectional size distribution.

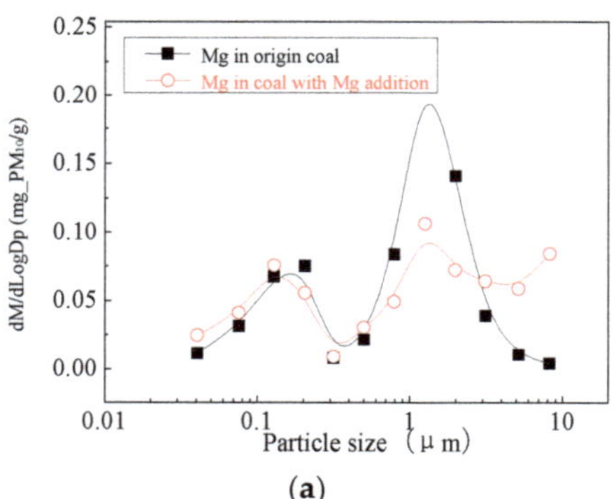

(a)

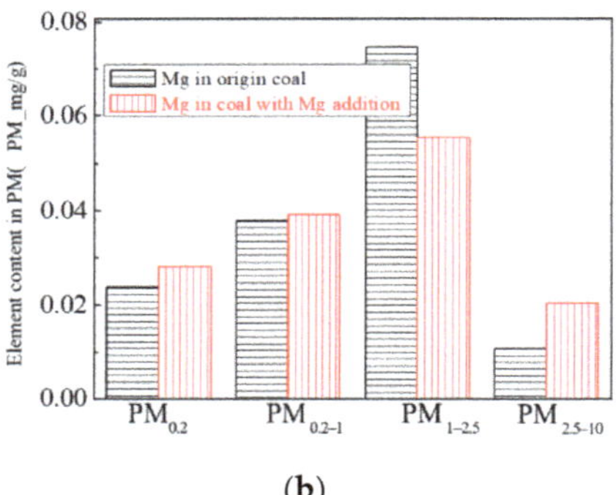

(b)

Figure 7. Mass particle size distribution of Mg (Mg addition). (**a**) The curves of mass particle size distribution. (**b**) Sectional size distribution.

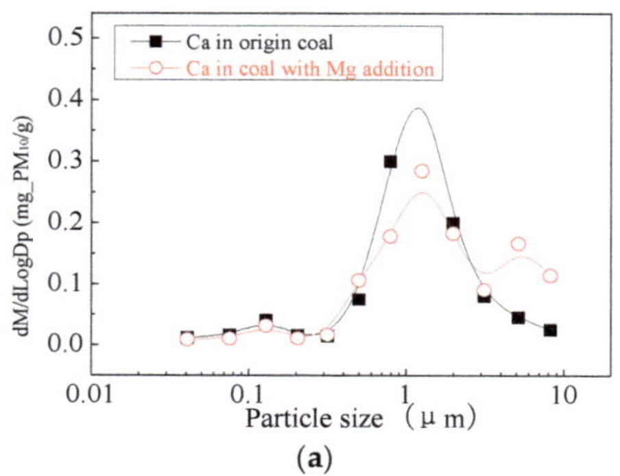

(a)

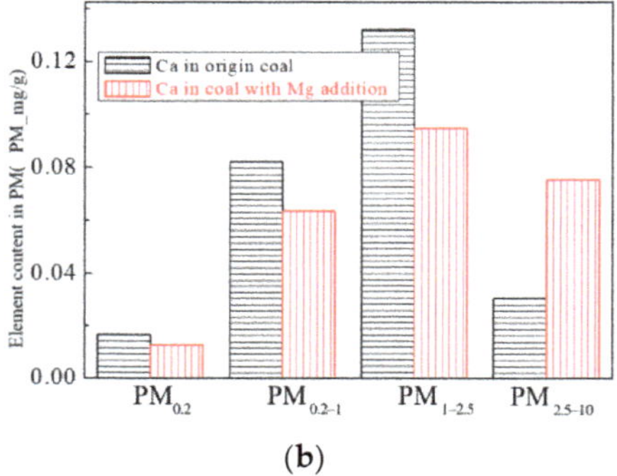

(b)

Figure 8. Mass particle size distribution of Ca (Mg addition). (**a**) The curves of mass particle size distribution. (**b**) Sectional size distribution.

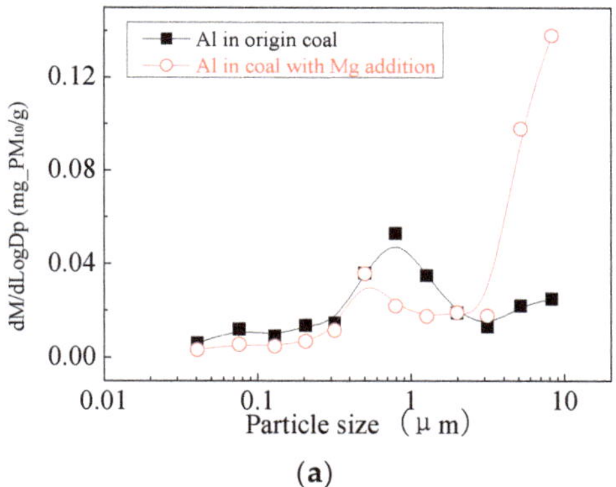

(a)

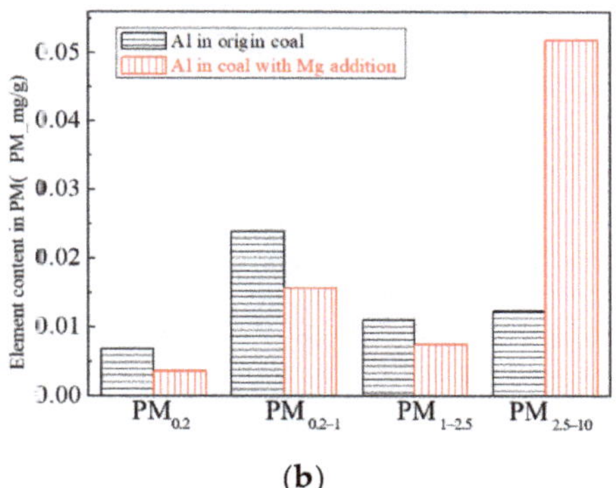

(b)

Figure 9. Mass particle size distribution of Al (Mg addition). (**a**) The curves of mass particle size distribution. (**b**) Sectional size distribution.

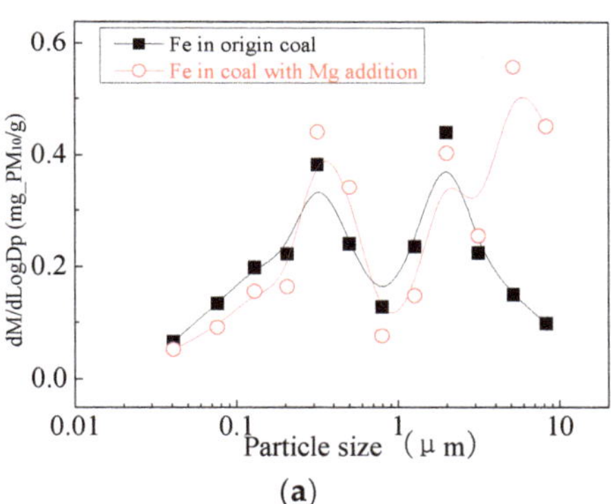

(a)

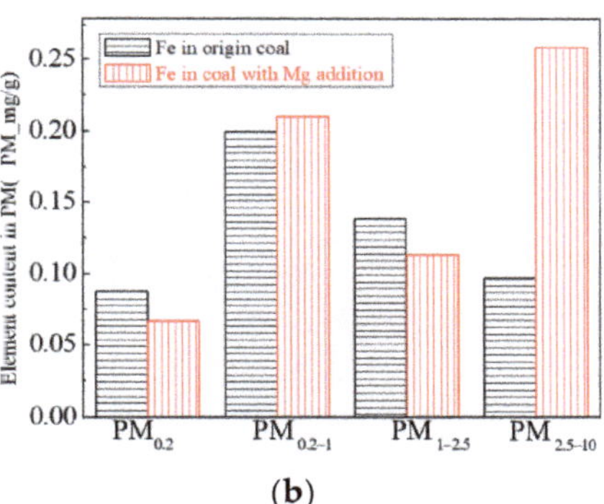

(b)

Figure 10. Mass particle size distribution of Fe (Mg addition). (**a**) The curves of mass particle size distribution. (**b**) Sectional size distribution.

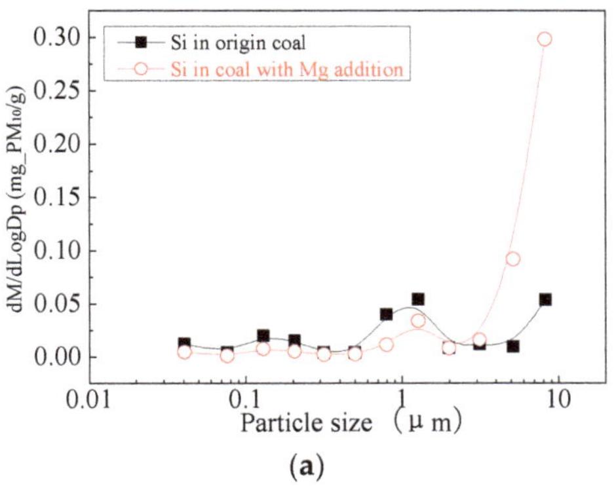

(a)

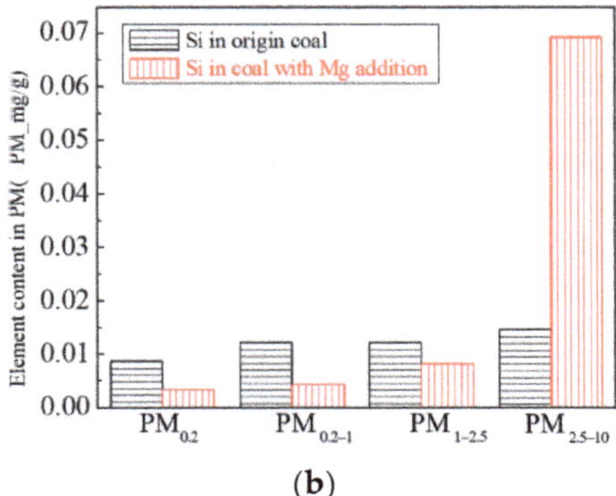

(b)

Figure 11. Mass particle size distribution of Si (Mg addition). (**a**) The curves of mass particle size distribution. (**b**) Sectional size distribution.

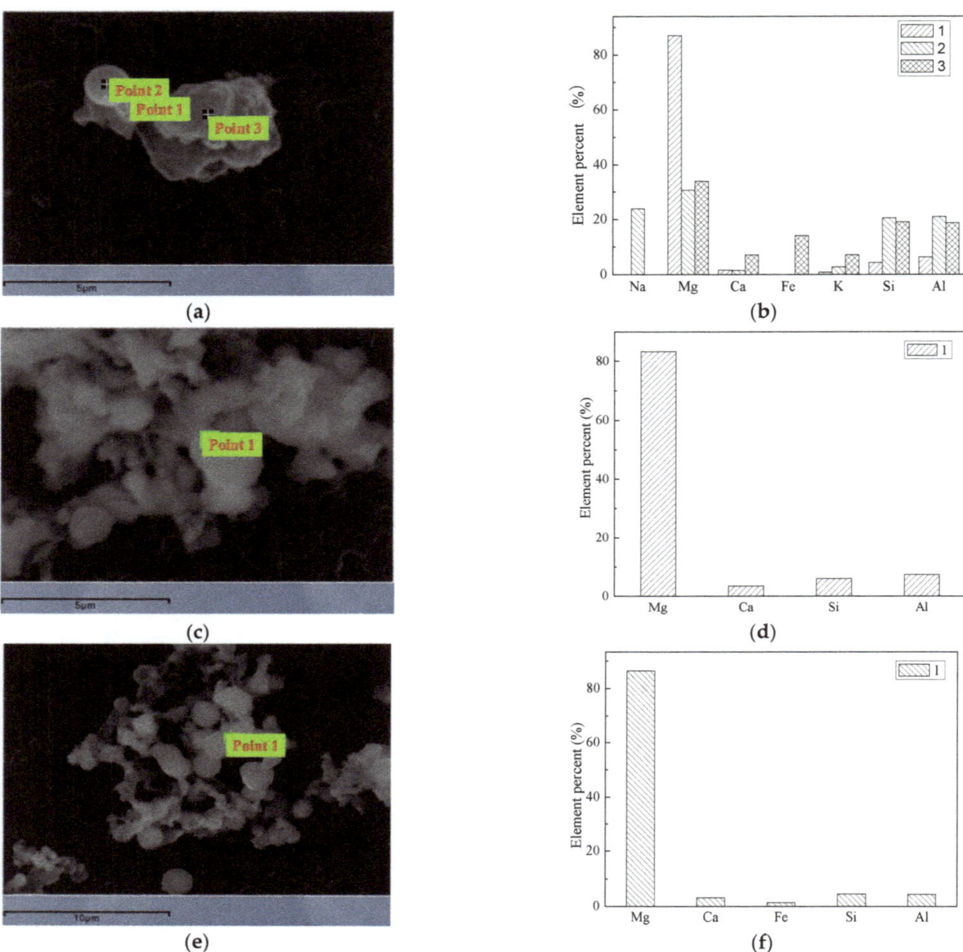

Figure 12. Morphology and Spectra of Particles after adding Mg. (**a**) Spectral Scanning Position. (**b**) Elements of Scan position. (**c**) Spectral Scanning Position. (**d**) Elements of Scan position. (**e**) Spectral Scanning Position. (**f**) Elements of Scan position.

Mg-based additives reduce the amount of PM_1 and $PM_{2.5}$, this is because the compositions of molten liquid during the combustion process of pulverized coal changes [30,39], which is an important condition to control the accumulation of mineral particles under high-temperature conditions. The addition of a soluble magnesium compound makes the magnesium element penetrate the coal particles and distribute evenly in the coal, and decompose into MgO forming submicron particles during combustion [40,41]. Submicron particles contained Al, Si in the coal blend and agglomerate with MgO to form large particles. Magnesium and refractory elements form co-materials like Mg-Ca-Fe-Al-Si, Mg-Ca-Al-Si, Mg-Al-Si and magnesium oxide. Its main mechanism is as follows:

$$MgO + SiO_2 \rightarrow Mg_2SiO_4 \tag{1}$$

$$MgO + 3Al_2O_3 \cdot 2SiO_2 \rightarrow MgO \cdot Al_2O_3 \cdot 2SiO_2 \tag{2}$$

$$MgO\cdot Al_2O_3\cdot 2SiO_2 + MgO \rightarrow Mg_3Al_2(SiO_4)_3 + Mg_2\,Al_4Si_5O_{18} \quad (3)$$

$$MgO + Ca_2SiO_4 \rightarrow CaMgSiO_4 \quad (4)$$

Yoshihiko [30] conducted a FactSage simulation of fly ash with different Mg contents. The results show that the adding magnesium significantly increases the amount of molten liquid in the fly ash. The combustion of coal with magnesium-containing additives compared to the direct combustion of coal shows a great increase in the accumulation of fine particles and a considerable part of the submicron particles attached to the coarse-mode particles surface. Magnesium-based additives promote ash accumulation, which shows a good potential to reduce particulate emissions during coal combustion [42], especially through the formation of the fused liquid form to promote the accumulation of fine-grain dust, thereby reducing the amount of PM_1 and $PM_{2.5}$.

3.2. Effect of Calcium on the Formation of Fine Particles

3.2.1. Effect of Calcium on the Particle Morphology and Particle Size Distribution

The morphology of the particulate matter produced by the combustion of raw coal-carrying Ca is shown in Figure 13. The SEM results show that the spherical particles formed by coal combustion indicate that they are transformed from droplets in the combustion process. Semi-molten irregular particles show that they are transformed from solid or incomplete liquid particles. Compared with the raw coal combustion, coals with Ca-based additives are burned to form accumulation state; a part of the submicron particles are accumulated to form larger particles.

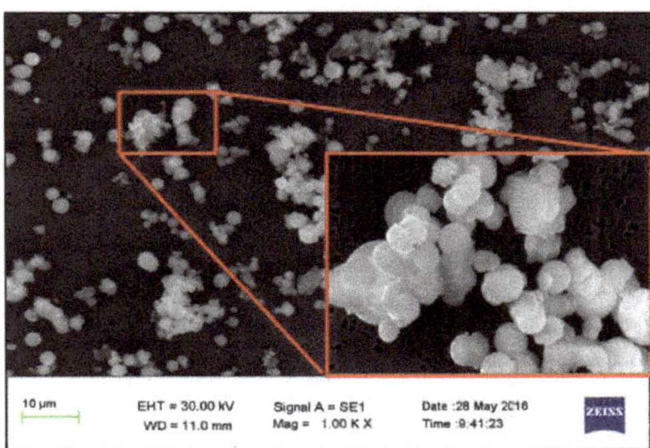

Figure 13. Morphology of particles after carrying calcium.

Figures 14 and 15 are the particulate matter mass size distributions for the particles generated by raw coal and raw coal with Ca-based addition [43]. From the figures, we can see the particle size changes after adding the Ca element. Raw minerals (including particle size and chemical composition) to some extent affect the effect of adding Ca minerals by comparing the effects of Ca minerals addition on the production of $PM_{0.2}$, $PM_{0.2\text{–}1}$ and $PM_{1\text{–}2.5}$.The production of $PM_{0.2}$, $PM_{0.2\text{–}1}$ and $PM_{1\text{–}2.5}$ were reduced from 0.32 mg/g, 0.31 mg/g, 0.45 mg/g to 0.20 mg/g, 0.21 mg/g, 0.26 mg/g, respectively. After adding Ca, the amount of $PM_{0.2}$, $PM_{0.2\text{–}1}$ and $PM_{1\text{–}2.5}$ decreased by 36.3%, 33.0% and 42.8%, and the total amount of $PM_{2.5}$ was reduced by 38%. While the $PM_{2.5\text{–}10}$ production increased from 0.17 mg/g to 0.35 mg/g. The production of PM_{10} generated by raw coal was 1.24 mg/g and 1.02 mg/g after the raw coal-carrying Ca, a decrease of 17.7%. The effect of adding Ca compounds on the particles is better than that of Mg.

3.2.2. Effect of Calcium on the Chemical Composition of Particles

The addition of Ca-containing compounds can significantly affect the conversion of minerals during pulverized coal combustion. The main element compositions of the particulate generated by pulverized coal combustion are shown in Figures 16–21. The composition analysis of particles with different particles show that the main reason for the decrease in PM_1 is the transfer of Ca, Mg, Al, Si and other elements from submicron particles to ultra-micrometer particles. The following figures show several major elements composition changes of the particles generated by original coal and original coal-carrying Ca. It can be seen that the total amount of Na and Mg elements on the submicron particles is reduced due to the decrease in the amount of submicron particulate formation. The changes of distribution on Al and Si in different particles indicate that the Al and Si in the submicron particles were transferred to the ultra-micron particles after carrying the Ca [44]. Especially the content of particulate matter around10 μm increased significantly.

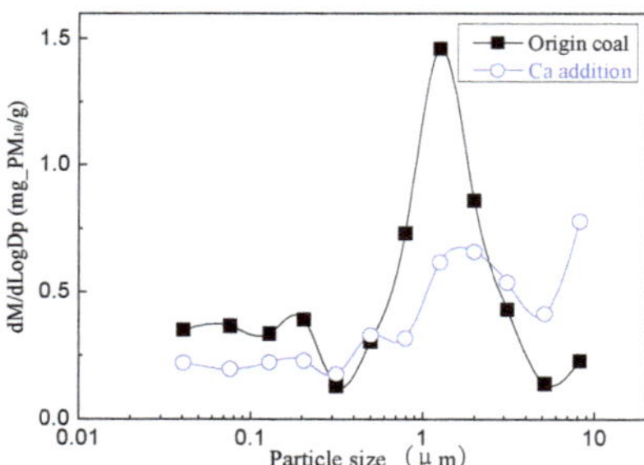

Figure 14. Particle size distribution (Ca addition).

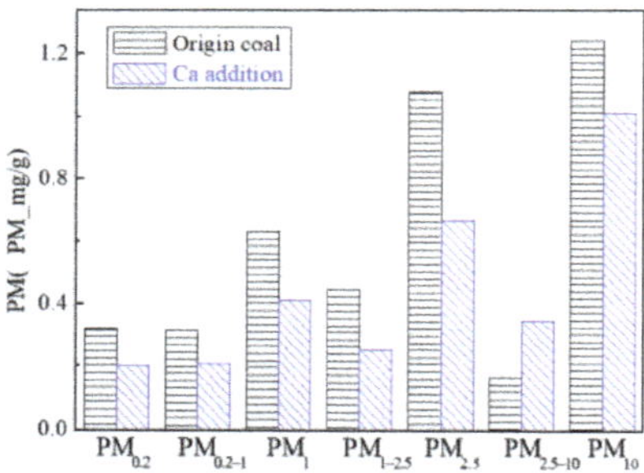

Figure 15. Particulate matter production (Ca addition).

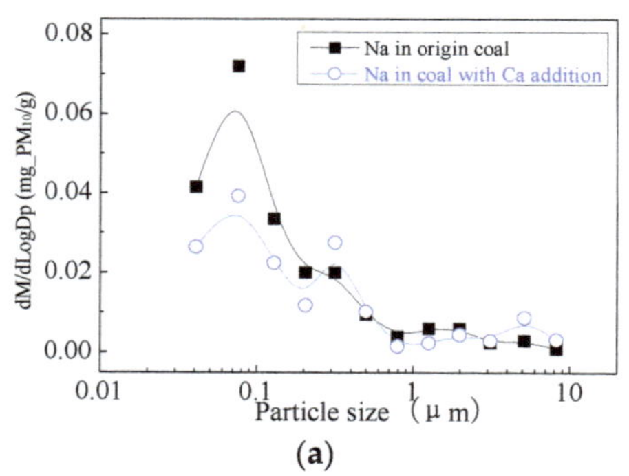

(a)

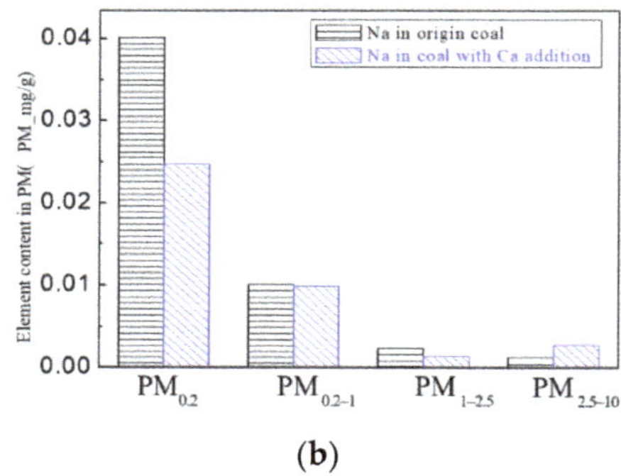

(b)

Figure 16. Mass particle size distribution of Na (Ca addition). (**a**) The curves of mass particle size distribution. (**b**) Sectional size distribution.

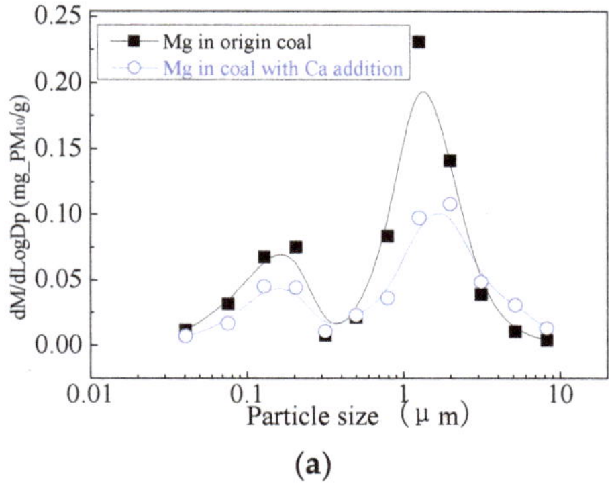

(a)

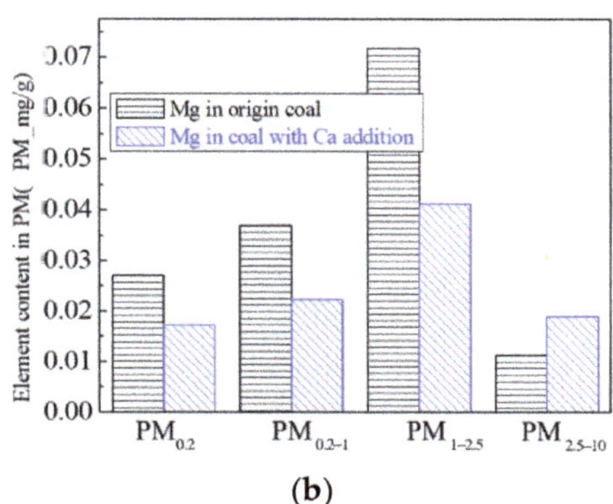

(b)

Figure 17. Mass particle size distribution of Mg (Ca addition). (**a**) The curves of mass particle size distribution. (**b**) Sectional size distribution.

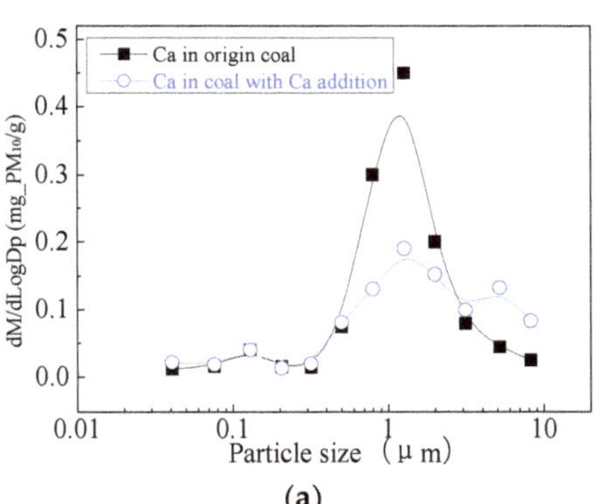

(a)

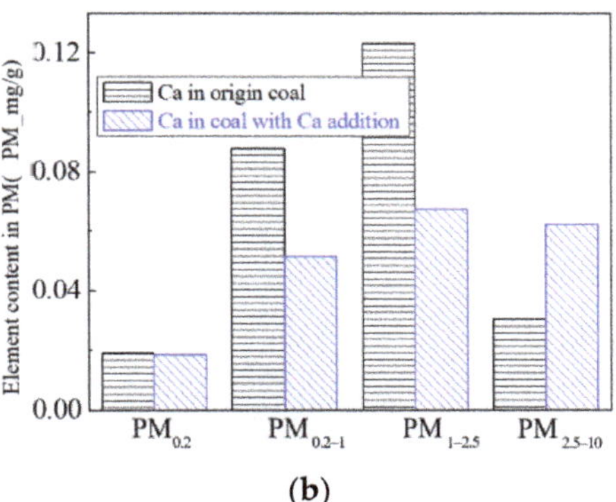

(b)

Figure 18. Mass particle size distribution of Ca (Ca addition). (**a**) The curves of mass particle size distribution. (**b**) Sectional size distribution.

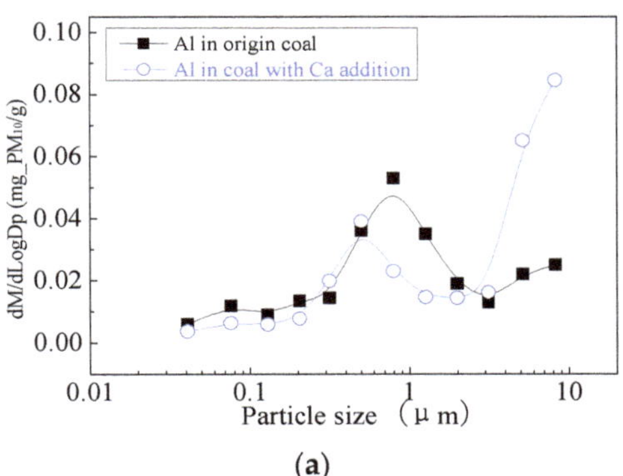

(a)

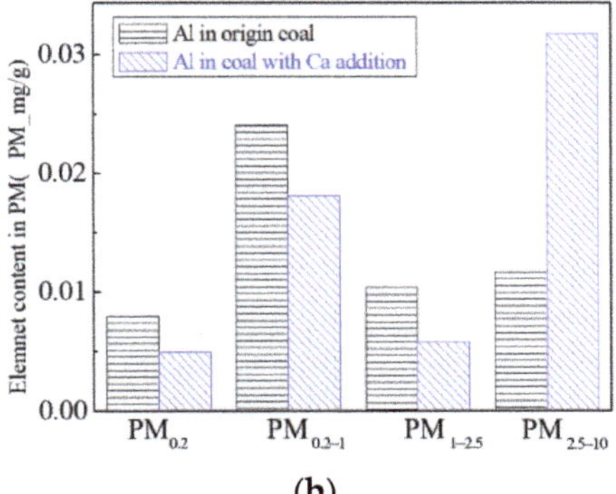

(b)

Figure 19. Mass particle size distribution of Al (Ca addition). (**a**) The curves of mass particle size distribution. (**b**) Sectional size distribution.

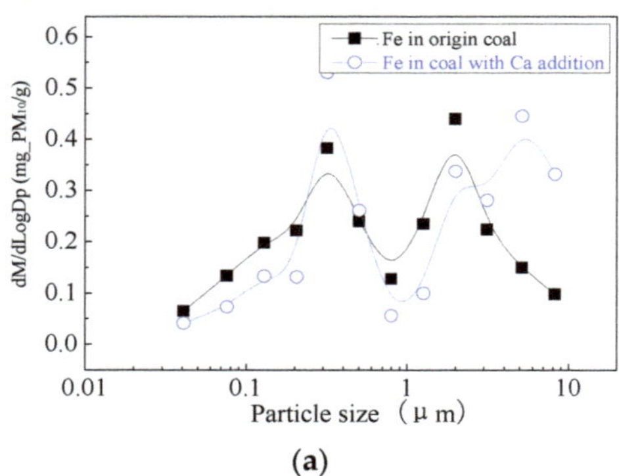

(a)

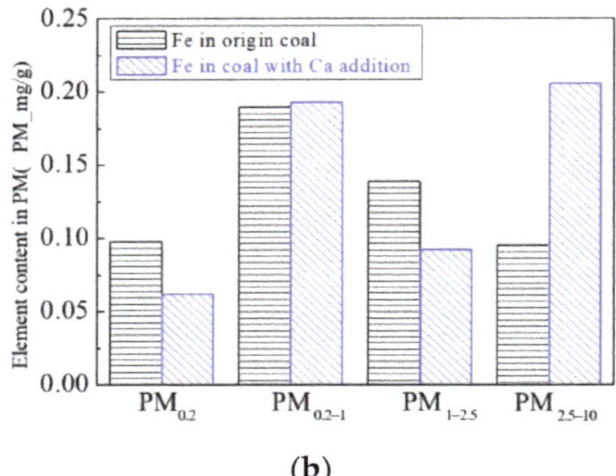

(b)

Figure 20. Mass particle size distribution of Fe (Ca addition). (**a**) The curves of mass particle size distribution. (**b**) Sectional size distribution.

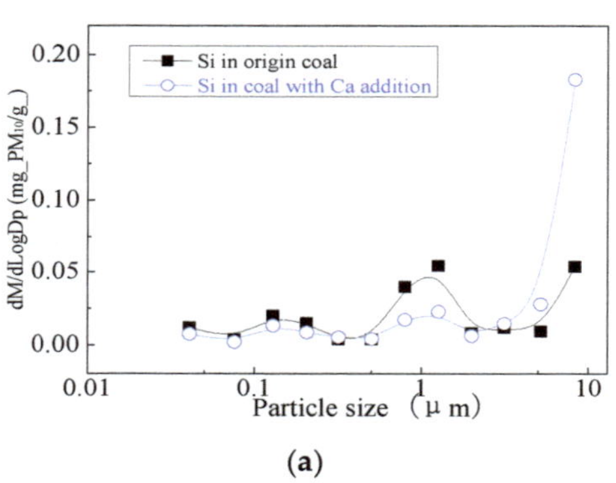

(a)

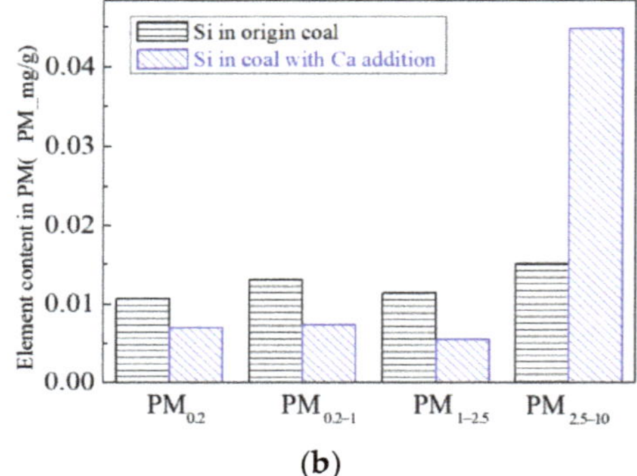

(b)

Figure 21. Mass particle size distribution of Si (Ca addition). (**a**) The curves of mass particle size distribution. (**b**) Sectional size distribution.

The morphology and the energy spectrum of the large particles are shown in Figure 22. The SEM results show that the shape of semi-molten irregular particle shows that they are the transition from the solid or incomplete liquid particles. The addition of Ca contributes to the formation of low-temperature eutectic materials, which can bind submicron particles or fine particles together to form coarse-mode particles. The mass percentage of the liquid material is the primary cause for the formation of suitable viscosities, and the submicron particles are bound to the surface of the coarse mode particles, which can capture submicron particles during the combustion process. The addition of the Ca-containing compound acts as a binder and coalesces the generated submicron particles together to form ultra-micron particles. The main points of the spectrum scanning are located near the bonding sites of different particles, and the morphology is semi-melting irregular. From the three points of the spectrum scanning, it can be seen that the Ca content is the highest at point 1 where the Mg content is 76%, the content of Mg and Ca are 38% and 30% at point 2, respectively. At point 3, the content of Ca can reach 70%. Mg-Ca-Fe-Al-Si, Mg-Ca-Al-Si and Ca-Al-Si are the main constituents in the Ca-containing particles. These low-temperature eutectics bond small submicron particles together to form ultra-micron particles. Studies have found that coal with more calcium and iron generated less PM_1 during combustion. The main reason is that the intrinsic minerals such as calcite ($CaCO_3$) and pyrite (FeS) in the coal in the initial stage of pulverized coal combustion broke down into small particles. The aluminosilicates and fine particles released from the char combustion phase form a low-melting compound by collision bonding and condensed into aluminosilicates containing calcium and iron [45]. In the early stages of combustion, these minerals decompose to CaO and Fe_2O_3, both of which react with the aluminosilicic acid to form calcium aluminosilicate and iron aluminosilicate or complex compounds containing both [46,47], which reduces the melting point of the aluminosilicate.

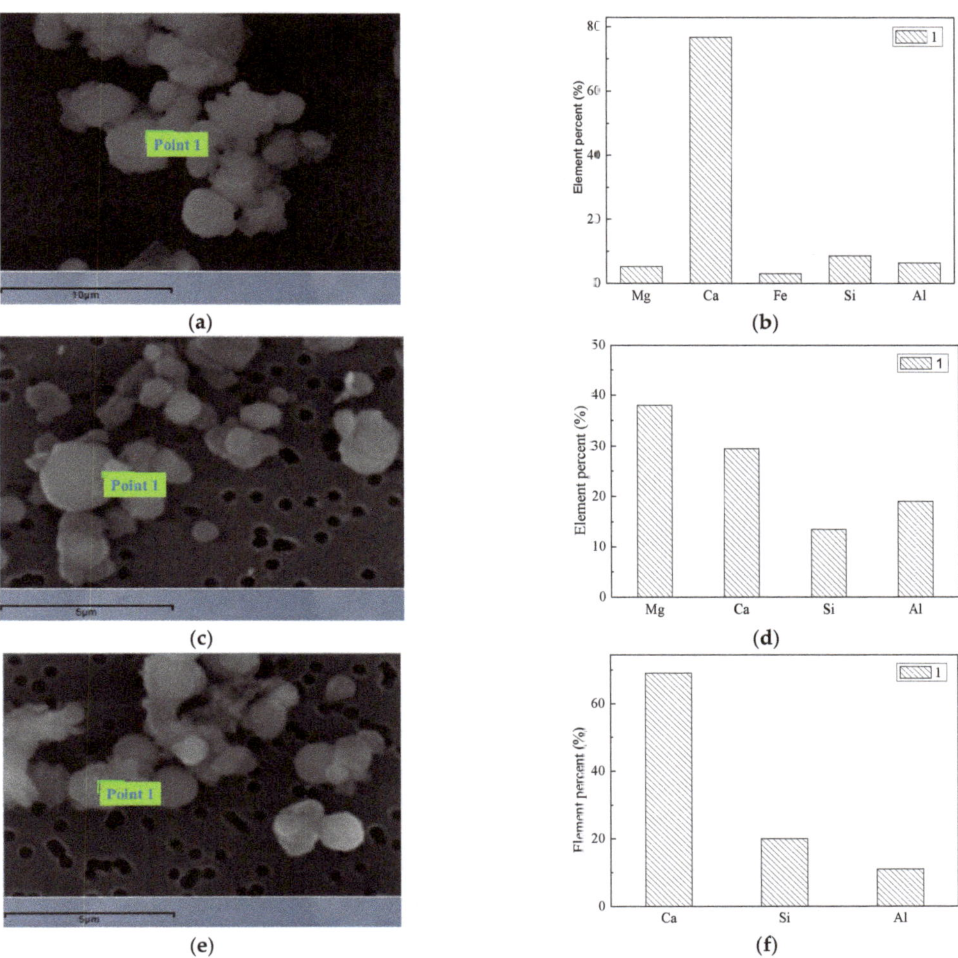

Figure 22. Morphology and spectra of particles after adding Ca. (**a**) Spectral scanning position. (**b**) Elements of scan position. (**c**) Spectral scanning position. (**d**) Elements of scan position. (**e**) Spectral scanning position. (**f**) Elements of scan position.

CaO is an alkaline-oxide and easy to react with SiO_2 to form a kind of silicate with a low melting point [48]. In addition, since the CaO monomer has a high melting point of 2590 °C, CaO cannot reduce the ash melting point when the CaO content is increased to a certain extent (40% to 50% or more), but will increase the melting point of ash instead. The results show that the mass ratio of SiO_2/Al_2O_3 in coal ash is above 3.0 for most coal species in China, and the melting point of ash is the lowest when the mass fraction of CaO in coal ash is 20~25%, the mass fraction of SiO_2/Al_2O_3 in coal ash is below 3.0, the mass fraction of CaO in coal ash is 30%~35%, the ash has the lowest melting point. When the mass fraction of CaO exceeds 30%~35%, CaO not only can reduce the melting point of ash, but play the role of increasing the ash melting point [49,50].

After carrying calcium acetate, calcium acetate changes in the coal combustion process [51]; minerals containing calcium may react with 10 μm or less Al-Si compound to form a molten Ca-Al-Si, then unite larger than 10 μm aluminosilicate salt. The internal mineral releases Al-Si, which carries minerals that release Ca, these elements collide under gaseous conditions to form widely dispersed melting droplets which adhere to large Al-Si surfaces or to each other to form ultra- particles or particles larger than 10 μm. Step 1. Adding minerals (calcium acetate) in the early stage of coal heat, when the temperature is 160 °C it began to decompose to $CaCO_3$, and $CaCO_3$ will decompose at about 825 °C to generate CaO. Step 2. Burned coke releases Al-Si (eg, mullite, quartz). Step 3. CaO reacts with the released Al-Si to form Ca-Al-Si, which either condenses into fine particles or is converted to PM_{1+}. Step 4. Most of the molten Ca-Al-Si aggregates together or agglomerate on the surfaces of unreacted Al-Si, CaO, or enhance their cohesive ability (unreacted Al-Si, CaO) to agglomerate into ultra-micron particles or particles bigger than 10 μm. The main chemical reaction is as following:

$$(CH_3COO)_2Ca \rightarrow CH_3COCH_3 + CaCO_3 \quad (5)$$

$$CaCO_3 \rightarrow CaO + CO_2 \quad (6)$$

$$3Al_2O_3 \cdot 2SiO_2 + CaO \rightarrow CaO \cdot Al_2O_3 \cdot 2SiO_2 \quad (7)$$

$$CaO \cdot Al_2O_3 \cdot 2SiO_2 + CaO \rightarrow 2CaO \cdot Al_2O_3 \cdot 2SiO_2 \quad (8)$$

$$CaO \cdot SiO_2 + SiO_2 \rightarrow 3CaO \cdot 2SiO_2 \quad (9)$$

$$SiO_2 + CaO \rightarrow CaO \cdot SiO_2 \quad (10)$$

By adding calcium in the coal minerals, the content of the calcium improves the proportion of liquid components, improving the surface adhesion of mineral elements Al, Si, Fe and others on the big particles during the combustion process, which makes submicron particles transform to PM_{1+}, and aggregated to form large particles of Ca-Fe-Al-Si and Ca-Al-Si components by collision. By changing the composition of ash, the ash melting point can be reduced, which makes $PM_{2.5}$ accumulate into bigger particles, increasing the production of $PM_{2.5+}$ and reducing $PM_{2.5}$ emissions [52].

3.3. *Effect of Iron on the Formation of Fine Particles*

3.3.1. Effect of Iron on the Particle Morphology and Particle Size Distribution

Figures 2 and 23 show the morphology of particles by the combustion of raw coal and raw coal-carrying Fe. Compared with the raw coal combustion, the particles produced by combustion of Fe-contained compounds showed more accumulation, and a part of the submicron particles adhered to the surface of the coarse-mode particles. The SEM results show that the spherical particles indicate that they are formed by droplets in the combustion process; semi-molten irregular particles show that they are from the transition of solid or incomplete liquid particles.

Figure 24 shows the particle size distribution of the particles by combustion of raw coal and raw coal-carrying Fe. It can be seen that, after adding Fe element, the mass particle size of the generated particle changes. Raw minerals (including particle size and chemical composition) affect the effect of adding Fe minerals to some extent [53]. As shown in Figure 25, by comparing the effects of raw coal and raw coal adding Fe on the formation of $PM_{0.2}$, $PM_{0.2–1}$ and $PM_{1–2.5}$; the production of $PM_{0.2}$, $PM_{0.2–1}$ and $PM_{1–2.5}$ were reduced from 0.32 mg/g, 0.31 mg/g, 0.45 mg/g to 0.27 mg/g, 0.26 mg/g and 0.31 mg/g. After adding Fe, the production of $PM_{0.2}$, $PM_{0.2–1}$ and $PM_{1–2.5}$ decreased by 15.6%, 16.2% and 31.1%, respectively, and the production of $PM_{2.5}$ was reduced by 22.4%. While the production of $PM_{2.5–10}$ increased from 0.17 mg/g to 0.34 mg/g, the production of PM_{10} was reduced from 1.24 m/g to 1.18 mg/g. It is clear that a significant fraction of the submicron particles produced during coal combustion are converted into ultra-micron particles after the addition of Fe minerals.

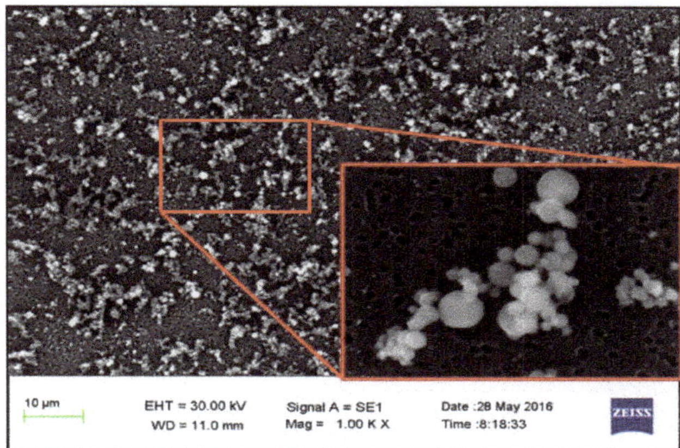

Figure 23. Morphology of particles after carrying Fe.

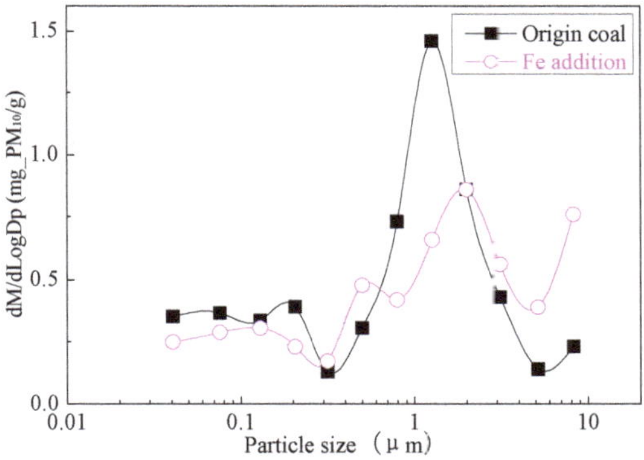

Figure 24. Particle size distribution (Fe addition).

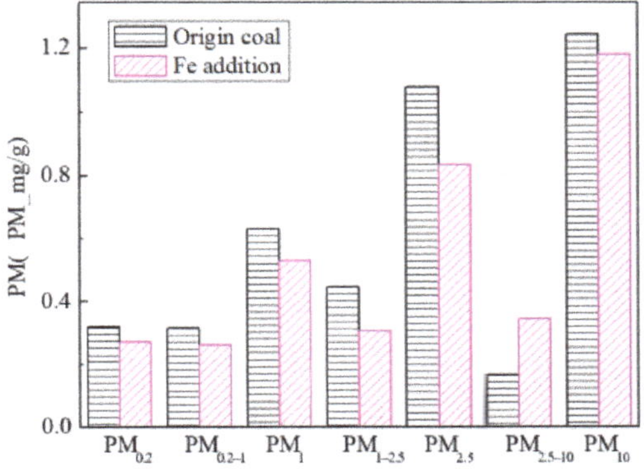

Figure 25. Particulate matter production (Fe addition).

3.3.2. Effect of Iron on the Chemical Composition of Particles

The addition of Fe-containing compounds can significantly affect the conversion of minerals during pulverized coal combustion. The main elemental compositions of pulverized coal combustion after carrying Fe are shown in Figures 26–31. The results show that the main reason for the decrease in PM1 is the transfer of Ca, Mg, Al and Si from submicron particles to ultra-micrometer particles. The following figure is the changes of several major elements composition in the particulate matter generated by the original coal and the original coal carrying Ca element. It can be seen that the total amount of Na and Mg elements on the submicron particles is reduced due to the decrease in the amount of submicron particulate formation. Distribution changes of Al and Si in different particles indicate that the Al and Si in the submicron particles are transferred to the ultra-micron particles after the addition of Fe; especially the particles in the vicinity of 10 μm have remarkably increased. The composition of particle generated by raw coal-carrying Fe is similar to that carrying Ca and Mg.

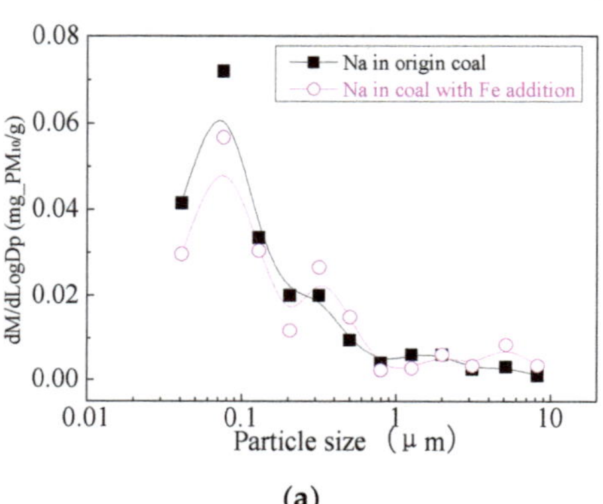

(a)

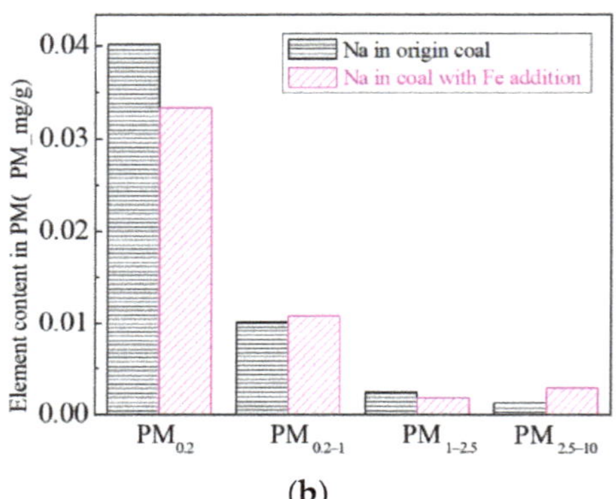

(b)

Figure 26. Mass particle size distribution of Na (Fe addition). (**a**) The curves of mass particle size distribution. (**b**) Sectional size distribution.

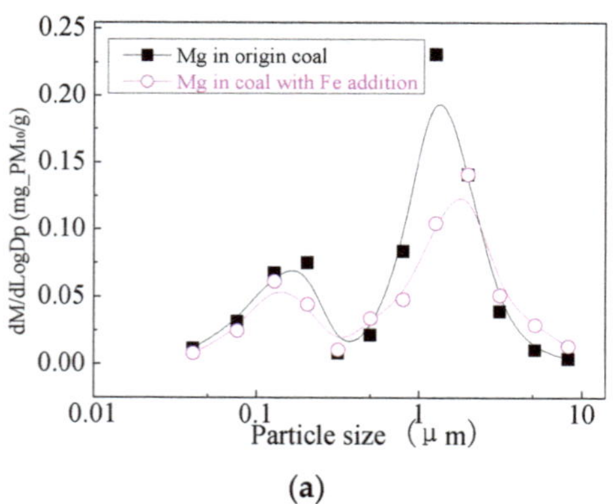

(a)

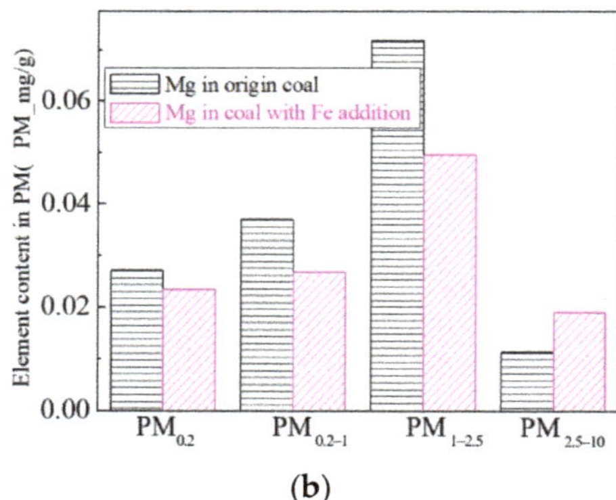

(b)

Figure 27. Mass particle size distribution of Mg (Fe addition). (**a**) The curves of mass particle size distribution. (**b**) Sectional size distribution.

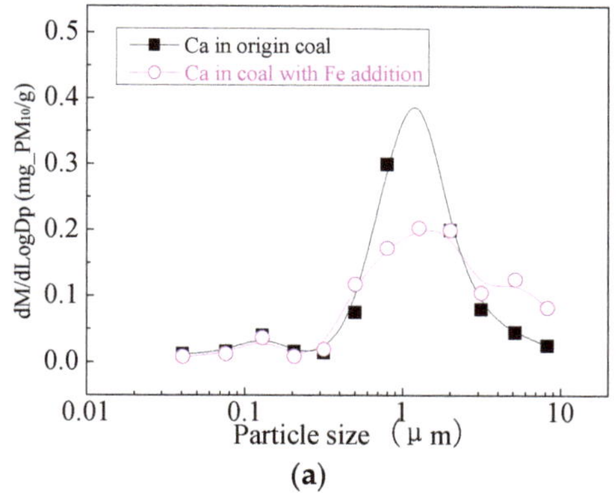

(a)

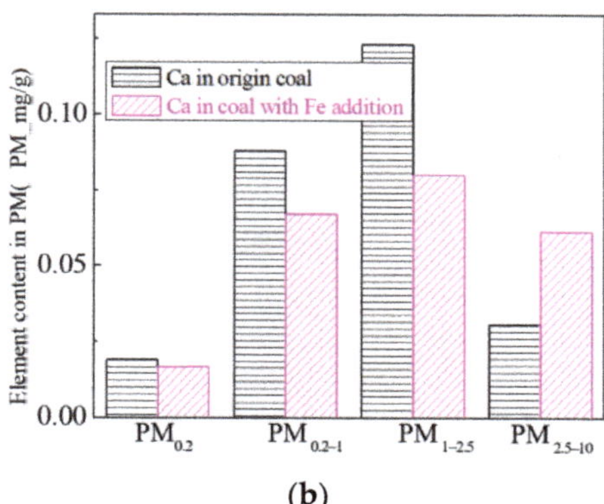

(b)

Figure 28. Mass particle size distribution of Ca (Fe addition). (**a**) The curves of mass particle size distribution. (**b**) Sectional size distribution.

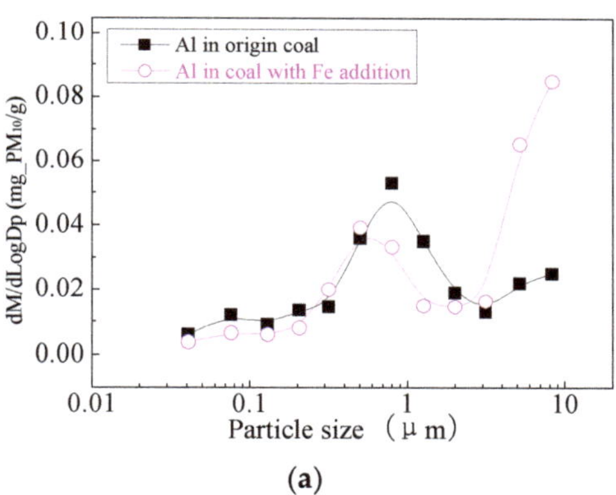

(a)

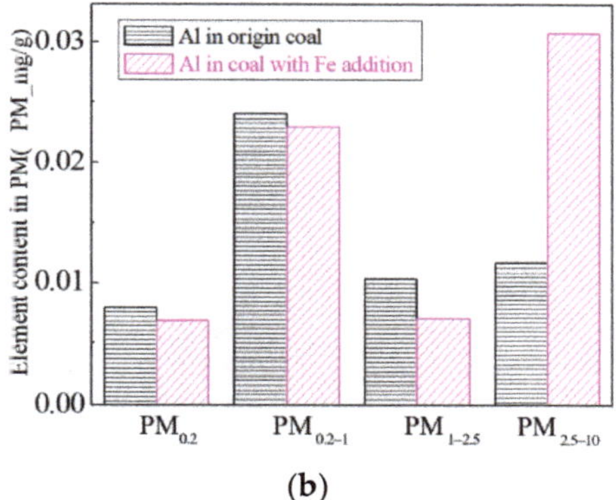

(b)

Figure 29. Mass particle size distribution of Al (Fe addition). (**a**) The curves of mass particle size distribution. (**b**) Sectional size distribution.

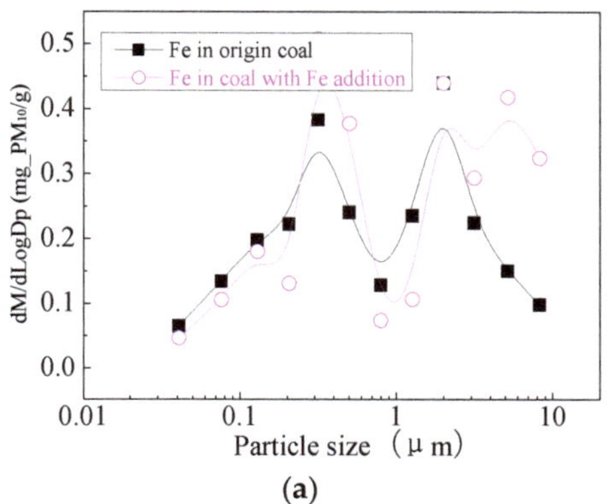

(a)

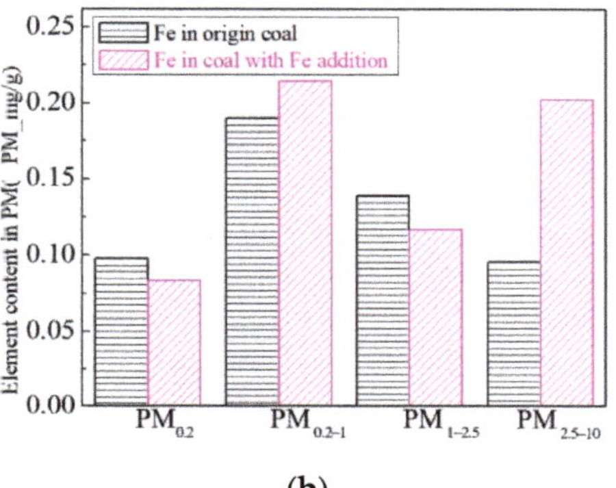

(b)

Figure 30. Mass particle size distribution of Fe (Fe addition). (**a**) The curves of mass particle size distribution. (**b**) Sectional size distribution.

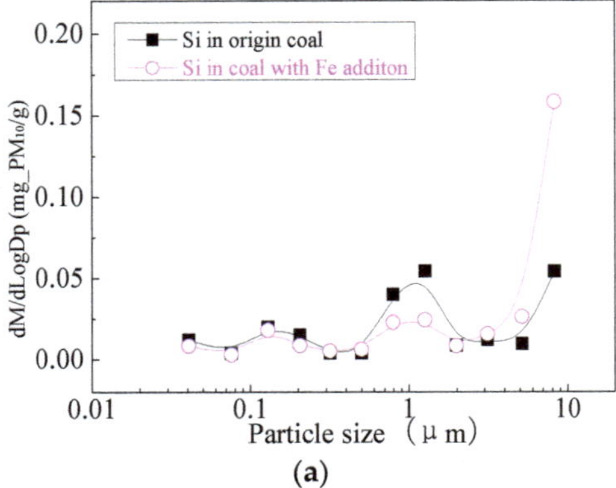

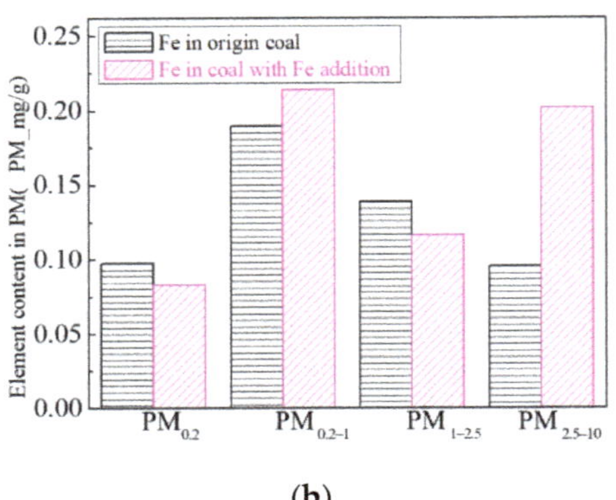

Figure 31. Mass particle size distribution of Si (Fe addition). (**a**) The curves of mass particle size distribution. (**b**) Sectional size distribution.

The main reason for the decrease in PM_1 generated by iron-added coal is that in the initial stage of pulverized coal combustion, the intrinsic minerals in coal are decomposed into fine Fe_2O_3 particles, the aluminosilicates released from the char combustion stage, Fe_2O_3 fine particles and aluminosilicates undergo a series of chemical reactions to form iron-containing minerals which can react with Al-Si compounds below 10 μm to form molten Fe-Al-Si, which aggregates into more than 10 μm aluminosilicates salt. The main processes are as follows: Step one. The carrying compound decomposes to produce Fe_2O_3; Step two. The burning coke releases fine particles such as Al-Si; Step three. Fe_2O_3 reacts with the released Al-Si to form Fe-Al-Si which either condenses into fine particles or is converted into PM_{1+}; Step four. Most of the molten Fe-Al-Si aggregates together or agglomerates on the surface of unreacted Al-Si, Fe_2O_3, or improves their cohesive ability (unreacted Al-Si, Fe_2O_3) to agglomerate to the particle more than 10 μm. By changing the content of the ash and reducing the melting point of the ash, $PM_{2.5}$ accumulated into bigger particles, increasing the production of $PM_{2.5+}$, which finally reduces the emission of $PM_{2.5}$.

Figure 32 is a diagram showing the morphology and the energy spectrum of particles formed after carrying Fe. At the two points of the first spectrum, the content of Fe is 11% and 22%, respectively. The content of Fe in the second spectrum is 27% and 47%, respectively. The main components compositions of Fe-containing particles are Fe-Al-Si, Fe-Ca-Al-Si and so on. The flux effect of Fe_2O_3 is largely affected by the atmosphere because Fe has different valence states in different atmospheres which have different effects on the silicate. The main reason is that Fe^{2+} is connected with an oxygen atom through octahedral structure, and Fe^{3+} is connected with oxygen through the tetrahedral structure. As a result of the charge balance, the tetrahedral structure of Fe^{3+} is easy to connect with four-sided structure of the SiO_2, which makes the Si-O bond brake, and part of the three-dimensional network structure decompose to form a new three-dimensional network structure, and the viscosity of the silicate at high temperatures decreased on the macro performance. The octahedral structure of Fe^{2+} is more likely to make the original three-dimensional network structure loose and disintegrate which leads to a significant reduction in viscosity. Due to the role of charge balance, Fe ions can participate in the formation of complex chelates. Fe^{3+} cations and O-containing anions (FeO^{2-} and $Fe_2O_5^{2-}$) can form complex $Fe_xO_y^{2-}$ anions, and this complex structure can increase the viscosity of coal ash.

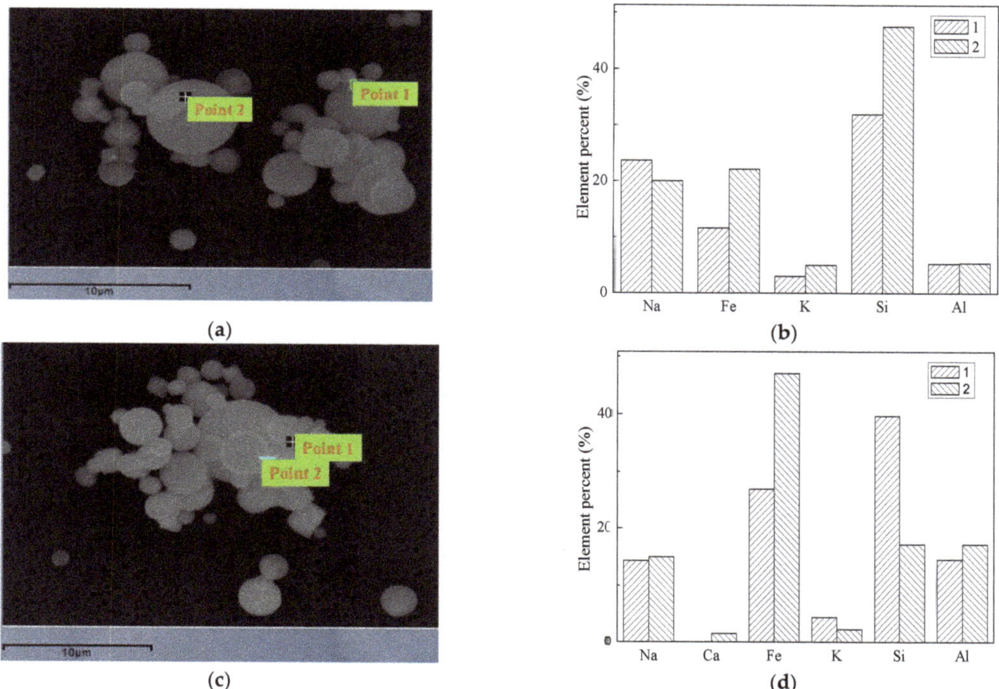

Figure 32. Morphology and spectra of particles after adding Fe. (**a**) The curves of mass particle size distribution. (**b**) Sectional size distribution. (**c**) Spectral scanning position. (**d**) Elements of scan position.

Fe_2O_3 in weak reduction or oxidation atmosphere systems can play a role in reducing ash melting point. Especially in the weak reducing atmosphere, Fe_2O_3 (1560 °C) can exist in the form of FeO (1420 °C), compared with Fe of other valence, it is easier to form low-temperature eutectic compounds with SiO_2 and other substances, which have the strongest flux effect. In the reducing atmosphere, when the temperature is 900 °C, hematite (Fe_2O_3) and siderite ($FeCO_3$) can decompose to FeO when they are heated. When the temperature is higher, 1000~1200 °C, FeO and other minerals began to react, such as quartz (SiO_2), mullite ($3Al_2O_3{\cdot}2SiO_2$, when the temperature is 1810 °C it decomposes to Al_2O_3, SiO_2 of liquid phase) and anorthite (The melting point of $CaO{\cdot}Al_2O_3{\cdot}2SiO_2$ is 1553 °C). After the reaction, iron olivine ($2FeO{\cdot}SiO_2$, The melting point is 1065 °C), iron spinel ($FeO{\cdot}Al_2O_3$) and so on with a lower melting point is produced. It is found that anorthite ($CaO{\cdot}Al_2O_3{\cdot}2SiO_2$) began to partially decompose at 1000~1100 °C and react with FeO to produce $2FeO{\cdot}SiO_2$ (melting point 1065 °C) with a low melting point. Studies have shown that iron-based minerals, especially those containing Al, Si, Fe minerals generally have a relatively low melting point. The flux effect of FeO is close to that of CaO. When the content of FeO exceeds 10%, the fluxing effect is better than that of CaO. Zhang [54] used frontier orbital theory from the view of quantum chemistry to show that the melting point of iron olivine ($2FeO{\cdot}SiO_2$) is 1065 °C lower than that of mullite ($3Al_2O_3{\cdot}2SiO_2$) and kaolinite ($Al_2O_3{\cdot}2SiO_2{\cdot}2H_2O$).

After the addition of the Fe compound, the following main reaction can reduce the formation of fine particles:

$$Fe(NO_3)_3 \rightarrow Fe_2O_3 + 12NO_2 + 3O_2 \quad (11)$$

$$Fe_2O_3 \rightarrow FeO + O_2 \quad (12)$$

$$SiO_2 + FeO \rightarrow FeO\ SiO_2 \quad (13)$$

$$FeO + Al_2O_3 \rightarrow Fe_2Al_2O_4 \quad (14)$$

$$FeO{\cdot}SiO_2 + FeO \rightarrow 2FeO{\cdot}SiO_2 \quad (15)$$

$$3Al_2O_3{\cdot}2SiO_2 + FeO \rightarrow 2FeO{\cdot}SiO_2 + FeO{\cdot}Al_2O_3 \quad (16)$$

$$CaO{\cdot}Al_2O_3{\cdot}2SiO_2 + FeO \rightarrow 2FeO{\cdot}SiO_2 + FeO{\cdot}Al_2O_3 + 3FeO{\cdot}Al_2O_3{\cdot}3SiO_2 \quad (17)$$

From these chemical reactions we can see that after the addition of Fe, the generated sub-micron particles aggregated and bonded to form big particles, thereby reducing the amount of submicron particulate matter.

3.4. Comparison of the Effect on Adding Several Inorganic Compounds

The addition of three minerals had some similarity to the control mechanism of the formation of submicron particles. Aiming to the control of PM_1, the amount of produced PM_1 was 0.63 mg/g coal. After adding Mg, Ca and Fe, the amount of PM_1 was 0.51 mg/g coal, 0.41 mg/g coal and 0.53 mg/g coal, respectively. The amount of PM_1 was reduced by 19%, 35% and 16%, respectively, when the raw coal was loaded with Mg, Ca and Fe. The effect of the three mineral elements in the control of PM_1 is Ca > Mg > Fe. Aiming to the control of $PM_{2.5}$, the amount of PM_1 was reduced by 19%, 35% and 16%, respectively, when the raw coal was loaded with Mg, Ca and Fe. The effect of the three mineral elements in the control of $PM_{2.5}$ is Ca > Fe > Mg, this is mainly due to Fe in the control of $PM_{1-2.5}$ is slightly better than the effect of Mg. After the addition of Mg, the amount of PM_{10} was almost unchanged, and the amount of PM_{10} decreased 17.7% and 5% after the addition of Ca and Fe, respectively, compared with the raw coal. These three kinds of minerals can make coal-generated PM_1 to some extent in the combustion process by aggregating and bonding ways to form bigger particles.

The amount of PM_1 generated by coal adding the mineral element is reduced, the main reason is that in the initial stage of pulverized coal combustion, the intrinsic minerals (carrying compounds containing Mg, Ca, Fe) in coal are decomposed into fine MgO, CaO and Fe_2O_3 particles. The aluminosilicates, fine MgO, CaO and Fe_2O_3 particles released from the coal combustion stage underwent a series of chemical reactions to form the corresponding aluminosilicates or complex materials containing the above three compounds, collide and bond to form a low melting point material, aggregate into aluminosilicate containing Mg, Ca and Fe.

The added minerals can react with the fine particles to form molten ash particles which can agglomerate into bigger aluminosilicates. The whole process is as follows: Step one. After the loading of mineral elements, the coal decompose combustion to release fine particles of MgO, CaO and Fe_2O_3 during combustion process; these substances collide under gaseous conditions to form the widely distribution of molten droplets. Step two. The burning coke releases fine particles containing Al-Si and so on. Step three. Mg-Ca-Al-Si, Ca-Al-Si and Fe-Al-Si in the molten state are formed by the reaction of MgO, CaO and Fe_2O_3 with the released fine particles of Al-Si, which either agglomerate into fine particles or convert to PM_{1+}.Step four. Most of the molten Mg-Ca-Al-Si, Ca-Al-Si and Fe-Al-Si aggregate together or aggregate on the surface of the unreacted Al-Si and oxide fine particles or improve their cohesive ability to make them condense into bigger particles. Therefore, by adding minerals to change the ash composition, reducing the ash melting point makes the sub-micron particles accumulate into larger particles, increasing the amount of ultra-microns and effectively reducing sub-micron particulate matter emissions.

These three mineral elements, Mg, Ca and Fe, can reduce the formation of submicron particles, mainly due to the decrease in the ash melting point [55] and contribute to the formation of low-temperature eutectic material which can bond the sub-micron particles or fine particles together to form coarse mode particulates. Scholars usually describe the composition of ash as 11 kinds of oxides: SiO_2, Al_2O_3, Fe_2O_3, CaO, MgO, TiO_2, Na_2O, K_2O, SO_3, MnO_2 and P_2O_5. In the study of the effect of ash composition on the melting point, the former several oxides are generally considered. These oxides are generally divided into two categories, one can increase the ash melting point of the acidic oxides, including SiO_2, Al_2O_3, TiO_2, etc., whereas the other can reduce the ash melting point of alkaline oxides, including Fe_2O_3, CaO, MgO, Na_2O, K_2O and so on.

It was proposed that the behavior of acidic and alkaline components is mainly affected by the chemical structure of ions [56]. The concept of "ion potential" is proposed. The ion potential is the ratio of ion valence to ionic radius. The ion potential of Mg^{2+}, Fe^{2+}, Ca^{2+}, Na^+ and K^+ are 3.0, 2.7, 2.0, 1.1 and 0.75, respectively. The ion potentials of Si^{4+}, Al^{3+}, $Ti^{4+}$and Fe^{3+} are 9.5, 5.9, 5.9 and 4.7, respectively. The ionic potential of the acidic oxide is high and the ionic potential of the basic oxide is low. Cations with high ionic potential are easily bound to O and form complexes or complex ions, and acidic components form a polymer in the ash. Alkaline components can prevent the formation of polymer, which play a role in reducing ash melting point. Ion potential of Na^+, K^+, Ca^{2+}, Mg^{2+} are the lowest, which can destroy the formation of polymer, which can help the role of melting. Some studies have shown that the total alkali content and ash fusion temperature have a good correlation (r = 0.84).When the content of Fe_2O_3 in coal ash is less than 20%, the average melting point of ash decreases by 18 °C for every 1% increase in Fe_2O_3 [56]. From the point of ionic potential, the ion potential of Fe^{3+} is 4.7 and the ion potential of Fe^{2+} is 2.7; Fe^{2+} is more effective than Fe^{3+} in reducing ash fusion temperature.

4. Conclusions

The production of PM_1 is mainly affected by the evaporation of mineral elements in the coal combustion process, which reduces the evaporation of the mineral elements or the generated PM_1, aggregating to form PM_{1-10} or particles bigger than 10 μm which can reduce the emission of PM_1.

(1) The amount of $PM_{0.2}$, $PM_{0.2-1}$ and $PM_{1-2.5}$ produced by the raw coal-carrying Mg are reduced by 36.7%, 17.4% and 24.6%, respectively, and the amount of $PM_{2.5}$ is reduced by 21.6%. The amount of PM_{10} is almost unchanged. Adding Mg can effectively increase the viscosity of submicron particles, making it easier to aggregate and bond together to form ultra-micron particles.

(2) The amount of $PM_{0.2}$, $PM_{0.2-1}$ and $PM_{1-2.5}$ produced by the raw coal-carrying Ca are reduced by 36.3%, 33.0% and 42.8%, respectively, and the amount of $PM_{2.5}$ is reduced by 38%. The production of PM_{10} is reduced by 17.7%. The effect of adding Ca compounds on the particles is better than that of Mg.

(3) The amount of $PM_{0.2}$, $PM_{0.2-1}$ and $PM_{1-2.5}$ produced by the raw coal-carrying Fe are reduced by 15.6%, 16.2% and 31.1%, respectively, and the amount of $PM_{2.5}$ is reduced by 22.4%. While the production of $PM_{2.5-10}$ increased from 0.17 mg/g to 0.34 mg/g, the production of PM_{10} is reduced by 5%. It is clear that a significant fraction of the submicron particles produced during the combustion of the raw coal are transformed into ultra-micron particles after the addition of Fe minerals.

(4) After comparing the particulate matter produced by raw coal after carrying Mg, Ca and Fe, it can be found that the addition of these three mineral elements can effectively reduce the ash melting point, so that part of the sub-microns generated during the process of coal combustion can be aggregated, bonded or adhered to other particles to form ultra-micron particles.

(5) In the process of coal combustion, some submicron particles form submicron particles, which are easy to be removed by dust removal equipment, which is conducive to simplifying the control process of coal-fired particles and reducing the cost of particle control. At the same time, compared with many particle removal technologies, the cost of mineral-element-modified coal is lower, the dedusting equipment is simpler and the dedusting efficiency is higher, which has a broader application prospect.

Author Contributions: Methodology, L.D.; Formal analysis, H.D. and Q.D.; Data curation, J.W.; Writing—original draft, S.W.; Writing—review & editing, Y.Z. and J.G. All authors have read and agreed to the published version of the manuscript.

Funding: This work is supported by the National Natural Science Foundation of China (52006047) and the Fundamental Research Funds for the Central Universities.

Data Availability Statement: Date openly available in a public repository.

Conflicts of Interest: The authors declare no conflict of interest.

References

1. Zhang, H.; Sun, W.; Li, W.; Wang, Y. Physical and chemical characterization of fugitive particulate matter emissions of the iron and steel industry. *Atmos. Pollut. Res.* **2022**, *13*, 101272. [CrossRef]
2. Yao, Q.; Li, S.; Xu, H.; Zhuo, J.; Song, Q. Studies on formation and control of combustion particulate matter in China: A review. *Energy* **2010**, *35*, 4480–4493. [CrossRef]
3. Yu, D.; Xu, M.; Yi, F.; Huang, J.; Li, G. A review of particle formation mechanisms during coal combustion. *Coal Convers.* **2004**, *27*, 7–12.
4. Yue, M.; Gu, X.; Zou, H.; Zhu, R.; Su, W. Killer of healthr polycyclic aromatic hydrocarbons. *J. Cap. Norm. Univ. (Nat. Sci. Ed.)* **2003**, *24*, 40–44.
5. Meij, R. Trace element behavior in coal fired power plants. *Fuel Process. Technol.* **1994**, *39*, 199–217. [CrossRef]
6. Yi, H.; Guo, X.; Hao, J.; Duan, L.; Li, X. Characteristics of inhalable particulate matter concentration and size distri bution from power plants in China. *J. Air Waste Manag. Assoc.* **2006**, *56*, 1243–1251. [CrossRef]
7. Xu, M.; Yan, R.; Zheng, C.; Qiao, Y.; Han, J.; Sheng, C. Status of trace element emission in a coal combustion process: A review. *Fuel Process. Technol.* **2004**, *85*, 215–237. [CrossRef]
8. Lu, B.; Kong, S.; Han, B.; Li, Z.; Bai, Z. Source profile of TSP and PM10 from coal-fired boilers. *J. China Coal Soc.* **2011**, *36*, 1928–1933.
9. Wu, Y.; Xu, Z.; Liu, S.; Tang, M.; Lu, S. Emission characteristics of PM2.5 and components of condensable particulate matter from coal-fired industrial plants. *Sci. Total Environ.* **2021**, *796*, 148782. [CrossRef]
10. Rönkkö, T.J.; Hirvonen, M.-R.; Happo, M.S.; Leskinen, A.; Koponen, H.; Mikkonen, S.; Bauer, S.; Ihantola, T.; Hakkarainen, H.; Miettinen, M.; et al. Air quality intervention during the Nanjing youth olympic games altered PM sources, chemical composition, and toxicological responses. *Environ. Res.* **2020**, *185*, 109360. [CrossRef]
11. Xu, M.; Yu, D.; Yao, H.; Liu, X.; Qiao, Y. Coal combustion-generated aerosols: Formation and properties. *Proc. Combust. Inst.* **2011**, *33*, 1681–1697. [CrossRef]
12. Buhre, B.; Elliott, L.; Sheng, C.; Gupta, R.; Wall, T. Oxy-fuel combustion technology for coal-fired power generation. *Prog. Energy Combust. Sci.* **2005**, *31*, 283–307. [CrossRef]
13. Buhre, B.; Hinkley, J.; Gupta, R.; Nelson, P.; Wall, T. Fine ash formation during combustion of pulverised coal–coal property impacts. *Fuel* **2006**, *85*, 185–193. [CrossRef]
14. Qiu, X.; Duan, L.; Duan, Y.; Li, B.; Lu, D.; Zhao, C. Ash deposition during pressurized oxy-fuel combustion of Zhundong coal in a lab-scale fluidized bed. *Fuel Process. Technol.* **2020**, *204*, 106411. [CrossRef]
15. Li, G. *Expermental Study on Formation, Evilution and Deposition Charctreistics of Pulverized Coal Combustion*; Tsinghua University: Beijing, China, 2014.
16. Galgani, L.; Tsapakis, M.; Pitta, P.; Tsiola, A.; Tzempelikou, E.; Kalantzi, I.; Esposito, C.; Loiselle, A.; Tsotskou, A.; Zivanovic, S.; et al. Microplastics increase the marine production of particulate forms of organic matter. *Environ. Res. Lett.* **2019**, *14*, 124085. [CrossRef]
17. Zhan, Z.; Chiodo, A.; Zhou, M.; Davis, K.; Wang, D.; Beutler, J.; Cremer, M.; Wang, Y.; Wendt, J.O.L. Modeling of the submicron particles formation and initial layer ash deposition during high temperature oxy-coal combustion. *Proc. Combust. Inst.* **2021**, *38*, 4013–4022. [CrossRef]
18. Oleschko, H.; Schimrosczyk, A.; Lippert, H.; Müller, M. Influence of coal composition on the release of Na$^-$, K$^-$, Cl$^-$, and S$^-$species during the combustion of brown coal. *Fuel* **2007**, *86*, 2275–2282. [CrossRef]
19. Thomson, L.L. Abstracts of Papers and Posters Presented at the Sixteenth Annual Education Conference of the National Society of Genetic Counselors (Baltimore, Maryand). *J. Genet. Couns.* **1997**, *6*, 433–512. [CrossRef]
20. Zhang, L.; Ninomiya, Y.; Yamashita, T. Formation of submicron particulate matter (PM 1) during coal combustion and influence of reaction temperature. *Fuel* **2006**, *85*, 1446–1457. [CrossRef]
21. Quann, R.J.; Sarofim, A.F. Vaporization of refractory oxides during pulverized coal combustion. *Symp. (Int.) Combust.* **1982**, *19*, 1429–1440. [CrossRef]
22. Liu, Y.; Guan, Y.; Zhang, Y.; Xiong, Y. Effects of atmosphere on mineral transformation of Zhundong coal during gasification in CO_2/H_2O conditions. *Fuel* **2022**, *310*, 122428. [CrossRef]

23. Li, M.; Yu, S.; Chen, X.; Li, Z.; Zhang, Y.; Song, Z.; Liu, W.; Li, P.; Zhang, X.; Zhang, M.; et al. Impacts of condensable particulate matter on atmospheric organic aerosols and fine particulate matter (PM2.5) in China. *Atmos. Chem. Phys.* **2022**, *22*, 11845–11866. [CrossRef]
24. Zhang, H.; Wu, S.; Yang, Y.; Cheng, J.; Lun, F.; Wang, Q. Mineral Distribution in Pulverized Blended Zhundong Coal and Its Influence on Ash Deposition Propensity in a Real Modern Boiler Situation. *ACS Omega* **2020**, *5*, 4386–4394. [CrossRef]
25. Feng, D.; Zhao, Y.; Zhang, Y.; Sun, S.; Meng, S.; Guo, Y.; Huang, Y. Effects of K and Ca on reforming of model tar compounds with pyrolysis biochars under H_2O or CO_2. *Chem. Eng. J.* **2016**, *306*, 422–432. [CrossRef]
26. Mlonka-Mędrala, A.; Dziok, T.; Magdziarz, A.; Nowak, W. Composition and properties of fly ash collected from a multifuel fluidized bed boiler co-firing refuse derived fuel (RDF) and hard coal. *Energy* **2021**, *234*, 121229. [CrossRef]
27. Tositti, L. Physical and Chemical Properties of Airborne Particulate Matter. In *Clinical Handbook of Air Pollution-Related Diseases*; Capello, F., Gaddi, A.V., Eds.; Springer International Publishing: Cham, Switzerland, 2018; pp. 7–32.
28. Ruan, R.; An, Q.; Tan, H.; Jia, S.; Wang, X.; Peng, J.; Li, P. Effect of calcined kaolin on PM0.4 formation from combustion of Zhundong lignite. *Fuel* **2022**, *319*, 123622. [CrossRef]
29. Zhao, J.; Zhang, Y.; Wei, X.; Li, T.; Qiao, Y. Chemisorption and physisorption of fine particulate matters on the floating beads during Zhundong coal combustion. *Fuel Process. Technol.* **2020**, *200*, 106310. [CrossRef]
30. Ninomiya, Y.; Wang, Q.; Xu, S.; Teramae, T.; Awaya, I. Evaluation of a Mg-Based Additive for Particulate Matter (PM)2.5 Reduction during Pulverized Coal Combustion. *Energy Fuels* **2010**, *24*, 199–204.
31. Wei, Y.; Wang, Q.; Zhang, L.; Awaya, I.; Ji, M.; Li, H.; Yamada, N.; Sato, A.; Ninomiya, Y. Effect of magnesium additives on PM 2.5 reduction during pulverized coal combustion. *Fuel Process. Technol.* **2013**, *105*, 188–194. [CrossRef]
32. Cao, X.; Zhang, T.-A.; Zhang, W.; Lv, G. Solvent Extraction of Sc(III) by D2EHPA/TBP from the Leaching Solution of Vanadium Slag. *Metals* **2020**, *10*, 790. [CrossRef]
33. Wang, H.; Zhang, X.; Wu, W.; Liaw, P.K.; An, K.; Yu, Q.; Wu, P. On the torsional and coupled torsion-tension/compression behavior of magnesium alloy solid rod: A crystal plasticity evaluation. *Int. J. Plast.* **2022**, *151*, 103213. [CrossRef]
34. Yang, Y.; Lin, X.; Li, S.; Luo, M.; Yin, J.; Wang, Y. Formation factors and emission characteristics of ultrafine particulate matters during Na-rich char gasification. *Fuel* **2019**, *253*, 781–791. [CrossRef]
35. Hu, W.; Wang, Y.; Wang, T.; Ji, Q.; Jia, Q.; Meng, T.; Ma, S.; Zhang, Z.; Li, Y.; Chen, R.; et al. Ambient particulate matter compositions and increased oxidative stress: Exposure-response analysis among high-level exposed population. *Environ. Int.* **2021**, *147*, 106341. [CrossRef]
36. Liu, J.; Banerjee, S.; Oroumiyeh, F.; Shen, J.; del Rosario, I.; Lipsitt, J.; Paulson, S.; Ritz, B.; Su, J.; Weichenthal, S.; et al. Co-kriging with a low-cost sensor network to estimate spatial variation of brake and tire-wear metals and oxidative stress potential in Southern California. *Environ. Int.* **2022**, *168*, 107481. [CrossRef]
37. Ninomiya, Y.; Wang, Q.; Xu, S.; Mizuno, K.; Awaya, I. Effect of Additives on the Reduction of PM2. 5 Emissions during Pulverized Coal Combustion†. *Energy Fuels* **2009**, *23*, 3412–3417. [CrossRef]
38. Zhu, N.Y.; Sun, C.Y.; Li, Y.L.; Qian, L.Y.; Hu, S.Y.; Cai, Y.; Feng, Y.H. Modeling discontinuous dynamic recrystallization containing second phase particles in magnesium alloys utilizing phase field method. *Comput. Mater. Sci.* **2021**, *200*, 110858. [CrossRef]
39. Zeng, K.; Yang, Q.; Zhang, Y.; Mei, Y.; Wang, X.; Yang, H.; Shao, J.; Li, J.; Chen, H. Influence of torrefaction with Mg-based additives on the pyrolysis of cotton stalk. *Bioresour. Technol.* **2018**, *261*, 62–69. [CrossRef]
40. Sezer, N.; Evis, Z.; Kayhan, S.M.; Tahmasebifar, A.; Koç, M. Review of magnesium-based biomaterials and their applications. *J. Magnes. Alloy.* **2018**, *6*, 23–43. [CrossRef]
41. Pi, S.; Zhang, Z.; He, D.; Qin, C.; Ran, J. Investigation of Y_2O_3/MgO-modified extrusion–spheronized CaO-based pellets for high-temperature CO_2 capture. *Asia-Pac. J. Chem. Eng.* **2019**, *14*, e2366. [CrossRef]
42. Seetharaman, S.; Jayalakshmi, S.; Arvind Singh, R.; Gupta, M. The Potential of Magnesium-Based Materials for Engineering and Biomedical Applications. *J. Indian Inst. Sci.* **2022**, *102*, 421–437. [CrossRef]
43. Yang, Y.; Jiang, C.; Guo, X.; Peng, S.; Zhao, J.; Yan, F. Experimental investigation on the permeability and damage characteristics of raw coal under tiered cyclic unloading and loading confining pressure. *Powder Technol.* **2021**, *389*, 416–429. [CrossRef]
44. Pallarés, S.; Gómez, E.T.; Martínez-Poveda, Á.; Jordán, M.M. Distribution Levels of Particulate Matter Fractions (<2.5 μm, 2.5–10 μm and >10 μm) at Seven Primary Schools in a European Ceramic Cluster. *Int. J. Environ. Res. Public Health* **2021**, *18*, 4922. [PubMed]
45. Shin, N.; Velmurugan, K.; Su, C.; Bauer, A.K.; Tsai, C.S.J. Assessment of fine particles released during paper printing and shredding processes. *Environ. Sci. Process. Impacts* **2019**, *21*, 1342–1352. [CrossRef] [PubMed]
46. Torres-Luna, J.A.; Carriazo, J.G. Porous aluminosilicic solids obtained by thermal-acid modification of a commercial kaolinite-type natural clay. *Solid State Sci.* **2019**, *88*, 29–35. [CrossRef]
47. Sun, J.; Guo, Y.; Yang, Y.; Li, W.; Zhou, Y.; Zhang, J.; Liu, W.; Zhao, C. Mode investigation of CO_2 sorption enhancement for titanium dioxide-decorated CaO-based pellets. *Fuel* **2019**, *256*, 116009. [CrossRef]
48. Hu, Y.; Lu, H.; Liu, W.; Yang, Y.; Li, H. Incorporation of CaO into inert supports for enhanced CO2 capture: A review. *Chem. Eng. J.* **2020**, *396*, 125253. [CrossRef]
49. Song, W.; Tang, L.; Zhu, X.; Wu, Y.; Rong, Y.; Zhu, Z.; Koyama, S. Fusibility and flow properties of coal ash and slag. *Fuel* **2009**, *88*, 297–304. [CrossRef]

50. Gao, N.; Chen, K.; Quan, C. Development of CaO-based adsorbents loaded on charcoal for CO_2 capture at high temperature. *Fuel* **2020**, *260*, 116411. [CrossRef]
51. Zhu, S.; Zhang, M.; Li, Z.; Zhang, Y.; Yang, H.; Sun, J.; Lyu, J. Influence of combustion temperature and coal types on alumina crystal phase formation of high-alumina coal ash. *Proc. Combust. Inst.* **2019**, *37*, 2919–2926. [CrossRef]
52. Wang, Q.; Zhang, L.; Sato, A.; Ninomiya, Y.; Yamashita, T. Interactions among inherent minerals during coal combustion and their impacts on the emission of PM10. 1. Emission of micrometer-sized particles. *Energy Fuels* **2007**, *21*, 756–765. [CrossRef]
53. Pongrac, P.; McNicol, J.W.; Lilly, A.; Thompson, J.A.; Wright, G.; Hillier, S.; White, P.J. Mineral element composition of cabbage as affected by soil type and phosphorus and zinc fertilisation. *Plant Soil* **2019**, *434*, 151–165. [CrossRef]
54. Zhang, Z.; Wu, X.; Zhou, T.; Chen, Y.; Hou, N.; Piao, G.; Kobayashi, N.; Itaya, Y.; Mori, S. The effect of iron-bearing mineral melting behavior on ash deposition during coal combustion. *Proc. Combust. Inst.* **2011**, *33*, 2853–2861. [CrossRef]
55. Guo, F.; Liu, Y.; Wang, Y.; Li, X.; Li, T.; Guo, C. Pyrolysis kinetics and behavior of potassium-impregnated pine wood in TGA and a fixed-bed reactor. *Energy Convers. Manag.* **2016**, *130*, 184–191. [CrossRef]
56. Dai, X.; Bai, J.; Yuan, P.; Du, S.; Li, D.; Wen, X.; Li, W. The application of molecular simulation in ash chemistry of coal. *Chin. J. Chem. Eng.* **2020**, *28*, 2723–2732. [CrossRef]

Article

Effects of Temperature and Chemical Speciation of Mineral Elements on PM10 Formation during Zhundong Coal Combustion

Qiaoqun Sun [1], Zhiqi Zhao [2], Shizhang Wang [2], Yu Zhang [2,*], Yaodong Da [3], Heming Dong [2], Jiwang Wen [3], Qian Du [2] and Jianmin Gao [2]

1 School of Aerospace and Construction Engineering, Harbin Engineering University, Harbin 150001, China
2 School of Energy Science and Engineering, Harbin Institute of Technology, Harbin 150001, China
3 China Institute of Special Equipment Inspection, Beijing 100029, China
* Correspondence: zhang.y@hit.edu.cn

Abstract: Particulate matter (PM) pollution from coal combustion is a leading contributor to the influence of atmospheric visibility, photochemical smog, and even global climate. A drop tube furnace was employed to explore the effects of temperature and chemical speciation of mineral elements on PM formation during the combustion of Zhundong coal. Chemical fractionation analysis (CFA), X-ray fluorescence (XRF), and inductively coupled plasma-atomic emission spectrometry (ICP-AES) were used to investigate the chemical and physical characteristics of the solid samples. It can be indicated that the combustion of similarly sized coal particles yielded more PM10 when the combustion temperature was increased from 1000 to 1400 °C. Zhundong coal is fractionated with deionized water, ammonium acetate, and hydrochloric acid, and pulverized coal, after fractionation, is burned to study the influence of mineral elements with different occurrence forms, such as water-soluble mineral elements, exchangeable ion elements, hydrochloric acid soluble elements and acid-insoluble elements, on the formation of particles. The results show that water-soluble salts play an important role in forming ultrafine particles (PM0.2); Fe, Ca, and other elements in organic form are distributed in flue gas through evaporation during pulverized coal combustion. When the flue gas temperature decreases, PM1 is formed through homogeneous nucleation and heterogeneous condensation, resulting in the distribution of these two elements on PM1. Different fractionation methods do not significantly affect the distribution of Si and Al in the PM1–10 combustion process.

Keywords: particulate matter (PM); Zhundong coal; PM10; PM1; PM0.2

Citation: Sun, Q.; Zhao, Z.; Wang, S.; Zhang, Y.; Da, Y.; Dong, H.; Wen, J.; Du, Q.; Gao, J. Effects of Temperature and Chemical Speciation of Mineral Elements on PM10 Formation during Zhundong Coal Combustion. *Energies* **2023**, *16*, 310. https://doi.org/10.3390/en16010310

Academic Editor: Fernando Rubiera González

Received: 27 October 2022
Revised: 30 November 2022
Accepted: 1 December 2022
Published: 27 December 2022

1. Introduction

Airborne particulate matter (PM) has become a major air pollutant in urban China in recent years. It is a leading factor contributing to decreased atmospheric visibility, global climate change, and photochemical smog [1–3]. PM is a complex mixture of organic substances and inorganic matter [4,5], including solid particles and liquid droplets. PM10 with an aerodynamic diameter of fewer than 10 μm can be breathed in through the nose/mouth and enter the human respiratory tract. Within the PM10 classification, PM can be further classified into submicron particulate matter (PM1) and ultra-micro particulate matter (PM1–10).

China accounts for 23% of the world's energy consumption, and, in the short term, at least, coal will remain the dominant source of Chinese energy [6]. Large reserves of Zhundong (ZD) coal exist in China [7], but the raw coal contains levels of alkali metals and alkaline earth metals (AAEMs) that are far higher than those found in other coals [8,9]. Combusting coals with significant AAEM species can lead to multi-particle formation during combustion, which can result in slagging and fouling problems and limit the utilization of ZD coal in large-scale applications [10,11].

Several variables impact the composition and formation characteristics of PM10. For example, increasing the combustion temperature would increase the number of PM10 produced [12–14]. Changing the mineral content in raw coal can influence how the mineral components react with each other during combustion, producing ash with different compositions that [15], in turn, influence the ash melting point. An increase in the ash melting point increases the formation of PM10, while an increase in the S content in the raw coal increases the production of PM1 [16]. A high silicate content in coal ash increases the melting temperature, while a high content of oxides and sulfate tends to lower its fusion temperature [17]. In addition, coal's mineral content and the division between organic and inorganic fractions can significantly influence PM formation [18,19]. Among others, Na, K, Ca, Al, and Fe can promote lower-temperature melting and polymerization of fine particles, forming larger particles [11,20,21]. PM1 mainly comprises alkali metal oxides, refractory oxides, and sulfides [16]. High-Na/K coals burn at 900–1100 °C, with the sulfate core forming the main component of PM1. At 1300 °C, Na and K elements are precipitated as hydroxides and easily combine with aluminate in coal to form larger particles (PM10+) [22]. The occurrence of Ca and Mg elements in low-rank coal is the primary determinant of their precipitation behavior [23,24]. Al is mainly present in the form of aluminosilicate, with less precipitation.

During combustion, various forms of different mineral elements precipitate at different stages, impacting PM formation with various sizes. Indeed, relatively few studies have been carried out addressing the impact of mineral matter composition on PM formation and composition characteristics. The work aims to provide helpful information that will aid the control of PM pollution from coal combustion.

2. Material and Methods

2.1. Sample Preparation

As shown in Figure 1, the drop tube furnace (DTF) was used for experiments with Chinese ZD coal (90–125 μm). The experiment was carried out in the air at 1000–1400 °C. The furnace was composed of a corundum tube of 80 mm diameter and 2000 mm length. The pulverized coal sample was pneumatically conveyed into the furnace chamber at a constant rate of 1.0 g/min with the primary air. The primary air flow is 1 L/min, and the secondary airflow is 9 L/min. A secondary air was mixed at the burner to ensure the burn-out rate in the furnace [25]. Fly ash was collected at the bottom of the reactor. The proximate and ultimate analyses of ZD coal are listed in Table 1. The compositional analysis of raw ZD coal ash is listed in Table 2.

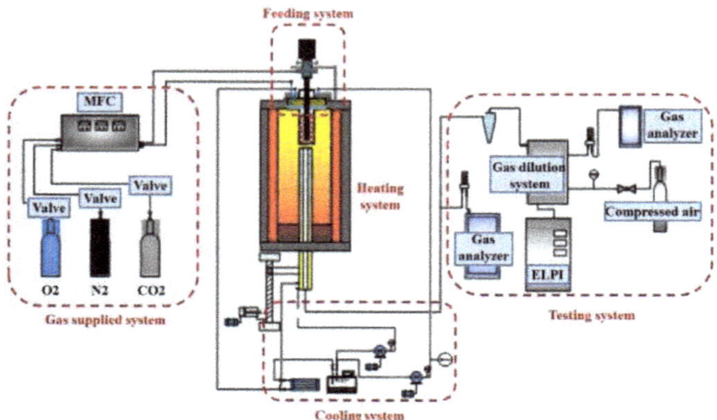

Figure 1. Schematic diagram of the experimental furnace.

Table 1. Proximate and ultimate analysis of ZD coal.

Proximate Analysis (wt.%, ad. = Air Dry Basis)				Ultimate Analysis (wt.%, ad. = Air Dry Basis)				
M(Moisture)	A(Ash)	V(Volatile)	FC(Fixed Carbon)	C	H	Odiff.	N	S
9.63	5.5	40.3	44.57	61.4	4.41	17.69	0.89	0.48

Note: Heating value = 18.5 MJ/kg. diff. = by difference.

Table 2. Compositional analysis of ZD coal ash.

Ash Compositions (wt. %)										
SiO_2	Al_2O_3	Fe_2O_3	CaO	MgO	TiO_2	SO_3	MnO_2	K_2O	P_2O_5	Na_2O
30.56	29.85	10.56	8.64	4.42	0.22	7.66	0.31	0.6	0.12	6.02

2.2. Chemical Fractionation Analysis

The chemical speciation of mineral elements in raw ZD coal was assessed by chemical fractionation analysis (CFA) [26–29]. Samples were dried overnight at 105 °C, and then each sample was successively extracted with deionized water (H_2O), 1.0 mol/L ammonium acetate (NH_4Ac), and 1.0 mol/L hydrochloric acid (HCl). In the first extraction, H_2O-soluble compounds, such as K and Na salts, would be dissolved [30]. Organically bound ion-exchangeable elements were removed by the NH_4Ac solution [31]. Finally, acid-soluble compounds, such as carbonates and sulfates, were extracted using an HCl solution. After each extraction, the solid residue was washed with deionized water until the pH of the leachate was constant. According to the procedures carried out during the CFA process, the samples were named raw coal, H_2O-coal, NH_4Ac-coal, and HCl-coal, respectively. These samples were then introduced to the DTF in the same way as the raw coal at 1200 °C.

2.3. Sampling System

PM samples were simultaneously sampled from the collected ash and classified for analysis. In order to study the morphology and composition, polycarbonate film was used for microscopic analysis [32]. A Teflon filter membrane was used during inorganic content analysis using ICP-AES and XRF.

An electrical low-pressure impactor (ELPI) (97 2E, NO.24423, Dekati LTD, Kangasala, Finland) was used to measure the ash particle size distribution. X-ray fluorescence (XRF, PW4400) further developed the compositional analysis. Major (content more than 1000 g/g) and minor (in 100~1000 g/g) ash components were then qualitatively and semi-quantitatively analyzed with accurate detection of elements in the range 9F to 92U. A vacuum of less than 100 Pa (minimum 1 Pa) and a temperature of approximately 60–70 °C were used; these conditions were assumed not to denature the sample.

The mineral matter that was retained in the ash particles was digested by a microwave digestion system (Ethos 1, Milestone, Sorisole, Italy) [29]. Each digested residue was re-dissolved, transferred to a 25 mL volumetric flask, and made up to a set volume with deionized H_2O. The mineral elements were then quantified by an inductively coupled plasma atomic emission spectrometer (ICP-AES).

3. Results and Discussion

3.1. Impact of Temperature on PM10 Formation

As shown in Figure 2, the impact of Raw-coal's combustion temperature on the formation of variously sized PM can be seen. Figure 2a shows a similar distribution for all temperatures. The amount of PM10 formed increased as the combustion temperature increased. The peak at approximately 2 μm was mainly affected by the formation of PM1–10 particles. At higher temperatures, the internal temperature gradient of the char particles is steeper, and the particles endure a relatively higher level of thermal stress, rendering them easily broken [33,34]. Figure 2b initially shows a gradual increase in the amount of PM0.2, from 0.29 to 0.32 mg/g, as the temperature was increased from 1000 to 1200 °C, with a more significant increase to 0.37 mg/g at 1400 °C. This might be driven by increased evaporation and condensation of mineral elements at higher temperatures [35]. The amount of PM0.2–1 produced is similar at 1000 and 1200 °C (0.30 and 0.31 mg/g, respectively), but it increases to 0.37 mg/g at 1400 °C. The amount of PM1 increases to 0.62 mg/g and 0.74 mg/g. A similar trend was observed for the different temperatures for particles in the PM10 and PM1 ranges. Results at 1000, 1200, and 1400 °C yielded 0.57, 0.61, and 0.71 mg/g for PM1–10 and 1.15, 1.23, and 1.46 mg/g for PM10, respectively. The precipitation and condensation of mineral elements were the primary mechanisms for forming PM1 [16,36,37]. The extent to which mineral elements volatilize is known to be positively correlated with the amount of PM1 [38], and, in general, their vapor-phase concentration is high. At the same time, the associated particle enjoys a slight temperature gradient across it, a long aggregation time, and a relatively large size [39].

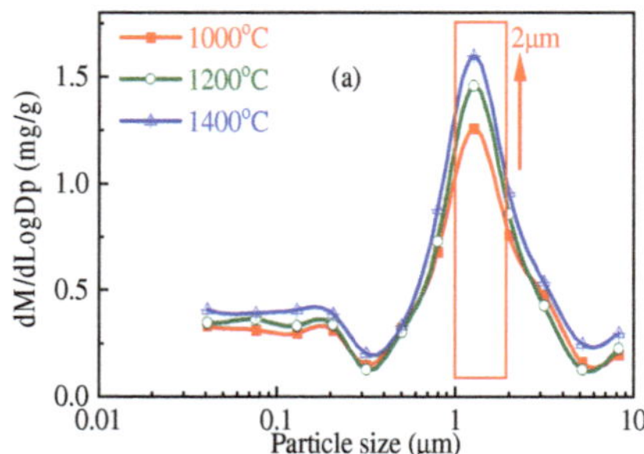

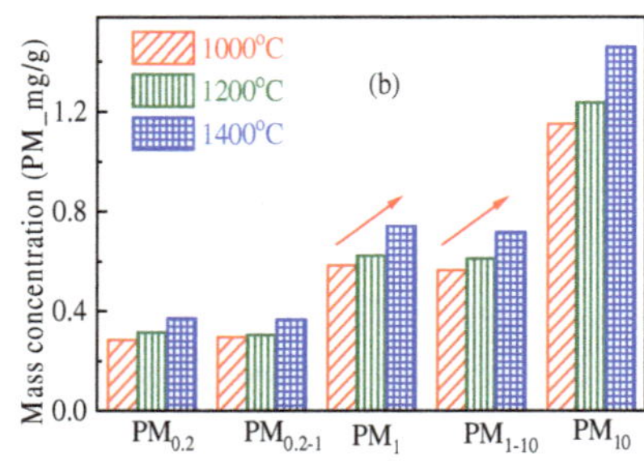

Figure 2. Continuous (**a**) and discrete (**b**) distribution of PM10 at different temperatures.

An analysis of the major mineral components in PM10 combusted at different temperatures is shown in Figure 3, while Figure 4 presents the bulk compositions of PM with various sizes. For PM0.2, an increase of the combustion temperature from 1000 to 1200 and then to 1400 °C increased the mass fraction of K and particularly Na, which increased from 9.0 to 12.5 and then to 20.6%. The change of element S is basically the same. Figure 4b shows that compared with the results for PM0.2, the PM0.2–1 fraction exhibited a significantly lower mass fraction of the volatile components but a much higher proportion of the non-volatile (Ca, Mg, and Fe) elements and Al-Si salts. (Under three temperature conditions, the proportion of volatile elements in PM0.2 is 26.28%, 29.40%, and 37.08%, respectively. The proportion of volatile elements in PM0.2–1 is 7.00%, 8.12%, and 12.66%, respectively). Because the mineral substances needed for particle nucleation come from two sources: the volatilization of minerals or their release during coal combustion [40]. This can be explained by considering that PM10 is more likely to be formed by homogeneous nucleation mechanisms and the condensation of volatile elements on the particle surface [41–44]. The main reason for the difference in mineral content in PM0.2 and PM0.2+ particles is that the formation mechanism of particles with different sizes is different. The content of volatile elements in PM0.2 is high. PM0.2 is mainly formed by steam condensation. The steam source is the decomposition of mineral elements in organic form in the coal powder pyrolysis stage, releasing mineral steam or the formation of inorganic minerals in the coke combustion stage. PM0.2+ is mainly generated in the coke combustion stage, mainly including volatile elements condensed on fine particles and particles rich in silicon and aluminum formed in the molten state, which accumulates to form irregular particles. As shown in Figure 4c, the content of volatile mineral elements in PM1–10 was less than that in the other samples. The fraction of volatile elements in the samples at 1200/1400 °C is very similar (6.2 and 6.9%, respectively), significantly greater than that observed at 1000 °C (1.7%). The most volatile elements reached the maximum value at 1400 °C, slightly increasing at 1200 °C.

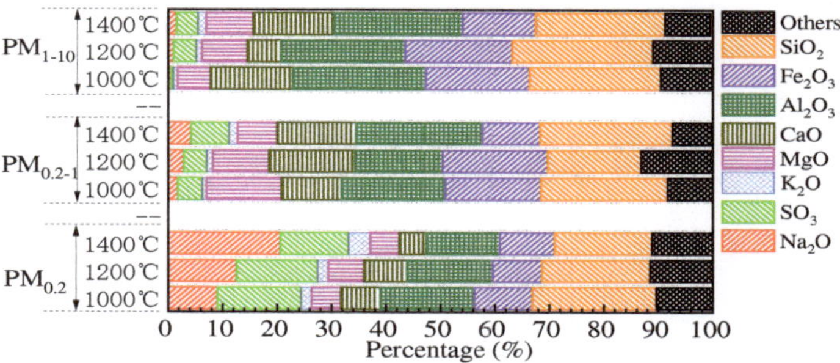

Figure 3. Composition of significant mineral components in PM10.

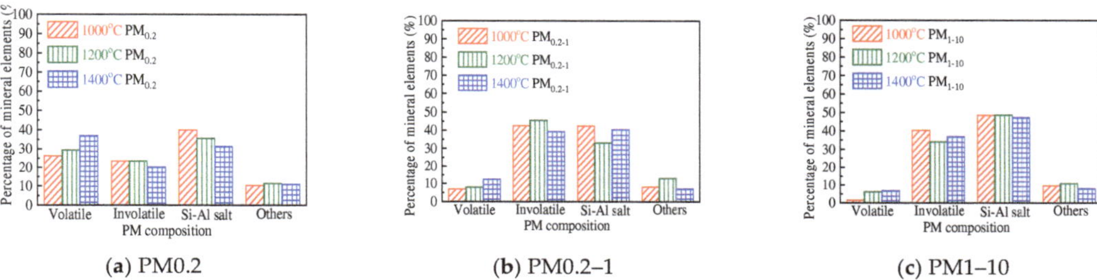

Figure 4. Composition of various size fractions of coal ash generated at different combustion temperatures. Volatile denotes Na, K, and S; non-volatile denotes Mg, Ca, and Fe.

3.2. *Effect of Chemical Speciation of Mineral Elements on PM10 Formation*

3.2.1. Chemical Speciation of Mineral Elements

Table 3 shows the composition of significant elements for the four stages of CFA. At the same time, Figure 5 displays the main speciation of these inorganic elements depending on whether they are soluble in H_2O, NH_4Ac, or HCl or insoluble in all the above solvents. The highest compositions observed in the raw sample are for Al and Si, followed by Fe, Ca, and Na. Na is mainly found as an H_2O-soluble salt. Washing with H_2O removes 61.1% of the actual Na, though this process removes little Mg, Ca, Fe, or Al. Washing with NH_4Ac removes 32.1% of the Mg and 20.2% of the Ca remaining after the H_2O wash. Mg and Ca species that then dissolve in HCl but are insoluble in NH_4Ac mainly comprised carbonates [45–48], such as calcite ($CaCO_3$) and dolomite ($CaMg(CO_3)_2$). Washing with HCl dissolves 39.2% of Al with the insoluble fraction assumed to be aluminosilicates, such as kaolin. For Fe, 17.9% is NH_4Ac-soluble, 44.7% is HCl-soluble, and the remainder consists of insoluble salts. With most Fe compounds insoluble in H_2O, compounds soluble in HCl or NH_4Ac may include siderite ($FeCO_3$) [49], ankerite ($(FeCaMg)CO_3$), or Fe(OH). Fe compounds that are insoluble in HCl may have been pyrite (FeS_2).

Table 3. Content of mineral elements in CFA (mg/g).

Coal	K	Na	Ca	Mg	Fe	Al	Si	Ash
Raw-Coal	0.2	1.8	2.6	1.2	2.8	6.9	6.8	5.3
H_2O-Coal	0.1	0.7	2.4	1.1	2.5	6.4	6.0	4.6
NH_4Ac-Coal	0.1	0.6	1.9	0.8	2.3	5.8	5.3	3.9
HCl-Coal	0.1	0.1	0.2	0.1	1.0	3.1	2.0	1.2

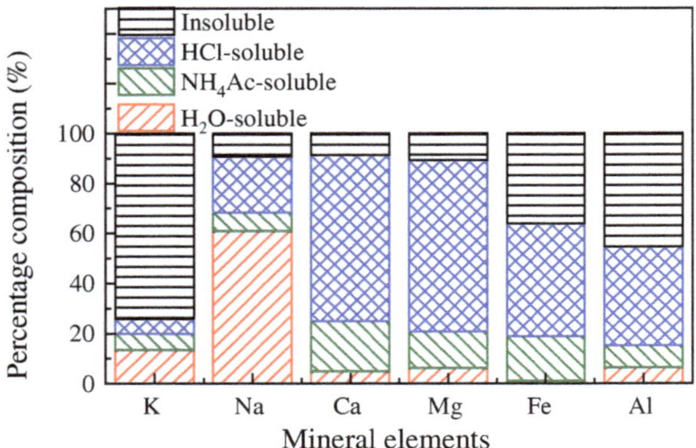

Figure 5. Fractionation of major elements.

3.2.2. Distribution of the Primary Mineral in PM10

CFA results showing the major mineral components for PM0.2, PM0.2–1, and PM1–10 are shown in Figure 6. The main components of PM0.2 were volatile elements such as S, Na, and K. The proportions of the insoluble elements—Mg, Ca, Fe, Al, and Si—were relatively small. The PM0.2 content of Na decreased significantly following H_2O-washing. At the same time, acid-washing caused an increase in the concentration of Si and Al elements as other mineral elements (such as Mg, Ca, and Fe) dissolved in HCl. The PM0.2–1 and PM1–10 fractions generated from the combustion of four different coal samples were very similar and mainly comprised refractory oxides (of Mg, Ca, and Fe) and Al and Si compounds. A bulk analysis of the composition of the three PM size groups is shown in Figure 7. These results suggest that the PM samples were mainly composed of Na_2O, MgO, CaO, SiO_2, Al_2O_3, Fe_2O_3, SO_3, and K_2O.

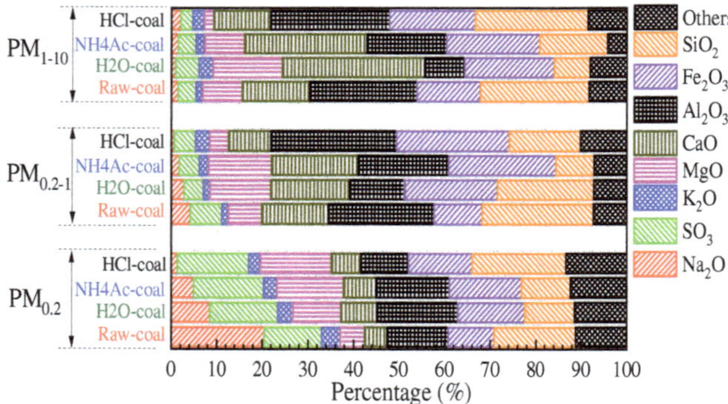

Figure 6. Major mineral in PM10 from CFA-treated coal samples.

3.2.3. Speciation of Mineral Elements in PM10

The impact of various treatments of the coal samples on the ash particle size distribution can be seen in Figure 8a. Notably, the acid-treated coal sample displayed no apparent peak in PM distribution, which was relatively uniform. Indeed, the ash content of the raw coal sample decreased from 5.3% to 1.2% after the acid-washing process. The PM distribution shows the two distinct peaks divided into three areas: ultrafine, intermediate, and coarse. The precipitation of mineral elements forms ultrafine particles. ZD coal has a high H_2O-soluble alkali content (especially Na), much of which was removed following H_2O washing. The number of ultrafine particles was constant, indicating that the ultrafine particles play an important part in the combustion process of coal with the H_2O-soluble minerals.

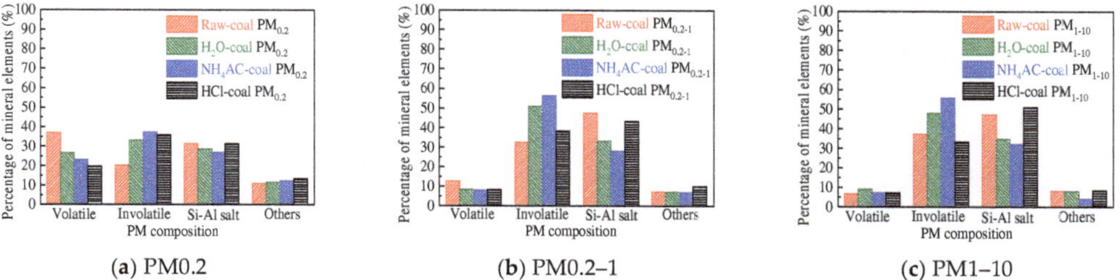

(**a**) PM0.2 (**b**) PM0.2–1 (**c**) PM1–10

Figure 7. Composition of various size fractions of coal ash generated from CFA-treated coal samples. Volatile denotes Na, K, and S; Non-volatile denotes Mg, Ca, and Fe.

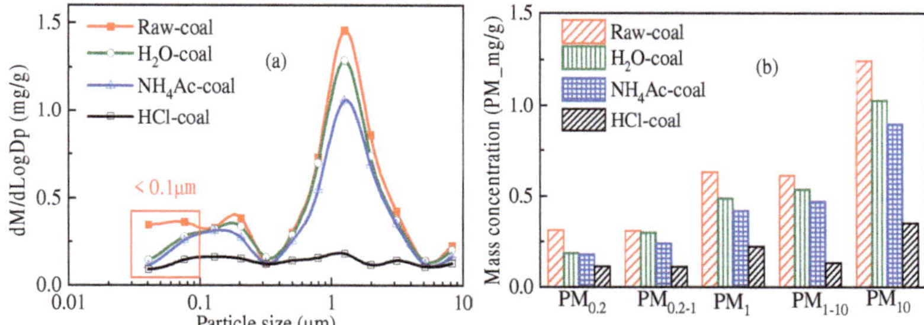

Figure 8. Continuous (**a**) and discrete (**b**) distribution of PM10 at different processing methods.

Na catalyzes the combustion process, accelerating the char burning rate and causing the particles to burn out more quickly. It decreases the accumulation time for melting ash and cannot adhere to the ash surface. Removing 61% of Na from the raw coal by H_2O-washing could decrease the reactivity of the char (compared with that of raw coal) during combustion. This decrease in reactivity may increase the accumulation and agglomeration time for the melting ash particles on the char surface, promoting the formation of larger particulates (>10 μm).

Figure 8 shows the particle size distribution for PM10 produced from the various coal samples. H_2O-washing caused the amount of PM0.2 to decrease from 0.32 to 0.19 mg/g, mainly because of the leaching of Na, which is key to the formation of particles via the evaporation–condensation mechanism in the PM0.2 range. Fine particulates mainly develop via two pathways: vapor condensation reactions at an inorganic surface of a nucleation point or the collision and coagulation of two existing fine particles [49]. The amount of PM0.2–1 generated from the combustion of raw coal, H_2O-coal, and NH_4Ac-coal was 0.31, 0.30, and 0.24 mg/g, respectively. From these results, it is possible to conclude that the role of H_2O-soluble mineral elements in the PM0.2–1 generation process is non-obvious and that the existence of the mineral elements in the H_2O-coal and raw coal (such as Ca, Fe, and Mg) suggests that organic minerals also impacted the formation of PM0.2–1. The fraction of ash produced as PM1–10 for the raw coal, H_2O-coal, NH_4Ac-coal, and HCl-coal samples was 0.61, 0.53, 0.47, and 0.13 mg/g, respectively, suggesting that the HCl-soluble minerals were essential in PM1–10 formations. These minerals may include carbonates, siderite, ankerite (Fe, Ca, Mg)O_3, or $Fe(OH)_2$. For the PM10 fraction, the raw coal produced 1.24 mg/g, which included 0.32 (25%), 0.31 (25%), and 0.61 (50%) mg/g of, respectively, PM0.2, PM0.2–1, and PM1–10. The fractional amounts of PM10 generated by the combustion of H_2O-coal, NH_4Ac-coal, and HCl-coal were all lower than that of raw coal (at 1.03, 0.90, and 0.35 mg/g, respectively). This analysis suggests that H_2O-soluble mineral elements mainly affected the formation of PM0.2; organically bound elements impacted the amount of PM0.2–1 generated and washing with HCl significantly decreased the amount of PM1–10 produced by coal combustion.

4. Conclusions

(1) When Zhundong pulverized coal with the same combustion atmosphere and particle size burns, the total amount of PM10 generated at high combustion temperature is generally more. At 1000 °C, 1200 °C and 1400 °C, the amount of PM10 generated by the unit mass of Zhundong pulverized coal is 1.15 mg/g, 1.23 mg/g and 1.46 mg/g, respectively. The reason may be that with the increase in combustion temperature, the steam concentration of mineral elements in coal during combustion is increased. When the temperature decreases, more mineral steam will condense on the surface of particles.

(2) The mineral elements in raw coal are divided into water-soluble mineral elements, organically bound mineral elements, acid-soluble mineral elements and acid-insoluble mineral elements by chemical fractionation. Water-soluble salts play an important role in the formation of ultrafine particles (PM0.2). The organic Fe and Ca elements do not react with the aluminosilicate in the pulverized coal during the combustion of pulverized coal but evaporate into steam and distribute in the flue gas. When the flue gas temperature decreases, they condense and form on PM1, resulting in these two elements on PM1; Mg, Ca, and Fe react with aluminosilicate to form molten particles, which makes particles PM1–10 contain these three elements. Different fractionation methods will not greatly affect Si and Al in the PM1–10 combustion process.

Author Contributions: Conceptualization, Y.Z.; Software, J.W.; Resources, Y.D., H.D., Q.D. and J.G.; Data curation, Z.Z.; Writing—original draft, Q.S.; Writing—review & editing, S.W. All authors have read and agreed to the published version of the manuscript.

Funding: This work is supported by the National Natural Science Foundation of China (52006047), Heilongjiang Provincial Natural Science Foundation (LH2022E065) and Foundation of State Key Laboratory of High-Efficiency Utilization of Coal and Green Chemical Engineering (2021-K45).

Data Availability Statement: Date openly available in a public repository.

Conflicts of Interest: No conflict of interest.

References

1. Yao, Q.; Li, S.Q.; Xu, H.W.; Zhuo, J.K.; Song, Q. Studies on formation and control of combustion particulate matter in China: A review. *Energy* **2009**, *34*, 1296–1309. [CrossRef]
2. Chung, W.S.; Chen, Q.; Osammor, O.; Nolan, A.; Zhang, X.H.; Sharifi, V.N.; Swithenbank, J. Characterisation of particulate matter on the receptor level in a city environment. *Environ. Monit. Assess.* **2012**, *184*, 1471–1486. [CrossRef] [PubMed]
3. Fiore, A.M.; Naik, V.; Leibensperger, E.M. Air Quality and Climate Connections. *J. Air Waste Manag. Assoc.* **2015**, *65*, 645–685. [CrossRef] [PubMed]
4. Zhao, T.; Yan, Y.; Zhou, B.; Zhong, X.; Hu, X.; Zhang, L.; Huo, P.; Xiao, K.; Zhang, Y.; Zhang, Y. Insights into reactive oxygen species formation induced by water-soluble organic compounds and transition metals using spectroscopic method. *J. Environ. Sci.* **2023**, *124*, 835–845. [CrossRef] [PubMed]
5. Feng, D.D.; Guo, D.W.; Zhang, Y.; Sun, S.Z.; Zhao, Y.J.; Shang, Q.; Sun, H.L.; Wu, J.Q.; Tan, H.P. Functionalized construction of biochar with hierarchical pore structures and surface O-/N-containing groups for phenol adsorption. *Chem. Eng. J.* **2021**, *410*, 127707. [CrossRef]
6. Li, L.; Karatzos, S.; Saddler, J. The potential of forest-derived bioenergy to contribute to China's future energy and transportation fuel requirements. *For. Chron.* **2012**, *88*, 547–552. [CrossRef]
7. Zhang, X.-P.; Zhang, C.; Tan, P.; Li, X.; Fang, Q.-Y.; Chen, G. Effects of hydrothermal upgrading on the physicochemical structure and gasification characteristics of Zhundong coal. *Fuel Processing Technol.* **2018**, *172*, 200–208. [CrossRef]
8. Dai, S.; Ren, D.; Chou, C.L.; Finkelman, R.B.; Seredin, V.V.; Zhou, Y. Geochemistry of trace elements in Chinese coals: A review of abundances, genetic types, impacts on human health, and industrial utilization. *Int. J. Coal Geol.* **2011**, *94*, 3–21. [CrossRef]
9. Zhou, X.; Huang, D.; Guo, J.; Ning, H. Hydrogenation properties of Mg17Al12 doped with alkaline-earth metal (Be, Ca, Sr and Ba). *J. Alloys Compd.* **2019**, *774*, 865–872. [CrossRef]

10. Li, G.Y.; Wang, C.A.; Yan, Y.; Jin, X.; Liu, Y.H.; Che, D.F. Release and transformation of sodium during combustion of Zhundong coals. *J. Energy Inst.* **2016**, *89*, 48–56. [CrossRef]
11. Feng, D.D.; Shang, Q.; Dong, H.M.; Zhang, Y.; Wang, Z.L.; Li, D.; Xie, M.; Wei, Q.Y.; Zhao, Y.J.; Sun, S.Z. Catalytic mechanism of Na on coal pyrolysis-derived carbon black formation: Experiment and DFT simulation. *Fuel Process. Technol.* **2021**, *224*, 107011. [CrossRef]
12. Linak, W.P.; Miller, C.A.; Seames, W.S.; Wendt, J.O.L.; Ishinomori, T.; Endo, Y.; Miyamae, S. On trimodal particle size distributions in fly ash from pulverized-coal combustion. *Proc. Combust. Inst.* **2002**, *29*, 441–447. [CrossRef]
13. Jia, Y.; Lighty, J.A.S. Ash Particulate Formation from Pulverized Coal under Oxy-Fuel Combustion Conditions. *Environ. Sci. Technol.* **2012**, *46*, 5214–5221. [CrossRef]
14. Zhu, S.; Zhang, M.; Li, Z.; Zhang, Y.; Yang, H.; Sun, J.; Lyu, J. Influence of combustion temperature and coal types on alumina crystal phase formation of high-alumina coal ash. *Proc. Combust. Inst.* **2019**, *37*, 2919–2926. [CrossRef]
15. Ke, X.; Li, D.; Zhang, M.; Jeon, C.-h.; Cai, R.; Cai, J.; Lyu, J.; Yang, H. Ash formation characteristics of two Indonesian coals and the change of ash properties with particle size. *Fuel Process. Technol.* **2019**, *186*, 73–80. [CrossRef]
16. Buhre, B.J.P.; Hinkley, J.T.; Gupta, R.P.; Nelson, P.F.; Wall, T.F. Fine ash formation during combustion of pulverised coal–coal property impacts. *Fuel* **2006**, *85*, 185–193. [CrossRef]
17. Vassilev, S.V.; Kitano, K.; Takeda, S.; Tsurue, T. Influence of mineral and chemical-composition of coal ashes on their fusibility. *Fuel Processing Technol.* **1995**, *45*, 27–51. [CrossRef]
18. Liu, X.; Xu, M.; Yao, H.; Yu, D.; Lv, D. Study on the effect of the occurrence of sodium elements in coal on the formation of submicron particles. *J. Eng. Therm. Phys.* **2009**, *30*, 1589–1592.
19. Su, X.B.; Ding, R.; Zhuang, X.G. Characteristics of Dust in Coal Mines in Central North China and Its Research Significance. *ACS Omega* **2020**, *5*, 9233–9250. [CrossRef]
20. Yu, D. *Modal Identification and Formation Mechanism of Fine Particle in Coal Combustion*; Huazhong University of Science and Technology: Huazhong, China, 2007.
21. Feng, D.D.; Guo, D.W.; Shang, Q.; Zhao, Y.J.; Zhang, L.Y.; Guo, X.; Cheng, J.; Chang, G.Z.; Guo, Q.J.; Sun, S.Z. Mechanism of biochar-gas-tar-soot formation during pyrolysis of different biomass feedstocks: Effect of inherent metal species. *Fuel* **2021**, *293*, 120409. [CrossRef]
22. Zhou, K. *Study on the Generation Characteristics and Furnace Control of Fine Particulate Matter in Coal Combustion*; Huazhong University of Science and Technology: Huazhong, China, 2011.
23. Quann, R.J.; Sarofim, A.F. Vaporization of refractory oxides during pulverized coal combustion. *Symp. Combust.* **1982**, *19*, 1429–1440. [CrossRef]
24. Sun, J.; Guo, Y.; Yang, Y.; Li, W.; Zhou, Y.; Zhang, J.; Liu, W.; Zhao, C. Mode investigation of CO2 sorption enhancement for titanium dioxide-decorated CaO-based pellets. *Fuel* **2019**, *256*, 116009. [CrossRef]
25. Zheng, W.; Zhang, M.; Zhang, Y.; Lyu, J.; Yang, H. The effect of the secondary air injection on the gas–solid flow characteristics in the circulating fluidized bed. *Chem. Eng. Res. Des.* **2019**, *141*, 220–228. [CrossRef]
26. Jordan, C.A.; Akay, G. Speciation and distribution of alkali, alkali earth metals and major ash forming elements during gasification of fuel cane bagasse. *Fuel* **2012**, *91*, 253–263. [CrossRef]
27. Zevenhoven-Onderwater, M.; Backman, R.; Skrifvars, B.J.; Hupa, M. The ash chemistry in fluidised bed gasification of biomass fuels. Part I: Predicting the chemistry of melting ashes and ash–bed material interaction. *Fuel* **2001**, *80*, 1489–1502. [CrossRef]
28. Pettersson, A.; Åmand, L.E.; Steenari, B.M. Chemical fractionation for the characterisation of fly ashes from co-combustion of biofuels using different methods for alkali reduction. *Fuel* **2009**, *88*, 1758–1772. [CrossRef]
29. Zhao, Y.; Feng, D.; Zhang, Y.; Huang, Y.; Sun, S. Effect of pyrolysis temperature on char structure and chemical speciation of alkali and alkaline earth metallic species in biochar. *Fuel Process. Technol.* **2016**, *141*, 54–60. [CrossRef]
30. Feng, D.D.; Guo, D.W.; Zhang, Y.; Sun, S.Z.; Zhao, Y.J.; Chang, G.Z.; Guo, Q.J.; Qin, Y.K. Adsorption-enrichment characterization of CO_2 and dynamic retention of free NH_3 in functionalized biochar with H_2O/NH_3 center dot H2O activation for promotion of new ammonia-based carbon capture. *Chem. Eng. J.* **2021**, *409*, 128193. [CrossRef]
31. Feng, D.D.; Zhao, Y.J.; Zhang, Y.; Sun, S.Z.; Gao, J.M. Steam Gasification of Sawdust Biochar Influenced by Chemical Speciation of Alkali and Alkaline Earth Metallic Species. *Energies* **2018**, *11*, 205. [CrossRef]
32. Wang, W.; Wang, Q.; Lyu, J.; Liu, M.; Yue, G.; Liu, J. The Effect of Inner Secondary Air on the Flow Field of a Swirl Burner. In Proceedings of the 2nd International Conference on Smart Power & Internet Energy Systems (SPIES), Bangkok, Thailand, 15–18 September 2020; pp. 452–456.
33. Lv, J.; Li, D. Effect of temperature on the characteristics of primary particles in pulverized coal combustion. *Chin. J. Electr. Eng.* **2007**, *27*, 24–29.
34. Nguyen, C.B.; Scherer, J.; Hartwich, M.; Richter, A. The morphology evolution of char particles during conversion processes. *Combust. Flame* **2021**, *226*, 117–128. [CrossRef]
35. Yaroshchuk, A. Evaporation-driven electrokinetic energy conversion: Critical review, parametric analysis and perspectives. *Adv. Colloid Interface Sci.* **2022**, *305*, 102708. [CrossRef]
36. Hao, Z.; Xie, G.; Liu, X.; Tan, Q.; Wang, R. The precipitation behaviours and strengthening mechanism of a Cu-0.4 wt% Sc alloy. *J. Mater. Sci. Technol.* **2022**, *98*, 1–13. [CrossRef]

37. Lu, Y.; Cao, R.; Huang, D.W.; Cui, Z.N.; Wang, Y.; Zhang, Y.F. Investigation on the ash characteristics and AAEM migration during co-combustion of Zhundong coal and shale char in a fixed bed. *Fuel* **2022**, *327*, 125214. [CrossRef]
38. Lighty, J.S.; Veranth, J.M.; Sarofim, A.F. Combustion aerosols: Factors governing their size and composition and implications to human health. *J. Air Waste Manag. Assoc.* **2000**, *50*, 1619–1622. [CrossRef]
39. Zhang, L.; Ninomiya, Y.; Yamashita, T. Formation of submicron particulate matter (PM 1) during coal combustion and influence of reaction temperature. *Fuel* **2006**, *85*, 1446–1457. [CrossRef]
40. Wood, B.J.; Smythe, D.J.; Harrison, T. The condensation temperatures of the elements: A reappraisal. *Am. Mineral. J. Earth Planet. Mater.* **2019**, *104*, 844–856. [CrossRef]
41. Wang, Y.B.; Tan, H.Z. Condensation of KCl(g) under varied temperature gradient. *Fuel* **2019**, *237*, 1141–1150. [CrossRef]
42. Zhang, Y.; Shang, Q.; Feng, D.D.; Sun, H.L.; Wang, F.H.; Hu, Z.C.; Cheng, Z.Y.; Zhou, Z.J.; Zhao, Y.J.; Sun, S.Z. Interaction mechanism of in-situ catalytic coal H_2O-gasification over biochar catalysts for H_2O-H_2-tar reforming and active sites conversion. *Fuel Process. Technol.* **2022**, *233*, 107307. [CrossRef]
43. Feng, D.D.; Zhang, Y.; Zhao, Y.J.; Sun, S.Z.; Wu, J.Q.; Tan, H.P. Mechanism of in-situ dynamic catalysis and selective deactivation of H2O-activated biochar for biomass tar reforming. *Fuel* **2020**, *279*, 118450. [CrossRef]
44. Hu, Y.; Lu, H.; Liu, W.; Yang, Y.; Li, H. Incorporation of CaO into inert supports for enhanced CO_2 capture: A review. *Chem. Eng. J.* **2020**, *396*, 125253. [CrossRef]
45. Pi, S.; Zhang, Z.; He, D.; Qin, C.; Ran, J. Investigation of Y_2O_3/MgO-modified extrusion–spheronized CaO-based pellets for high-temperature CO_2 capture. *Asia-Pac. J. Chem. Eng.* **2019**, *14*, e2366. [CrossRef]
46. Zhang, Y.; Wang, S.Z.; Feng, D.D.; Gao, J.M.; Dong, L.H.; Zhao, Y.J.; Sun, S.Z.; Huang, Y.D.; Qin, Y.K. Functional Biochar Synergistic Solid/Liquid-Phase CO_2 Capture: A Review. *Energy Fuels* **2022**, *36*, 2945–2970. [CrossRef]
47. Feng, D.D.; Zhao, Y.J.; Zhang, Y.; Xu, H.H.; Zhang, L.Y.; Sun, S.Z. Catalytic mechanism of ion-exchanging alkali and alkaline earth metallic species on biochar reactivity during CO_2/H_2O gasification. *Fuel* **2018**, *212*, 523–532. [CrossRef]
48. Jiang, S.; Xu, H.; Sun, Y.; Song, Y. Performance analysis of Fe-N compounds based on valence electron structure. *J. Alloys Compd.* **2019**, *779*, 427–432. [CrossRef]
49. Helble, J.J. Mechanisms of Ash Formation and Growth During Pulverized Coal Combustion. Ph.D. Thesis, Department of Chemical Engineering, Massachusetts Institute of Technology, Cambridge, MA, USA, 1987.

Article

Simulation of Micron and Submicron Particle Trapping by Single Droplets with Electrostatic Fields

Qiaoqun Sun [1], Wei Zhang [2], Yu Zhang [2,*], Yaodong Dan [3], Heming Dong [2], Jiwang Wen [2], Qian Du [2] and Jianmin Gao [2]

1 School of Aerospace and Construction Engineering, Harbin Engineering University, Harbin 150001, China
2 School of Energy Science and Engineering, Harbin Institute of Technology, Harbin 150001, China
3 China Institute of Special Equipment Inspection, Beijing 100029, China
* Correspondence: zhang.y@hit.edu.cn

Abstract: Wet electrostatic precipitators have problems such as uneven water distribution and poor economy in applying ultra-clean particulate matter emissions from coal-fired boilers. Upgrading the droplets in wet dust removal to charged mobile collectors can effectively compensate for these shortcomings. In this paper, the effects of particle sphericity, particle size, and charge on the capture efficiency of a single droplet for capturing micron and submicron particles are qualitatively studied by simulating the process of particle capture by charged droplets in a turbulent flow field. The simulation results show that the trapping efficiency of charged droplets is positively correlated with the sphericity and the amount of charge. The particle size significantly impacts the capture efficiency, and the increase in size increases the capture efficiency, and the capture efficiency of 5.49 μm particles reaches 100%. The effect of particle movement speed on the capture efficiency needs to be considered in combination with particle size. For micron particles, the capture efficiency is close to 100% when the movement speed is 0.3 m/s and 0.5 m/s. For submicron particles, the aggregation morphology is lower at lower speeds. Simple non-spherical particles have greater capture efficiency.

Keywords: charged droplets; sphericity; numerical simulation; trapping efficiency

Citation: Sun, Q.; Zhang, W.; Zhang, Y.; Dan, Y.; Dong, H.; Wen, J.; Du, Q.; Gao, J. Simulation of Micron and Submicron Particle Trapping by Single Droplets with Electrostatic Fields. *Energies* **2022**, *15*, 8487. https://doi.org/10.3390/en15228487

Academic Editor: Dumitran Laurentiu

Received: 18 October 2022
Accepted: 10 November 2022
Published: 14 November 2022

1. Introduction

With the massive consumption of coal, particulate matter emission severely impacts the environment. However, in recent years, China has experienced rapid industrialization and the electricity demand is still in the increasing stage. In particular, the production and operation of coal-fired power plants generate a large amount of particulate emissions. A large amount of particulate matter in their soot emissions can cause serious pollution to the environment and affect people's health. To control the environmental damage of particulate matter, waste gas generated in the industrial production process must be purified before being discharged into the atmosphere. The wet dust collector has a better effect than various dust removal technologies [1]. Wet electrocoagulation technology is widely used due to its advantages of low energy consumption, pressure reduction, and the ability to remove soluble gas, but it also has some shortcomings, such as serious corrosion and the easy formation of corona inhibition [2,3]. The study of particle capture by single droplet is the basis for improving wet electrocoagulation. Studying the capture of particles by single droplet can deeply understand and use the wet deposition mechanism in the atmospheric environment to remove particles [2].

At present, scholars have studied the particle capture process of charged single droplets, most of which are based on the assumption of spherical particles, and the research on non-spherical particles is still in the qualitative research stage. However, Jiang [4] and Yao [5] and other studies found that the shape of the particles affects the humidity and thus the charging capacity. If the particle shape is not considered, the charged single droplet captures the particles with a large deviation, so the particle shape affects the capture. The effect of efficiency cannot be ignored, and the "sphericity" of the particles is proposed as an important characteristic parameter [6]. Wadell [7] defined, by the ratio of the spherical particle surface area Av to the non-spherical surface area Ap with the same volume of non-spherical particles, a highly accurate relationship between sphericity and drag force established through experimental measurements.

According to the current research progress, the factors affecting the capture efficiency in the process of capturing particles by a single charged droplet are not only particle sphericity but also particle size, particle and droplet charge amount, and particle motion speed [8]. Zuo Ziwen [9] and others designed an experimental device for particle capture by charged droplets and found that the particle capture capacity of charged droplets was more than an order of magnitude higher than that of non-charged droplets, and the capture capacity of droplets was similar to that of non-charged droplets. The amount of charge is linearly proportional. Li Lin [10] conducted an experimental study on the process of electrostatic spraying to remove particulate matter from flue gas. The results show that electrostatic spraying has a significant effect on agglomerating particulate matter and small particles can aggregate to form larger particles, which improves the capture efficiency. Wang Junfeng and others [11] carried out an experiment of trapping particles by charged droplets. The research results show that charged droplets can capture more particles, and particles stay on the surface of the droplets due to surface tension, which affects the capture of the particles' set efficiency. Zuo and others [12] developed an experimental setup for trapping particles by charged droplets and studied the entire trapping process. The experimental results show that the number of particles captured by charged droplets is much larger than that of non-charged droplets, and the particles on the surface of charged droplets obviously aggregate together. Zhang Yaowen [13] developed an electrostatic spray cyclone dust removal system and carried out research on the influence of various factors on the dust reduction rate under different dust concentrations. The experimental results show that the increase in electric field strength is beneficial to improve the dust reduction rate of the device. In the thermo-dynamic characterization of free water and surface water of colloidal monomolecular polymeric particles using DSC, Peng Geng et al. [14] found that the effect of Manning condensation occurs when the charge on the surface of the particles is very high, and the effective charge density decreases, reducing the effective charge on its surface.

In terms of simulation, the core is the construction of the model and the definition of the capture efficiency. From the 2D model to the 3D model, the definition of the capture efficiency needs to be rewritten. In terms of numerical simulation, Wang [15], Shapiro and Laufer [16], and others found that if the droplets and particles are not charged or the charge is low, the particles can be driven to collide with the droplets by adding an electric field, but if the droplets and particles are two, all of them are charged, and the addition of an electric field weakens the capture of the particles by the droplets. Wang Junfeng [17] and Xie Liyu [18] carried out numerical simulation research on particles trapped by charged droplets. The research results show that when the direction of airflow movement and the direction of droplet deformation and projection are the same, although the collision efficiency is reduced, it can be improved and improve capture efficiency. Zhao Haibo and Zheng Chuguang [19–21] numerically simulated the process of removing boiler flue gas by electrostatic spraying, and discussed the effects of inertial collision, interception, Brownian diffusion, and electrostatic force on the capture efficiency. Wang Ao [22] used a three-dimensional model to study and considered the combined effect of inertial and thermophoretic mechanisms. The study found that in the range of the Reynolds number involved in spraying, the flow boundary layer near the droplets was separated and showed non-steady state and non-axisymmetric morphology. Based on the study of particle inertial motion behavior, Slinn and Davenport [23–26] established an inertial capture efficiency formula.

Weber et al. [27] studied the inertial trapping efficiency of droplets flowing around particles under the assumptions of the Stokes flow model and the potential flow model and estimated the inertial trapping efficiency at each Reynolds number (Re) between the two flow assumptions using the interpolation method, and the calculated results have a large deviation from the experimental and numerical simulation results. Bauer and others [28] assumed a steady-state axisymmetric flow field around the flow droplet and numerically simulated the inertial trapping of particles with St = 0.1 to 100 by the droplet at Re = 1 to 400. Slinn and others [25] fitted this numerical simulation result to obtain a formula for calculating inertial trapping efficiency at different Reynolds numbers (Re) and Stokes numbers (St), which is widely used to calculate inertial collision kernels to predict The removal coefficient of particulate matter during rainfall. The formula fitted by Slinn is in good agreement with the experimentally obtained efficiency in the range of Re < 300, but the relative deviation from the experimental data at higher Reynolds number (Re), such as in the study by Horn [29], is more than 20%.

To sum up, the research on wet electrocoagulation technology is extremely important, but the related analysis methods are not yet mature, and more in-depth research is needed. Therefore, a single droplet captures spherical and non-spherical particles under the action of an electrostatic field, which is simulated and studied to establish a numerical model and a physical model of the particle and study the particle size, particle charge, droplet charge, particle velocity, and particle size. Due to the effects of aggregation behavior and other factors on the capture efficiency, the research results are of great significance for improving the wet electrocoagulation technology and improving the particle capture efficiency.

2. Numerical Model

2.1. Force Analysis Equation

The main forces on the particles during their motion are shown in Table 1.

Table 1. Force analysis equations for particle motion.

Action Force	Force Analysis Equation	Serial Number
Gravity	$F_g = \frac{1}{6}\pi {d_p}^3 \rho_p g$	(1)
Buoyancy	$F_g = \frac{1}{6}\pi {d_p}^3 \rho g$	(2)
Inertial force	$F_g = -\frac{1}{6}\pi {d_p}^3 \rho_p \frac{\mathrm{d}u_p}{d_t}$	(3)
Resistance force	$C_D = \frac{F_\gamma}{\pi {r_p}^2\left[\frac{1}{2}\rho\left(u-{u_p}^2\right)\right]}$	(4)
	$F_r = \frac{\pi^2}{2} C_D \left\| u - u_\gamma \right\| (u - u_\gamma)$	(5)
Basset force	$F_B = \frac{3}{2}{d_\gamma}^2\sqrt{\pi\rho\mu}\int_{-\infty}^{t}\frac{\frac{du}{dt}-\frac{du_\gamma}{dt}}{\sqrt{t-\tau}}d\tau$	(6)
Saffman lift force	$F_s = 1.61(\mu\rho)^{\frac{1}{2}} d_\rho^2 (u - u_\rho)\left\|\frac{du}{dy}\right\|^{\frac{1}{2}}$	(7)
Additional mass force	$F_{VM} = \frac{1}{2}\rho V\left(\frac{du}{dt} - \frac{du_\rho}{dt}\right)$	(8)
Magnus lift force	$F_1 = \frac{1}{3}\pi {d^3}_p \rho u \varpi$	(9)

In addition to the main forces mentioned above, the traction forces on the particles should not be neglected. For the traction of spherical particles, the particle velocity (v) is the translational velocity of the particle center of mass, and the continuous fluid velocity (u) is usually defined in the region without particles. The continuous fluid velocity can also be extrapolated to the particle center of mass and expressed as $u_{@p}$, called the "unimpeded velocity." The relative velocity of particles (w) based on the "unobstructed velocity" can be expressed as [30]:

$$w(t) = v(t) - u_{@p}(t) \tag{10}$$

Assuming that the particle and fluid velocities are stable and uniformly distributed over the space away from the particles, $\nabla - u_{@p} = 0$ [31]. At this point, the magnitude of the trapping force is mainly determined by the particle Reynolds number (Rep).

$$\mathrm{Re}_{\rho} = \frac{\rho_f \left| w \right| d}{\mu_g} \tag{11}$$

where d denotes the particle diameter, ρ_f denotes the fluid density, and μ_f denotes the fluid viscosity. Nedelcu [32] derived the trajectory of spherical particles under the condition that the convection term (Rep $\ll 1$) can be neglected:

$$F_D = -3\pi d \mu_f w \tag{12}$$

When the particle Reynolds number increases, the flow behind the particle is initially an attached laminar flow with Rep < 22, then a separate laminar flow zone, which becomes an unstable transition zone at 22 < Rep < 130, and then a turbulent zone at 130 < Rep < 1000 [31–35]. At 2000 < Rep < 300,000, the boundary layer starting in front of the particles ($\theta = 0°$) is laminar and separates at about 80° [35], creating a fully turbulent wake behind the particles. The total drag force is defined according to the drag coefficient (CD) as:

$$F_D = -\frac{\pi}{8} d^2 \rho_f C_D w w \tag{13}$$

For smaller values of Rep, the Stokes traction equation is appropriate. The traction coefficient (CD) measured in the range 2000 < Rep < 300,000 is almost constant, from about 0.4 to 0.45, which is also commonly referred to as the "Newtonian zone" [36]. The Stokes correction can be obtained by normalizing the traction by the creep flow solution.

$$f_{\mathrm{Re}} = \frac{F_D(\mathrm{Re}_\gamma)}{F_D(\mathrm{Re}_\gamma \to 0)} = \frac{C_D(\mathrm{Re}_\gamma)}{24/\mathrm{Re}_\gamma} \tag{14}$$

Equation (14) is uniform for Rep $\ll 1$ and is proportional to Newton's law for Rep (about 3000 < Rep < 200,000). A transition occurs in between due to the appearance and growth of the wake separation bubble [35].

An empirical formula including a traction coefficient can be derived from a series of assumptions. An empirical curve consisting of 10 components was given by Clift et al. [37] for extending the study of spheres to Rep as 106. The Stokes correction in this calculation gives

$$f_{\mathrm{Re}} = 1 + 0.15 {\mathrm{Re}_\gamma}^{0.687} \quad \text{for Rer} \; < \; 800 \tag{15}$$

Regularly shaped non-spherical particles do not have an analytical solution for the trajectory even in creeping flows with large flow velocities; their shape and the corresponding correction for the trajectory can be approximated as ellipsoids by determining the effective aspect ratio (E). As shown in Figure 1, the cylinder can also be corrected for shape (fshape) to maintain sufficient accuracy [38–40]. As with the sphere shape factor, fshape is inversely proportional to the change in terminal velocity (for d = constant), as the trajectory correlation is linear in creeping flows.

$$f_{shape} = \frac{C_{D,shape}}{C_{D,shaere}} \bigg| \mathrm{Re} \ll 1 \& const.wol = \frac{W_{term,sphere}}{W_{term,sphape}} \tag{16}$$

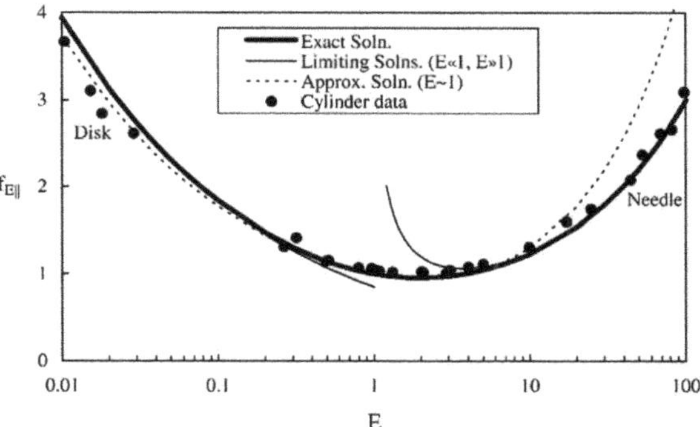

Figure 1. Effect of eccentricity on sphere trajectory correction [28].

In order to estimate the shape factor of non-spherical regular particles, two dimensionless area parameters, the surface area ratio and the projected area ratio, are usually considered [30]:

$$A^*_{surf} = \frac{A_{surf}}{\pi l^2}, A^*_{proj} = \frac{A_{proj}}{\frac{1}{4}\pi l^2} \tag{17}$$

The surface area ratio and the projected area ratio are usually considered. The surface area ratio is always greater than 1, i.e., $A^*_{surf} \geq 1$. The reciprocal of the surface area ratio is more commonly defined as the "sphericity ratio" [7], "sphericity" [24] or "shape factor". For cylinders with length-to-diameter ratio (Acyl), ratio (Acyl), the surface area ratio and equivalent volume diameter can be determined from geometric relationships [30]:

$$E_{cyl} = \frac{L_{cyl}}{d_{cyl}}, A^*_{surf} = \frac{2E_{cyl} + 1}{(18E^2_{cyl})^{\frac{1}{3}}}, d = d_{cyl}\left(\frac{3E_{cyl}}{2}\right)^{\frac{1}{3}} \tag{18}$$

The projection area ratio depends mainly on the direction of projection of the particles and their shape. Under non-isometric conditions, Clift et al. [38] gave the A^*_{surf} and A^*_{proj} equations for biconical and rectangular equations. Although there are other ways to describe the non-spherical properties of particles, these are the two most commonly used parameters for symmetric particles, and both are also the most effective in correlating the traction [25,37].

Leith [41] proposed a correlation between the Stokes shape correction factor and these two area ratios:

$$f_{shape} = \frac{1}{3}\sqrt{A^*_{porj}} + \frac{2}{3}\sqrt{A^*_{surf}} \text{ for Rep } \ll 1 \tag{19}$$

From Equation (19), 1/3 of the sphere resistance is shape resistance (related to the projected area), 2/3 is friction resistance (related to the surface area), and the shape resistance and friction resistance are proportional to the particle size. A review by Ganser [39] shows that if the surface area ratio of a particle can be reasonably made to approach that of a sphere, the following relation can be used for a given aspect ratio.

$$\begin{aligned} A^*_{surf} &= \frac{E^{\frac{-2}{3}}}{2} + \frac{E^{\frac{4}{3}}}{4\sqrt{1-E^2}} \ln\left(\frac{1+\sqrt{1-E^2}}{1-\sqrt{1-E^2}}\right) \\ A^*_{surf} &= \frac{1}{2E^{\frac{2}{3}}} + \frac{E^{\frac{1}{3}}}{2\sqrt{1-E^{-2}}} \sin^{-1}\left(\frac{1+\sqrt{1-E^2}}{1-\sqrt{1-E^2}}\right) \end{aligned} \tag{20}$$

2.2. Governing Equation

2.2.1. Drag Equation

The trajectory of the particles in the vicinity of the adsorbed droplet is determined by Newton's equation [42].

$$m\frac{d\vec{w}}{dt} = \frac{C_d \mathrm{Re}_p}{24}\vec{F_r} + \vec{F_\tau} + \vec{F_g} \tag{21}$$

where w represents particle velocity, m represents particle mass, C_d represents traction coefficient, Re_p represents particle Reynolds number, F_r represents drag, F_g represents gravity, and F_ε represents electrostatic force.

Although the forces on the particles change during their motion and their trajectory changes, the whole physical process still follows the conservation of mass, momentum, and energy. The conservation of mass equation [43]:

$$\frac{\partial \rho}{\partial t} + \frac{\partial(\rho v_x)}{\partial x} + \frac{\partial(\rho v_y)}{\partial y} + \frac{\partial(\rho v_z)}{\partial z} = 0 \tag{22}$$

Conservation of momentum equation.

$$\frac{\partial(\rho v_x)}{\partial t} + div(\rho v_x v) = \frac{\partial \rho}{\partial t} + \frac{\partial \tau_{xx}}{\partial x} + \frac{\partial \tau_{yx}}{\partial y} + \frac{\partial \tau_{zx}}{\partial z} + F_x \tag{23}$$

$$\frac{\partial(\rho v_y)}{\partial t} + div(\rho v_y v) = \frac{\partial \rho}{\partial y} + \frac{\partial \tau_{xy}}{\partial x} + \frac{\partial \tau_{yy}}{\partial y} + \frac{\partial \tau_{zy}}{\partial z} + F_y \tag{24}$$

$$\frac{\partial(\rho v_z)}{\partial t} + div(\rho v_z v) = \frac{\partial \rho}{\partial z} + \frac{\partial \tau_{xz}}{\partial x} + \frac{\partial \tau_{yz}}{\partial y} + \frac{\partial \tau_{zz}}{\partial z} + F_z \tag{25}$$

Conservation of energy equation.

$$\frac{\partial}{\partial t}(\rho h) + div(phu) = div[(k + k_i)divT] + S_k \tag{26}$$

2.2.2. Capture Efficiency

The trapping efficiency of a moving particle in a laminar flow varies depending on the position of release. Here, by denoting y_0 as the maximum value of the trajectory of the trapped particles, the trapping efficiency can be defined as [44]:

$$E_T(d_d, d_p) = \frac{N}{N_0} = \left(\frac{2y_0}{d_d}\right)^2 \tag{27}$$

where d_d denotes droplet diameter, d_p denotes particle diameter, N denotes the number of particles trapped and N_0 denotes the number of particles released over the projected area of the droplet.

In turbulent flow, the flow field is unstable and the trajectory of motion is not unique. For this reason, the collision probability e(d_d,d_p,Y) is proposed to characterize the collision of particles with droplets under turbulent conditions. For simple micron spherical particle capture, the effect of each factor on the capture efficiency can be obtained by modelling the probability of collision for a given position.

The complexity of turbulent pulsations can be found from Figure 2, the previously defined trapping efficiency Equation (27) is not applicable to turbulent flows [45]. As the relative position of the particles, Y, affects the collision efficiency, the trapping efficiency can be synthesized by fitting the collision probability, as shown in Figure 3.

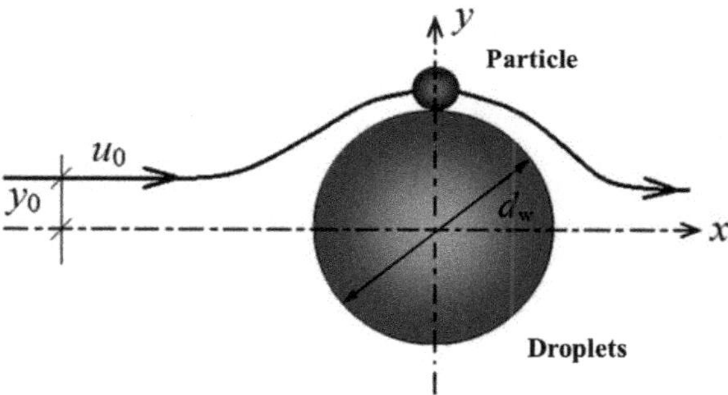

Figure 2. Schematic diagram of droplet trapping particles [44].

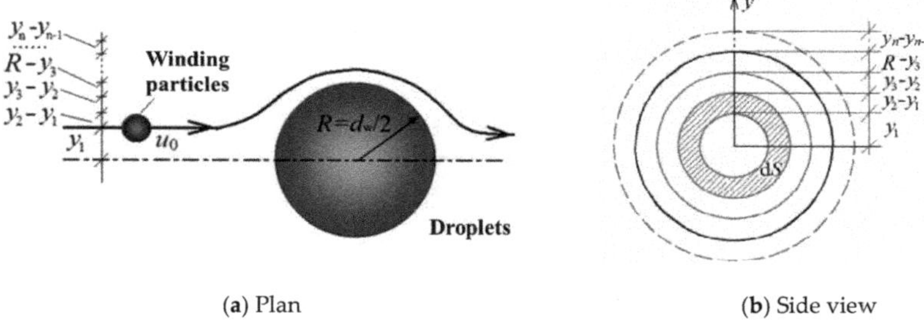

Figure 3. Schematic diagram of turbulence fitting efficiency [44].

We can define the trapping efficiency η_s of droplet trapping particles as [44]

$$\begin{aligned}\eta_s &= \int_0^R e(d_d, d_p, Y)\left[\pi(y+dy)^2 - \pi y^2\right]/\pi R^2 \\ &= \int_0^R 2\pi e(d_d, d_p, Y) y dy/\pi R^2\end{aligned} \tag{28}$$

For ease of fitting, the above equation can be expanded as

$$\begin{aligned}\eta_s &= \sum_{n-1}^{n} e_n \pi(y_n^2 - y_{n-1}^2)/\pi R^2 \\ &= \left[y_1^2 e_1 + (y_2^2 - y_1^2)e_2 + \ldots + (y_n^2 - y_{n-1}^2)e_n\right]/R^2\end{aligned} \tag{29}$$

where y_1, y_2 ... y_n are the 1st, 2nd ... n outer circle radii of the micro-element circle, e_1, e_2 ... e_n are the collision efficiencies of the released particles at the corresponding positions mentioned above, respectively. Using this equation in combination with numerical simulations at different locations, the capture efficiency of a single charged droplet can be fitted to calculate the capture of particulate matter.

The effect of particle sphericity on the capture efficiency is investigated by fitting the collision probability to the capture efficiency, using a model of spherical particles and non-spherical particles of different sphericity of the same volume and varying parameters such as particle size, particle charge, droplet charge, and particle motion velocity to investigate their effect on the capture efficiency and the positive and negative correlation with particle shape.

2.3. Solving Algorithm

Simulation studies are carried out to validate the computational models by building them and determining the physical parameters and boundary conditions. Simulation studies are then carried out first to investigate the collision probability of trapped micron particles. The trapping efficiency of submicron particles is then explored. The effects of particle size, droplet charge, and particle motion velocity are also discussed, and methods to improve the trapping efficiency are analyzed.

The fluid in the simulation is air, the simulation temperature is room temperature, The physical parameters are shown in Table 2 and the particle density is set to 2200 kg/m^3. The model inlet condition is a velocity–inlet boundary. The flow at the outlet is assumed to be fully developed and the model outlet boundary condition is the outflow boundary. A standard k-ω turbulence model is selected and the SIMPLE algorithm is chosen as the method for coupling pressure and velocity.

Table 2. Simulated air physical properties parameters [44].

Materials	Pressure (Mpa)	Temperature (°C)	Density (kg/m^3)	Viscosity Factor (Pa·s)
(air)	0	26.75	1.225	1.79×10^{-5}

2.4. Model Building and Validation

The commercial finite volume software ANSYS FLUENT 18.0 was used to solve the 3D Navier–Stokes equations and the particle equations of motion directly, with the computational domain shown in Figure 4. The single spherical droplet is fixed in the computational domain, and the droplet boundary is a wall boundary condition without considering the variation of the physical and chemical properties of the droplet. The air is fed from the left side of the computational domain in the plane (y–z), and the flow direction is the same as the x-axis. The particles are fed uniformly from the projection of the droplet on the inlet plane. Three hundred particles are fed at once, and the droplet diameter is taken to be 250 μm. The particle charge is set to 7.89E–17C, the particle velocity is set to 1.1 m/s, the electrostatic force is compiled and imported using UDF, and the particle motion and position are tracked using the DPM model. After each particle is tracked, the number of particles deposited onto the droplet (divided by the number of particles flowing since it is the capture efficiency) and the deposition position are counted.

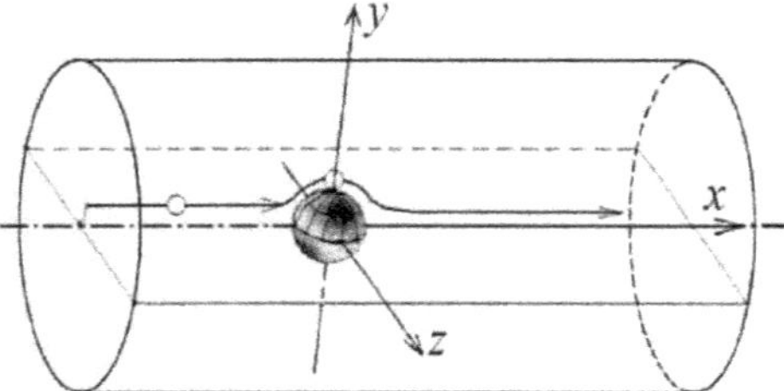

Figure 4. Geometry of the calculation area.

Considering the characteristics of the droplet boundary layer and the influence on the trajectory of the particles, the grid around the droplet was treated in an encrypted manner when dividing the grid, with a minimum grid of 9.4×10^{-13} m^3; see Figure 5.

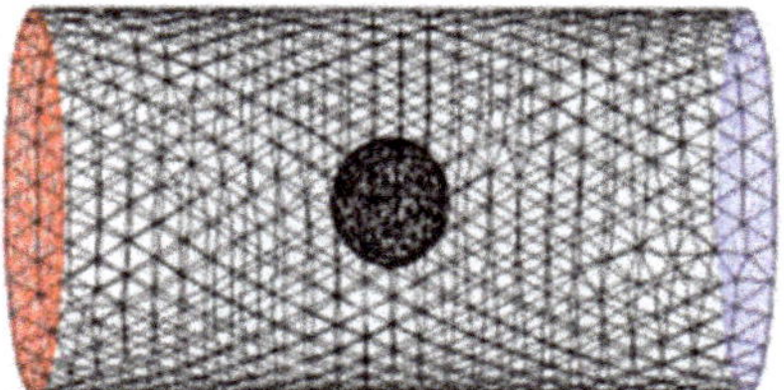

Figure 5. Grid division.

3. Results and Discussion

3.1. Effect of Particle Sphericity on Trapping Efficiency

A preliminary investigation was first carried out to obtain the effect of sphericity on droplet trapping micron particle trapping efficiency by investigating the effect of different particle sphericity on particle collision probability. Particles with sphericity of 0.1, 0.2, 0.3, 0.4, 0.5, 0.6, 0.7, 0.8, 0.9, and 1 were taken to simulate the collision probability of particles with different sphericity. The particle motion velocity was set at 1.1 m/s. Two particle sizes, 1.37 μm and 3.50 μm, were chosen for the simulations, and the results are shown in Figure 6.

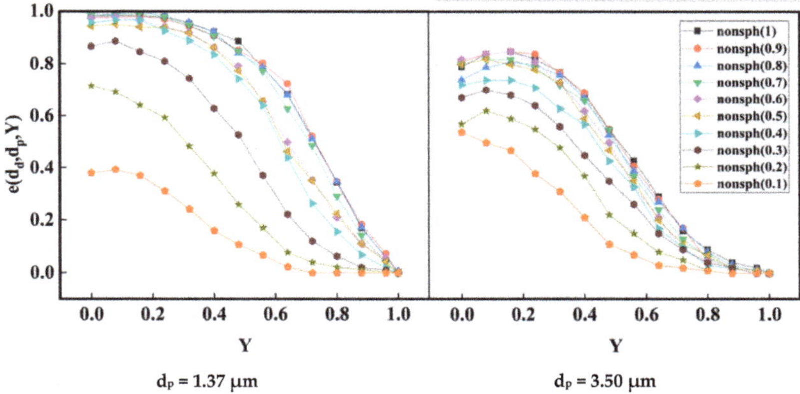

Figure 6. Collision probabilities at different sphericity values.

Analysis of Figure 6 shows that for particles of 1.37 μm, the collision probability is similar for sphericity of 0.6–1, with a significant reduction in particle impact on droplets occurring below 0.5. For particles of 3.50 μm, the difference in collision probability is slight. It overlaps more for sphericity of 0.5–1. Still, the difference in collision probability occurs at a sphericity of 0.4 or less, and particles of size 3.50 μm have a greater collision probability than particles of 1.37 μm under the same conditions. The analysis shows that particle sphericity affects the collision probability, and the effect is more significant for smaller particles. For both the 1.37 μm and 3.50 μm particle sizes, the probability of collision with a droplet decreases as the particle sphericity decreases. It can be assumed that particle shape affects the capture efficiency and that the more complex the particle shape, the lower the capture efficiency of single droplet capture particles.

3.2. Effect of Particle Size on Trapping Efficiency

For simple spherical micron particles, different particle sizes mainly affect their collision behavior, so the effect of particle size on collision probability is explored. Thus its effect on the trapping efficiency is analyzed. From the results in Figure 7a, it can be seen that particle size has a significant effect on the collision probability. As the particle size increases, the collision probability between the particle and the droplet also increases, and the collision efficiency for the 5.49 μm particles can reach 100%. As the particle release position gradually moves away from the droplet center, the particle–droplet collision probability decreases, reducing the capture efficiency.

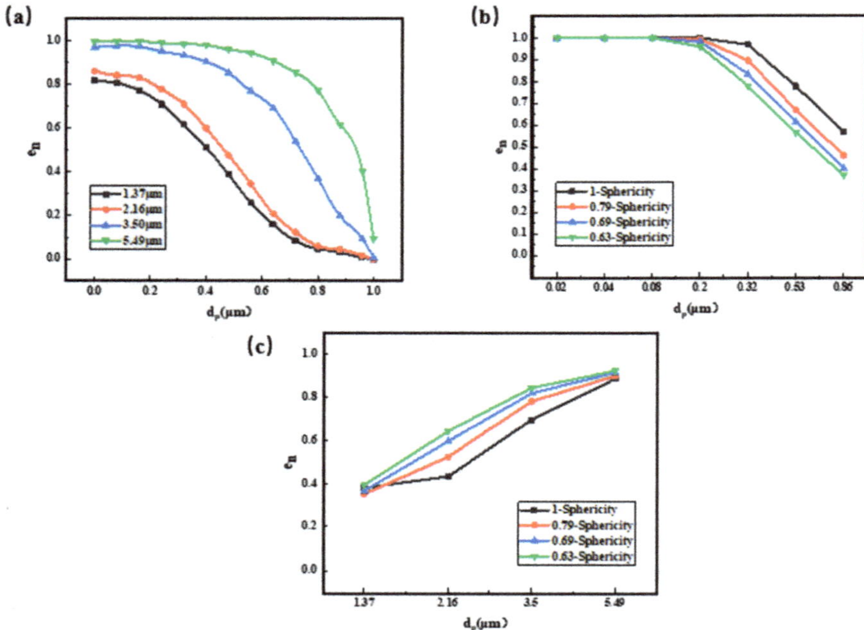

Figure 7. Trapping efficiency for different particle sizes (The sphericity is (**a**) 1 (**b**) 0.79 (**c**) 0.63).

For more complex, non-spherical particles, there is a positive correlation between size, charge, and capture efficiency. The four-micron particles of 1.37 μm, 2.16 μm, 3.50 μm, and 5.49 μm and seven submicron particles of 0.02 μm, 0.04 μm, 0.08 μm, 0.20 μm, 0.32 μm, 0.53 μm, and 0.86 μm given above were used to form linearly arranged aggregates of sphericity 1, 0.79, 0.69, and 0.63, respectively. The aggregates were formed with the sphericity of 1, 0.79, 0.69, and 0.63, respectively. The results in Figure 7b show that the trapping efficiency of the smaller non-spherical particles aggregated at 1.37 μm is less affected by aggregation behavior and mainly by electrostatic force. The larger non-spherical particles aggregated at 5.49 μm do not clearly show the effect of aggregation behavior because the non-spherical particles have a more significant inertial force and the trapping force affected by the particle shape is more diminutive. The electrostatic force mainly affects the non-spherical particles aggregated at 2.16 μm and 3.50 μm. The aggregation behavior of spherical particles with a size of 2.16 μm had the most significant effect on the capture efficiency, with an increase of around 25% at a sphericity of 0.79 and around 62.5% at a sphericity of 0.63. Although non-sphericality reduces the capture efficiency, the charge and size of the particles increase as they aggregate, which contributes to the increase in capture efficiency. The results in Figure 7c show that the capture efficiency of submicron spherical particles decreases as the particle size increases. The trapping efficiency is high for smaller-sized submicron particles, which are subject to less traction, and the non-spherical particles under this condition are mainly subject to electrostatic forces. At this point, the particles are primarily deposited towards the droplet surface, so the trapping efficiency is high. As the degree of particle aggregation increases, the effect of traction is more pronounced than the effect of inertia, and the trapping efficiency decreases.

3.3. Effect of Droplet Charge on Trapping Efficiency

For non-spherical particles, the droplet charge also affects the efficiency of droplet capture of particles. Droplets with charges of 3.95E–13C, 2.36E–13C, 1.38E–13C, and 6.78E–14C and non-spherical micron particles with the sphericity of 0.79 and 0.69, respectively, were selected for the simulations, and the results are shown in Figure 8a. The efficiency of droplet trapping of particles varies significantly with the droplet charge when the particle charge is a certain amount. As the droplet charge increases, the electrostatic force on the particles increases, the particles are pulled towards the droplet, and the efficiency of trapping non-spherical particles increases significantly. As the electrostatic forces on the particles are relatively smaller than the inertial and traction forces on the particles, the main parameter influencing the capture efficiency, in this case, is the particle size. For the same "equivalent diameter," the efficiency of trapping non-spherical particles is smaller than that of trapping spherical particles. For non-spherical particles formed by aggregation of submicron particles, the electrostatic force on the particles increases as the droplet charge increases for a given particle charge, making it easier for a droplet with a higher charge to trap more particles. When the droplet charge is 0C and 3.95E–13C, the capture efficiency is not affected by the aggregation behavior of the particles because the electrostatic force on the particles is 0. At a droplet charge of 3.95E–13C, the submicron particles are mainly affected by the electrostatic force and are less affected by the aggregation behavior of the particles. Analysis of Figure 8b, for the non-spherical particles formed by the aggregation of 0.20 μm spherical particles, shows a different trend when the non-spherical particles are four-particle aggregates, which is due to the larger particle inertial forces at this point coming into play. Analysis of Figure 8c shows that when the particle size is large enough, the particle traction force plays a significant role in trapping. The particle aggregation behavior still affects the trapping efficiency. In contrast, at a droplet charge of 0C, the submicron particles do not affect the trapping efficiency during aggregation due to their small size and low inertial forces. For submicron particles, the trapping efficiency decreases as the degree of particle aggregation increases.

3.4. Effect of Particle Motion Velocity on Trapping Efficiency

For spherical micron particles, the velocities of the above four different sizes of particles were selected as 0.3 m/s, 0.5 m/s, 0.8 m/s, 1.0 m/s, 1.1 m/s, 1.2 m/s, 1.5 m/s, 1.8 m/s, etc. The droplet charge was taken as 3.95E–13C, and the particle charge was taken as 7.89E–17C for the effect of particle motion velocity on the collision efficiency. In turn, its effect on the trapping efficiency was obtained, and the results are shown in Figure 9. When the particle velocity is small, the time the particle spends impacting the droplet is also long. At this time, the particle is subjected to less traction, accompanied by electrostatic force attraction, inertia force, and impact on the droplet. As can be seen from Figure 9a, the probability of collision for particles of size 1.37 μm is close to 100% when the velocity of motion is 0.3 m/s and 0.5 m/s. The impact of particle release position on the collision efficiency is almost non-existent. In contrast, for particles of larger size 5.49 μm, the analysis in Figure 9b shows that the collision probability changes as the particle release position changes.

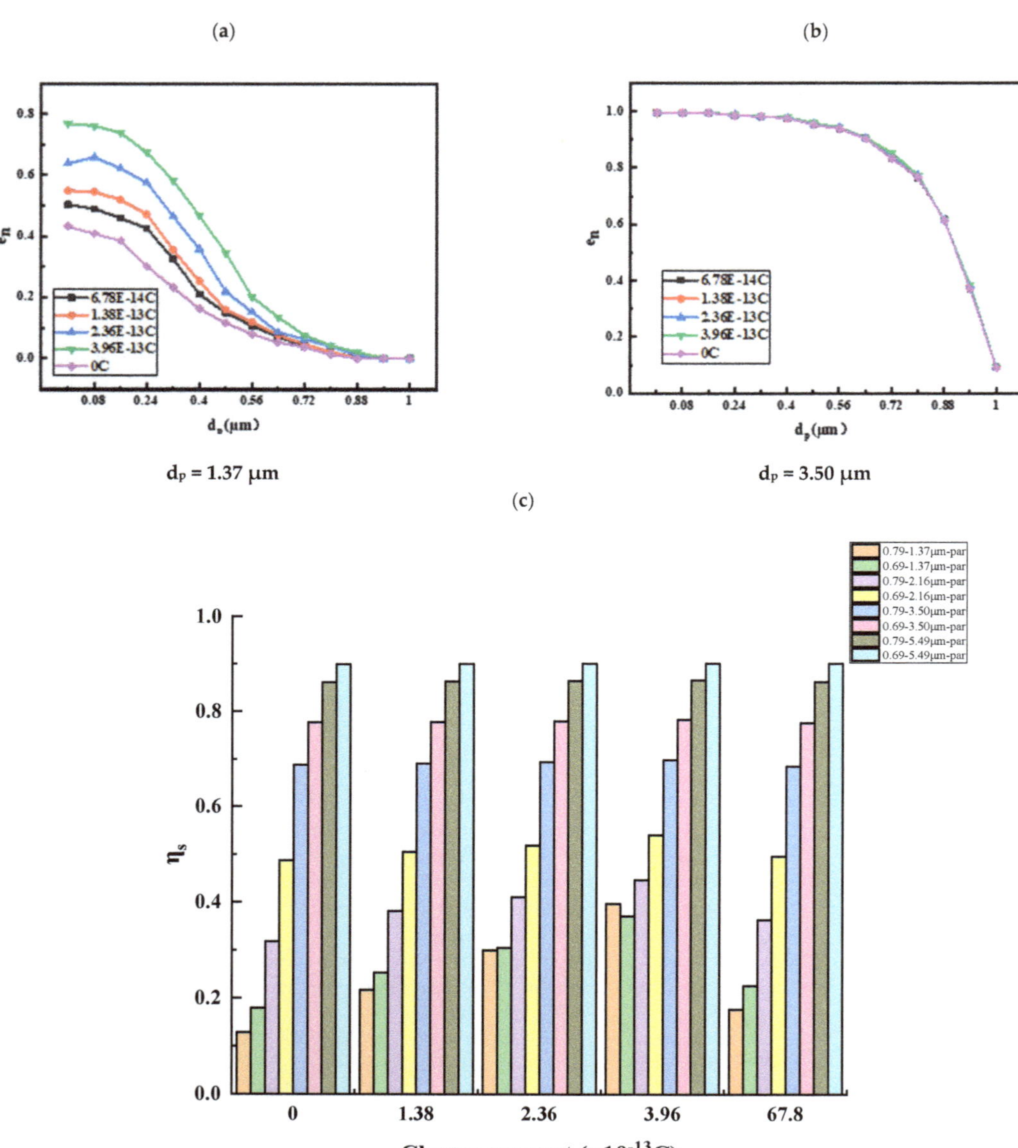

Figure 8. Trapping efficiency for different droplet charges.((**a**) micron particles (**b**) micron particles (**c**) submicron particles).

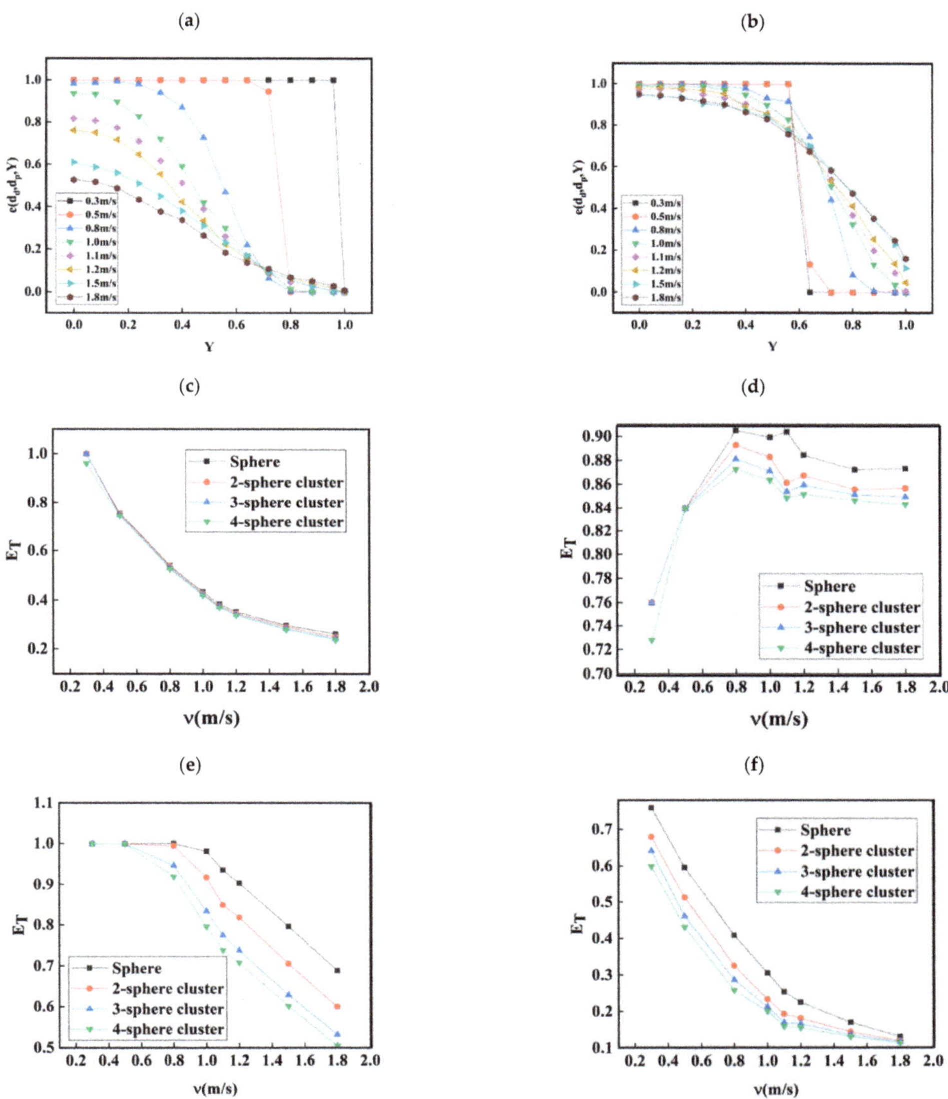

Figure 9. Trapping efficiency for different particle motion velocities ((**a**) 0.3 m/s, (**b**) 0.5 m/s, (**c**) 0.8 m/s, (**d**) 1.0 m/s, (**e**) 1.1 m/s, (**f**) 1.2 m/s).

For non-spherical particles, four different sizes and shapes of micron particles and submicron particle aggregates were selected above, and the velocities were set to 0.3 m/s, 0.5 m/s, 0.8 m/s, 1.0 m/s, 1.1 m/s, 1.2 m/s, 1.5 m/s, 1.8 m/s, etc. The droplet charge was set to 3.95E−13C, and the particle charge was set to 7.89E−17C for the simulation, and the results are shown in Figure 9c. For smaller micron particles of size 1.37 μm, the electrostatic force plays a dominant role when the velocity is small, and the droplets mostly trap the particles along with the electrostatic attraction, so their trapping efficiency is high. When the velocity of particle movement increases, the traction force also increases, which makes the particles deviate from the trajectory, and the trapping efficiency of non-spherical particles decreases. As seen from Figure 9d, for larger particles such as 5.49 μm, the effect of the electrostatic force is not obvious. The inertia force plays a dominant role when the particle motion speed is low, causing some particle motion trajectories to deviate from the droplet direction, leading to a decrease in the capture efficiency. As the velocity of particle motion increases, the trapping force on the particles increases, and the efficiency of non-spherical particles hitting the droplet increases under the combined effect of the forces on the particles. For particles of 1.37 μm size, the effect of particle shape is nearly absent. However, for particles of 5.49 μm size, the efficiency of capturing non-spherical particles is less than that of capturing spherical particles for the same "equivalent diameter." For non-spherical particles formed by aggregation of submicron particles of 0.08 μm, the electrostatic force plays a dominant role when the velocity is small. The droplets mostly trap the particles along with the electrostatic attraction. The trapping efficiency is not affected by the aggregation behavior of the particles when the velocity is 0.3 m/s and 0.5 m/s.

When the velocity of particle movement gradually increases, the particle is subjected to increased traction, making the particle gradually deviate from the trajectory. The efficiency of trapping non-spherical particles decreases, and the higher the degree of aggregation, the lower the trapping efficiency. The analysis in Figure 9e shows that the non-spherical particles formed by the aggregation of spherical particles of 0.20 μm are not affected by the aggregation behavior when the velocity is 0.3 m/s. However, as the velocity increases, the traction force on the non-spherical particles increases, and the efficiency of the non-spherical particles hitting the droplets decreases. The higher the degree of aggregation, the lower the trapping efficiency. As can be seen from Figure 9f, the aggregation of spherical particles of size 0.53 μm forms non-spherical particles because of the larger particle size, and the particles are subject to greater inertial forces and traction forces, so the aggregation behavior of particles has an impact on the capture efficiency, and the higher the degree of aggregation, the lower the capture efficiency. At an enormous particle movement speed, around 1.8 m/s, when the inertial force is more significant, particle aggregation behavior does not affect the capture efficiency. For non-spherical particles formed by the aggregation of 0.86 μm spherical particles, the effect on the trapping efficiency at 1.2 m/s is mainly absent; with the higher the degree of aggregation, the lower the trapping efficiency until then, after which the higher the degree of aggregation, the higher the trapping efficiency is demonstrated. At lower velocities, the capture efficiency is more significant for non-spherical particles with simpler aggregation patterns. When the velocity increases, the capture efficiency is smaller for non-spherical particles with more complex aggregation patterns, and the capture efficiency decreases as the particle aggregation increases.

4. Conclusions

(1) Using the standard k-ω turbulence model and SIMPLE algorithm, the trapping efficiency of charged droplets is positively correlated with the sphericity and the amount of charge. For micron particles, the efficiency of capturing spherical particles is greater than that of capturing non-spherical particles of equal volume. The aggregation behavior of submicron particles with low gravity is not conducive to the improvement of capture efficiency, and the capture efficiency can be reduced by up to 21.1% when non-spherical particles with a sphericity of 0.636 are formed.

(2) Particle size has a significant effect on the capture efficiency, increasing size increases the capture efficiency, and the effect of particle velocity on the capture efficiency needs to be considered in conjunction with particle size. Particle size less than or equal to 2.16 μm has a higher capture efficiency in the range of particle motion velocity less than 0.5 m/s. In comparison, particle size greater than or equal to 3.50 μm has a higher capture efficiency in the range of particle motion velocity greater than 0.8/s.

(3) Increasing the charge of particles and droplets could increase the Coulomb force on particles and improve the trapping efficiency of particles; as the gravity of submicron particles is minimal, the aggregation behavior is not conducive to the improvement of

trapping efficiency, and the trapping efficiency can be reduced by up to 21.1% when tetramers are formed.

Author Contributions: Conceptualization, Q.S. and J.G.; data curation, W.Z.; investigation, Q.S. and Y.D.; methodology, H.D.; project administration, Y.Z. and J.W.; visualization, Q.D.; writing—original draft, W.Z.; writing—review & editing, Y.Z. All authors have read and agreed to the published version of the manuscript.

Funding: This work was supported by the National Natural Science Foundation of China (52006047) and the National Key Research and Development Program of China (Gas Boiler Energy Efficiency Emission Online Inspection and Intelligent Diagnosis Platform Development 2021YFF0600605).

Conflicts of Interest: The authors declare no conflict of interest.

References

1. Luo, R.S.; Wei-Quan, Y.U.; Lin, G.X.; Cheng, G.H.; Chen, Y.C.; Lin, X.; Zheng, Y.F. Analysis on Reform of Wet ESP Technology Utilized in 700MW Generating Set of Power Plant. *China Environ. Prot. Ind.* **2018**, 110–116.
2. Liu, X.; Xu, Y.; Zeng, X.; Zhang, Y.; Xu, M.; Pan, S.; Zhang, K.; Li, L.; Gao, X. Field Measurements on the Emission and Removal of PM2.5 from Coal-Fired Power Stations: 1. Case Study for a 1000 MW Ultrasupercritical Utility Boiler. *Energy Fuels* **2016**, *30*, 6547–6554. [CrossRef]
3. Pettersson, J.; Andersson, S.; Bafver, L.; Strand, M. Investigation of the Collection Efficiency of a Wet Electrostatic Precipitator at a Municipal Solid Waste-Fueled Combined Heat and Power Plant Using Various Measuring Methods. *Energy Fuels* **2019**, *33*, 5282–5292. [CrossRef]
4. Jiang, S.; Feng, H.; Jxa, B.; Hu, Y.B.; Zga, B.; Xl, C. Exploring the influences of Li_2O/SiO_2 ratio on $Li_2O—Al_2O_3—SiO_2—B_2O_3—BaO$ glass-ceramic bonds for vitrified cBN abrasives. *Ceram. Int.* **2019**, *45*, 15358–15365.
5. Yao, J.; Wu, J.; Zhao, Y.; Lim, E.; Cao, P.; Fang, Z.; Wang, C.H. Experimental investigations of granular shape effects on the generation of electrostatic charge. *Particuology* **2018**, *15*, 82–89. [CrossRef]
6. Pendar, M.R.; Pascoa, J. Numerical Simulation of the Electrostatic Coating Process: The Effect of Applied Voltage, Droplet Charge and Size on the Coating Efficiency. *SAE Int. J. Adv. Curr. Pract. Mobil.* **2021**, *22*, 189–205.
7. Wadell, H. The Coefficient of Resistance as a Function of Reynolds Number for Solids of Various Shapes. *J. Frankl. Inst.* **1934**, *217*, 459–490. [CrossRef]
8. Yan, Z.X. Experimental Study of the Effect of Chemical Coagulation on the Performance of a Wet Type Electric Precipitator. Master's Thesis, Yanshan University, Qinhuangdao, China, 2021.
9. Ziwen, Z.; Junfeng, W.; Yuanping, H. Study of the particle trapping characteristics of charged droplets. *J. Eng. Thermophys.* **2018**, *39*, 355–360.
10. Cui, R.; Liu, H.; Gao, H.; Gao, Z. Research status and prospect of abatement of ultrafine particles in flue gas. *China Energy Environ. Prot.* **2018**, *40*, 123–126.
11. Liying, S. Experimental Study on the Charging Performance of Water Mist and Its Dust Reduction Effect. Masters' Thesis, Liaoning University of Engineering and Technology, Liaoning, China, 2019.
12. Yao, G.M.; Wei, B.; Ren, Y.C. Numerical simulation of charged droplets trapping particles. *Energy Conserv.* **2015**, *11*, 41–44.
13. Wen, Z.Y. Experimental Study on the Removal of Fine Particulate Matter from an Electrostatic Spray Cyclone Dust Collector. Master's Thesis, Jiangsu University, Jiangsu, China, 2017.
14. Geng, P.; Zore, A.; Van De Mark, M.R. Thermodynamic Characterization of Free and Surface Water of Colloidal Unimolecular Polymer (CUP) Particles Utilizing DSC. *Polymers* **2020**, *12*, 1417. [CrossRef] [PubMed]
15. Wang, P.K. Collection of aerosol particles by a conducting sphere in an external electric field—Continuum regime approximation. *J. Colloid Interface Sci.* **1983**, *94*, 301–318. [CrossRef]
16. Shapiro, M.; Laufer, G. The effect of electrostatic field direction on the diffusion of charged dust particles from the flowstream to a sphere. *J. Colloid Interface Sci.* **1984**, *99*, 256–269. [CrossRef]
17. Junfeng, W.; Liyu, X.; Yuanping, H. Effect of droplet deformation on the adsorption of fine particles by charged water mists. *Chin. Sci. Technol. Pap.* **2015**, *10*, 1303–1308.
18. Junfeng, W.; Jin, L.; Huibin, X. Advances in wet desulphurisation for the synergistic removal of fine particulate matter. *Chem. Prog.* **2019**, *38*, 3402–3411.
19. Su, L.; Zhang, Y.; Du, Q.; Dai, X.; Wang, H. An experimental study on the removal of submicron fly ash and black carbon in a gravitational wet scrubber with electrostatic enhancement. *RSC Adv.* **2020**, *10*, 5905–5912. [CrossRef]
20. Zhao, H.B.; Zheng, C.G. Operation Optimization of Electrostatically Enhanced Wet Scrubbers. *J. Power Eng.* **2007**, *27*, 954–959.
21. Li, H.; Zhang, M.; Zhu, L.; Yang, J. Stability of mercury on a novel mineral sulfide sorbent used for efficient mercury removal from coal combustion flue gas. *Environ. Sci. Pollut. Res.* **2018**, *25*, 28583–28593. [CrossRef]
22. Ao, W. Study of the Behaviour and Mechanism of Single Droplet Trapping of Fine Particles. Ph.D. Thesis, Tsinghua University, Beijing, China, 2016.

23. Chang, H.J.; Yong, P.K.; Lee, K.W. A moment model for simulating raindrop scavenging of aerosols. *J. Aerosol Sci.* **2003**, *34*, 1217–1233.
24. Feng, D.; Shang, Q.; Dong, H.; Zhang, Y.; Wang, Z.; Li, D.; Xie, M.; Wei, Q.; Zhao, Y.; Sun, S. Catalytic mechanism of Na on coal pyrolysis-derived carbon black formation: Experiment and DFT simulation. *Fuel Process. Technol.* **2021**, *224*, 107011. [CrossRef]
25. Slinn, W.G.N. Some approximations of the wet and dry removal of particles and gases from the atmosphere. *Water Air Soil Pollut.* **1977**, *7*, 513–543. [CrossRef]
26. Davenport, H.M.; Peters†, L. Field studies of atmospheric particulate concentration changes during precipitation. *Atmos. Environ.* **1978**, *12*, 997–1008. [CrossRef]
27. Weber, M.E.; Paddock, D. Interceptional and gravitational collision efficiencies for single collectors at intermediate Reynolds numbers. *J. Colloid Interface Sci.* **1983**, *94*, 328–335. [CrossRef]
28. Bauer, S.E.; Bausch, A.; Nazarenko, L.; Tsigaridis, K.; Xu, B.; Edwards, R.; Bisiaux, M.; Mcconnell, J. Historical and future black carbon deposition on the three ice caps: Ice core measurements and model simulations from 1850 to 2100. *J. Geophys. Res. Atmos.* **2013**, *118*, 7948–7961. [CrossRef]
29. Horn, H.G.; Bonka, H.; Gerhards, E.; Hieronimus, B.; Kalinowski, M.; Kranz, L.; Maqua, M. Collection efficiency of aerosol particles by raindrops. *J. Aerosol Sci.* **1988**, *19*, 855–858. [CrossRef]
30. Loth, E. Drag of non-spherical solid particles of regular and irregular shape. *Powder Technol.* **2008**, *182*, 342–353. [CrossRef]
31. Ma, L.; Qin, C.; Pi, S.; Cui, H. Fabrication of efficient and stable Li_4SiO_4-based sorbent pellets via extrusion-spheronization for cyclic CO_2 capture. *Chem. Eng. J.* **2020**, *379*, 122385. [CrossRef]
32. Nedelcu, D.; Iancu, V.; Gillich, G.R.; Bogdan, S.L. Study on the effect of the friction coefficient on the response of structures isolated with friction pendulums. *Vibroeng. Procedia* **2018**, *19*, 6–11. [CrossRef]
33. Leal, L.G. Bubbles, drops and particles. *Int. J. Multiph. Flow* **1979**, *5*, 229–230. [CrossRef]
34. Dallavalle, J.M. *Micrometrics*; Pitman Publishing: New York, NY, USA, 1948; pp. 55–60.
35. White, F.M. *Fluid Mechanics*; McGraw-Hill: New York, NY, USA, 1991.
36. Feng, D.; Zhao, Y.; Zhang, Y.; Xu, H.; Zhang, L.; Sun, S. Catalytic mechanism of ion-exchanging alkali and alkaline earth metallic species on biochar reactivity during CO_2/H_2O gasification. *Fuel* **2018**, *212*, 523–532. [CrossRef]
37. Schiller, L. Uber die grundlegenden Berechungen bei der Schwerkraftaufbereitung. *Z. Ver. Dtsch. Inge* **1933**, *77*, 318–321.
38. Clift, R.; Gauvin, W. The motion of particles in turbulent gas streams. *Rev. Argent. Cardiol.* **1970**, *82*, 14–28.
39. Dressel, M. Dynamic Shape Factors for Particle Shape Characterization. *Part. Part. Syst. Charact.* **1985**, *2*, 62–66. [CrossRef]
40. Feng, D.; Zhang, Y.; Zhao, Y.; Sun, S.; Wu, J.; Tan, H. Mechanism of in-situ dynamic catalysis and selective deactivation of H2O-activated biochar for biomass tar reforming. *Fuel* **2020**, *279*, 118450. [CrossRef]
41. Leith, D. Drag on Nonspherical Objects. *Aerosol Sci. Technol.* **1987**, *6*, 153–161. [CrossRef]
42. Ganser, G.H. A rational approach to drag prediction of spherical and nonspherical particles. *Powder Technol.* **1993**, *77*, 143–152. [CrossRef]
43. Lu, X. Conservation of Energy, Entropy Identity, and Local Stability for the Spatially Homogeneous Boltzmann Equation. *J. Stat. Phys.* **1999**, *96*, 765–796. [CrossRef]
44. Dai, X. Wet Electrocoagulation and Removal of Fine Particulate Matter from Coal Combustion. Master's Thesis, Harbin Institute of Technology, Harbin, China, 2018.
45. Fengqiao, H.; Yanming, K.; Ke, Z. Influence of turbulence effects on particle trapping processes on raindrop surfaces. *China Environ. Sci.* **2017**, *37*, 13–20.

Review

A Mini-Review on CO_2 Photoreduction by MgAl-LDH Based Materials

Changqing Wang [1], Jie Xu [1,2,3,*] and Zijian Zhou [1]

[1] State Key Laboratory of Coal Combustion, Huazhong University of Science and Technology, Wuhan 430074, China
[2] School of Chemical Engineering Technology, Hebei University of Technology, 5340 Xiping Road, Tianjin 300401, China
[3] Hebei Engineering Research Center of Pollution Control in Power System, Tianjin 300401, China
* Correspondence: j_xu@hebut.edu.cn

Abstract: In recent years, the rapid consumption of fossil fuels has brought about the energy crisis and excess CO_2 emission, causing a series of environmental problems. Photocatalytic CO_2 reduction technology can realize CO_2 emission reduction and fuel regeneration, which alleviates the energy crisis and environmental problems. As the most widely used LDH material in commercial application, MgAl-layered double hydroxide (MgAl-LDH) already dominates large-scale production lines and has the potential to be popularized in CO_2 photoreduction. The adjustable component, excellent CO_2 adsorption performance, and unique layer structure of MgAl-LDH bring specific advantages in CO_2 photoreduction. This review briefly introduces the theory and reaction process of CO_2 photocatalytic reduction, and summarizes the features and drawbacks of MgAl-LDH. The modification strategies to overcome the drawbacks and improve photocatalytic activity for MgAl-LDH are elaborated in detail and the development perspectives of MgAl-LDH in the field of CO_2 photoreduction are highlighted to provide a guidance for future exploration.

Keywords: MgAl-LDH; carbon oxide; photoreduction; modification strategy

Citation: Wang, C.; Xu, J.; Zhou, Z. A Mini-Review on CO_2 Photoreduction by MgAl-LDH Based Materials. *Energies* **2022**, *15*, 8117. https://doi.org/10.3390/en15218117

Academic Editor: Francesco Frusteri

Received: 27 September 2022
Accepted: 26 October 2022
Published: 31 October 2022

1. Introduction

The development of human society is closely related to the utilization of fossil energy, and the consumption of fossil energy has increased sharply accompanied by the advancement of industrialization. The massive consumption of fossil energy has triggered an energy crisis and brought about a large amount of CO_2 emissions, which leads to the serious environmental problems, such as climate change [1] and ocean acidification [2]. Therefore, the reduction of CO_2 emissions has become one of the common challenges faced by mankind [3], and developing an efficient and sustainable technology for CO_2 emissions reduction is of great significance. Pacala and Socolow [4] have proposed the technologies to stabilize the global CO_2 concentration at 500 ppm for the next 50 years, including the improvement in energy utilization efficiency, carbon capture, and storage (CCS) technology, the development of nuclear power and renewable energy, etc. The improvement in energy utilization efficiency is limited by technological breakthroughs, and the development of energy technology has currently hit a bottleneck. The development of nuclear power is conducive to the control of CO_2 and gas pollutant emissions, but radioactive and thermal pollutions are quite serious. The utilization of renewable energy, including wind, hydrogen, and biomass energy, is mainly limited by the high cost, which is derived from economic or technical deficiencies. Carbon capture and storage (CCS) technology is especially suitable for the control of CO_2 emissions from stationary sources and is considered one of the most effective methods for large-scale CO_2 emission control in the short term, but it cannot fundamentally solve the CO_2 emission problem. In addition, CO_2 is the most abundant and economical carbon source on earth, and the above technologies only focus on CO_2 emissions reduction without considering the sustainable utilization of CO_2. CO_2 photocatalytic reduction can utilize solar energy to convert CO_2 into carbon-containing fuels or valued chemical products, which achieves solar energy storage and CO_2 cyclic utilization simultaneously.

Limited by the stable and inert structure of CO_2, CO_2 reduction reaction is quite hard to attain [5]. Developing the efficient photocatalysts to promote the reaction process has attracted much attention, and an ideal photocatalyst should be able to overcome the energy loss as far as possible in the process of light absorption, charge separation and reduce the energy barrier of surface reaction. Traditional metal oxides or sulfides are always limited in practical application by their poor solar utilization efficiency due to the wide bandgap, such as TiO_2 (~3.2 eV) [6], SnO_2 (~3.6 eV) [7] and ZnS (~3.6 eV) [8], which can only utilize the UV light in sunlight. Traditional 2D materials, such as graphene [9] and graphitic carbon nitride [10], have been reported in photocatalysis due to the outstanding optical, electrical and mechanical performance, but the fast charge recombination always limits practical application [11]. Therefore, more novel materials have been explored and various strategies have been proposed to improve catalytic activity, such as surface loading [12], element doping [13], morphology control [14] and adsorption strengthening [15], etc.

Layered double hydroxides (LDHs), a class of 2D anionic inorganic materials, have been wildly reported in heterogeneous catalysis due to their unique layer structure, exclusive physical and chemical properties, and optical characteristics [16]. Due to the abundant hydroxy groups on the surface of LDHs, the superior affinity to CO_2 molecule strengthens CO_2 adsorption and promotes the photocatalytic reaction [17]. The synthetic methods of LDHs are quite facile, which can expediently regulate and control the components, morphologies, defects, grain size and electrical properties. Generally, the synthetic methods of LDHs can be divided into one-step methods and multistep methods. One-step methods always achieve the synthesis of LDHs by the direct preparation process, including coprecipitation [18] and hydrothermal [19] methods, etc. However, many specific LDHs cannot be synthesized by one-step methods directly, and hence multistep methods have been developed. Multistep methods always include ion exchange [20], structure reconstruction [21], and a sol–gel process with specific colloid as the precursor [22]. Among various LDH materials, MgAl-LDH is the most widely used commercially, and has been used in PVC stabilizer, polymer or asphalt additives and retardant [23,24]. Meanwhile, MgAl-LDH already has scaled up production lines [25], and it has the potential to be promoted in CO_2 photoreduction due to the low cost and mature market. Moreover, the synthesis of nanostructured LDHs has been developed in recent years, and the separate nucleation and aging method is the only industrial method for large-scale synthesis of nanoscale LDHs at present [26], which has the advantages of simplicity, speed, and yield amplification. Zhao et al. [27] prepared MgAl-LDH by the separate nucleation and aging method with uniform particle size distribution (60–80 nm), and found that the control of particle size could be simply achieved by adjusting the metal ratio, aging time and temperature. This method has been adapted to the production line of LDHs in China for industrial-scale preparation. Mature industrial production and the development of synthetic methods have indicated that MgAl-LDH has the most promising prospects as a photocatalyst for CO_2 photoreduction.

Numerous reviews have been published about the application of LDH materials for CO_2 photoreduction, and the universal structural features and development prospects of LDH materials have been introduced comprehensively [28–31]. However, systematic reviews of specific MgAl-LDH in the field of CO_2 photoreduction are still scarce. In this review, the theory and reaction process of CO_2 photocatalytic reduction based on LDH materials are introduced and the key steps are summarized. The properties and drawbacks of MgAl-LDH are introduced and summarized. Research on the application of MgAl-LDH in CO_2 photoreduction is reviewed and future development trends pointed out. We believe that this review provides guidance for further development of MgAl-

LDH in CO_2 photoreduction and a direction for the design and construction of MgAl-LDH-based photocatalysts.

2. Theory of CO_2 Photoreduction

CO_2 is a linear molecule with a stable structure, and the bond energy of C=O reaches 750 kJ/mol (the bond energy of C-C is 336 kJ/mol and the bond energy of C-O is 327 kJ/mol) [32]. The direct decomposition reaction of CO_2 usually requires a temperature above 2000 K [33,34], and the hydrogenation reaction of CO_2 can occur at a relatively low temperature with the assistance of hydrogen. Kattel et al. [35] realized the hydrogenation of CO_2 to methanol in the temperature range of 525–575 K. CO_2 photocatalytic reduction can achieve CO_2 reduction under mild conditions, while originally needing harsh conditions [36,37]. The main products of CO_2 photocatalytic reduction in current reports include CO [38], CH_4 [39], HCOOH [40], HCHO [41] and CH_3OH [42], but the conversion efficiency is far from the requirement for practical application. The potential advantages of CO_2 photocatalytic reduction in CO_2 emission control, carbon recycling utilization and solar energy storage are attractive and the exploration of this promising technology is precious and worthy.

2.1. Basic Concepts of Photocatalysis

Since Inoue et al. [43] first reported the use of TiO_2 and other semiconductors in photocatalytic CO_2 conversion into formaldehyde, formic acid and methanol, CO_2 photocatalytic reduction over semiconductor catalysts has attracted much attention. As shown in Figure 1a, the reaction process of CO_2 photocatalytic reduction utilizes the photogenerated electrons to achieve CO_2 reduction, which is inspired by the photosynthesis of plants in Figure 1b.

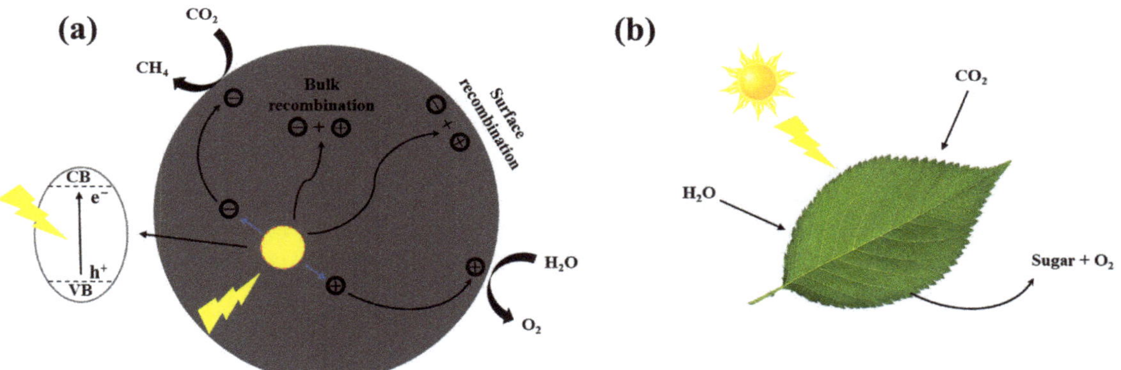

Figure 1. Schematic diagram of (**a**) photocatalytic reaction and (**b**) photosynthesis [44].

When the photon energy is enough to overcome bandgap, the semiconductor catalysts are activated under light illumination. The electrons transfer to the conduction band (CB) and the holes remain in the valence band (VB), and then the electron–holes pairs overcome the Coulomb force to achieve the separation and transfer to surface reaction sites. The redox reaction of adsorbed reactants occurs on the surface of semiconductor catalysts due to strong redox ability of photogenerated charge. In the process of photocatalysis, the interspace between VB and CB represents the energy state that electrons cannot occupy (forbidden band) and the bandgap of forbidden band determines the difficult degree of electron transfer from VB to CB. The recombination of electron–hole pairs always occurs in the process of charge transfer inside or on the surface of catalysts, which greatly limits the catalytic activity. Meanwhile, the adsorption of reactants on the surface is helpful to break the stable structure of CO_2 molecules, which influences the product yield and selectivity.

2.2. Performance Evaluation and Influencing Factors of CO_2 Photoreduction

Photocatalysts play an important role in the photocatalytic CO_2 reduction and the performance evaluation of photocatalysts is of great significance for estimating the photocatalytic activity and revealing the reaction mechanism. Product evolution rate is a common parameter in the performance evaluation of catalysts, which directly reflects the catalytic activity. The unit of product evolution rate is generally $\mu mol \cdot g^{-1} \cdot h^{-1}$ or $\mu mol \cdot h^{-1}$. Photocatalytic CO_2 reduction is a reaction involving multiple electrons, and different products will be generated due to the various intermediates. Product selectivity is also an important parameter for performance evaluation of photocatalysts, and the regulation of photocatalytic product, to targeting product, is of great significance for practical application. Generally, the competitive H_2 evolution reaction is an important factor affecting product selectivity. External quantum efficiency (EQE_λ) and turnover frequency (*TOF*) are always used for performance evaluation of photocatalysts, which reflect the ability of photocatalysts to capture the photons with a particular wavelength and frequency of reaction at specific active sites. In addition, the stability of the catalytic reaction is an important index to evaluate the catalytic performance of catalysts. The long-term photocatalytic reaction is the most direct method to evaluate the reaction stability and the cyclic experiment is also used widely to evaluate the reaction stability. At present, there is still no unified standard for performance evaluation of photocatalysts in CO_2 photoreduction. The performance evaluation methods and the defined evaluation parameters in various studies are different, and a unified standard for performance evaluation of photocatalysts is urgently needed.

$$EQE_\lambda = \frac{the\ number\ of\ reacted\ electrons}{the\ number\ of\ incident\ photons} \times 100\% \tag{1}$$

$$TOF = \frac{the\ number\ of\ reacted\ electrons\ per\ second}{the\ number\ of\ active\ sites} \tag{2}$$

In addition to the properties of catalysts, the experimental conditions also have an influence on the catalytic performance of photocatalysts, including light intensity, pressure, temperature, reactor structure and sacrificial reagents. Generally, the photocatalytic activity is proportional to light intensity, but the recombination of photogenerated electrons and holes will be aggravated and the catalytic activity will be inhibited when the light intensity is too high [45]. The increase of pressure in the reaction system can improve the solubility of CO_2 in the aqueous solution, which accelerates the catalytic reaction rate and affects the product selectivity. Theoretically, the photocatalytic reaction is not sensitive to the temperature change, but the electron collision frequency and diffusion rate will increase at the high temperature and the catalytic activity will also be improved. Reactor structure affects the light absorption and the contact between the catalysts and CO_2 molecules, and the catalytic activity is also affected. The addition of sacrificial reagents can effectively trap holes and inhibit the recombination of electron–hole pairs, and the catalytic activity can be greatly improved. Due to the different reaction systems and experimental conditions, it is very difficult to compare the catalytic activities of photocatalysts strictly in different reports, and a unified standard for experimental method also needs to be developed.

2.3. Reaction Process of CO_2 Photoreduction

Photocatalytic CO_2 reduction involves many complex physical and chemical processes, including the adsorption and activation of CO_2 molecules, the generation of electron–hole pairs, the separation and transfer of charge carriers and the surface redox reaction. Electrons will transfer from VB to CB of semiconductor catalysts after absorbing the energy of photons under light irradiation, which is a necessary step for photocatalytic reactions. Only when the energy of incident photon is higher than the bandgap of the semiconductor catalyst can electrons be excited from VB to CB and the photogenerated electron–hole pairs can generate. After the generation of electron–hole pairs, electrons and holes are still bound together due to the effect of Coulomb force and a bound electron–hole pair is called an exciton. The process of overcoming Coulomb force and achieving the separation of electron–hole pairs has an important influence on the photocatalytic activity. The separation efficiency of electron–hole pairs is closely related to the structure of the semiconductor catalysts. For example, layered materials usually have an advantage in the separation of electron–hole pairs because the presence of interlayer electric field can promote the transfer of electrons [46]. Photogenerated electrons and holes will diffuse to the surface of semiconductor catalysts and the recombination of electrons and holes always occurs inside or on the surface of catalysts. The recombination of charge carriers leads to the annihilation of electrons and holes and releases energy through radiation simultaneously. For the same material, the diffusion distance of electrons and holes is fixed and reducing the particle size can reduce the diffusion length, which is beneficial to reduce the recombination rate of electrons and holes.

After the electrons and holes migrate to the active sites on the surface of semiconductor catalysts, the redox reaction will occur between the adsorbed CO_2 or H_2O molecules and electrons or holes. Generally, H_2O molecules will be oxidized by holes to O_2 and CO_2 molecules will be reduced to different products depending on the number of electrons involved in the reaction. In this process, the electrons in the CB should provide higher energy (more negative) than the reduction potential of CO_2 to targeting products and the holes in the VB should provide higher energy (more positive) than the oxidation potential. The standard redox potentials of reactions are shown in Table 1 (vs. NHE, pH = 7). The first step of redox reaction is the activation of CO_2 molecule by a single-electron reaction, and CO_2 molecule obtain an electron to form $^{\bullet}CO_2^{-}$ and break the stable linear structure. The redox potential of $^{\bullet}CO_2^{-}$ formation is quite high, and this step is very difficult and needs large energy to promote the progression of subsequent reactions. In view of thermodynamics, the targeting product can be obtained when the electrons and holes in the CB and VB can provide sufficient energy to overcome the redox potential of corresponding product. However, the multiple-electrons-involved reaction is quite hard to achieve compared with the few-electrons-involved reaction in view of kinetics. For example, only two electrons are required for CO production, but eight electrons are required for CH_4 production, and more electrons are required for C2+ product production. Therefore, CO is the most readily available product. The selectivity of products can be regulated by improving the band structure, charge separation and surface adsorption.

Table 1. Standard redox potentials for photocatalytic CO_2 reduction to different products [47].

No.	Reaction	Standard Redox Potential (V, NHE, pH = 7)
1	$CO_2 + e^- \rightarrow {}^{\bullet}CO_2^-$	−1.90
2	$CO_2 + 2H^+ + 2e^- \rightarrow HCOOH$	−0.61
3	$CO_2 + 2H^+ + 2e^- \rightarrow CO + H_2O$	−0.53
4	$CO_2 + 4H^+ + 4e^- \rightarrow HCHO + H_2O$	−0.48
5	$CO_2 + 6H^+ + 6e^- \rightarrow CH_3OH + H_2O$	−0.38
6	$CO_2 + 8H^+ + 8e^- \rightarrow CH_4 + 2H_2O$	−0.24
7	$2CO_2 + 12H^+ + 8e^- \rightarrow C_2H_5OH + 3H_2O$	−0.16
8	$H_2O + 2h^+ \rightarrow 1/2O_2 + 2H^+$	+0.82

CO_2 adsorption is also an important step that affects the catalytic activity and product selectivity. Strengthening the adsorption of CO_2 is beneficial to overcome the energy barrier for CO_2 activation to $^{\bullet}CO_2^{-}$, and the metal or defect sites on the surface can facilitate the adsorption of CO_2 and promote the reduction reaction [48]. Product selectivity is closely related to the adsorption model of the CO_2 molecule. If CO_2 is absorbed on the surface of catalysts by only bonding with O atoms, it is inclined to form formic acid ions by combining with hydrogen atoms. Formic acid may be the final product if the C-O bond is not further broken. If CO_2 is absorbed on the surface of catalysts by bonding with C atom, it is inclined to form carboxyl groups by combining with hydrogen atoms preferentially and C-O bond is broken early to form CO. The adsorbed CO can be further hydrogenated to form CH_4 [32]. Zhou et al. [49] found that the introduction of Ru atoms into TiO_2 could convert the main product from CO to CH_4 in CO_2 photoreduction due to the synergy of Ru and oxygen defects. The competitive adsorption of CO_2 and H_2O also has an influence on the product selectivity, which can be adjusted by changing the surface metal sites. Han et al. [50] found that the adsorption energies of CO_2 and H_2O could be adjusted by changing the metal species over metal–organic framework monolayers and the product selectivity towards CO was enhanced. Previous research indicated that the metal species in LDHs also greatly affected product selectivity in CO_2 photoreduction, while Zn in LDHs is beneficial for CO generation [51] and Cu in LDHs is beneficial for CH_3OH generation [52].

3. Structures and Properties of MgAl-LDH

3.1. Component and Characteristic of LDH Materials

Layered double hydroxides (LDHs), a class of anionic inorganic materials, is a hydrotalcite-like compound with the chemical formula of $[M^{2+}{}_{1-x}M\_{}^{3+}{}_{x}(OH)_2]^{x+}(A^{n-}{}_{x/n})$ •yH_2O, where M^{2+} and M^{3+} represent divalent and trivalent metal cations, x is the molar ratio of trivalent cations/total cations, and y is the number of H_2O molecules between layers [53]. As shown in Figure 2, the layered structure of LDHs is composed of the arranged divalent and trivalent metal cations coordinated by OH-octahedrons, while the layered structure is positively charged and the charge-balancing anions (A^{n-}) exist in the interlayer. Common divalent metal cations in LDHs include Mg^{2+}, Zn^{2+}, Ni^{2+}, Mn^{2+}, and trivalent metal cations include Al^{3+}, Fe^{3+}, Cr^{3+}, while anions include F^-, Cl^-, $NO_3{}^-$, $CO_3{}^{2-}$. The basic structural parameters of MgAl-LDH are shown in Table 2. The chemical stability of LDHs is related to the metal species in LDHs. Generally, the metal elements with ionic radii similar to Mg^{2+} can form stable layer structure [54], which is close to $Mg(OH)_2$, but it is not completely absolute. For example, the ionic radius of Ca^{2+} (100 pm) has a huge difference compared with that of Mg^{2+} (72 pm), but it can form stable layer structure with Al^{3+} [55]. In addition, the properties of specific metal elements also influence the chemical stability. Although the ionic radius of Cu^{2+} (73 pm) is similar to that of Mg^{2+}, the Jahn–Teller effect causes the distortion of octahedron and the stable layer structure cannot be formed [56]. Pt^{2+} and Pd^{2+} have strong tendency to form planar tetra-coordination structures and cannot participate in the construction of stable LDH layers. The chemical stability of photocatalysts is quite important for practical application and it should be considered in the design of LDHs.

Table 2. Structural parameters of MgAl-LDH [57].

Structural Parameters	Value
Interlayer distance	0.76 nm
Bandgap	4.631 eV
Work function	5.052 eV
CB referred to vacuum level	−0.617 eV
VB referred to vacuum level	−5.248 eV

Due to the positive charges of layer structure, anions (A^{n-}) in the interlayer play the role of charge compensation for overall electric neutrality. Generally, the volume, number, valence state of anions and the bonding intensity with layer structure influence the interlayer spacing [58]. The neighboring layers are connected with each other by hydrogen bonds originated from H_2O molecules and anions. Therefore, the interrelation between neighboring layers can be weakened by changing anions and the monolayer LDH can be exfoliated. LDHs will lose surface OH, interlaminar anions and H_2O molecules during the calcining process and the highly dispersed metal oxides can be obtained. When metal oxides derived from the calcination at relatively low temperature get in touch with water or anion solution, it will restore the layer structure of LDHs due to the memory effect of LDHs and the absorbed H_2O molecules will convert to OH [59].

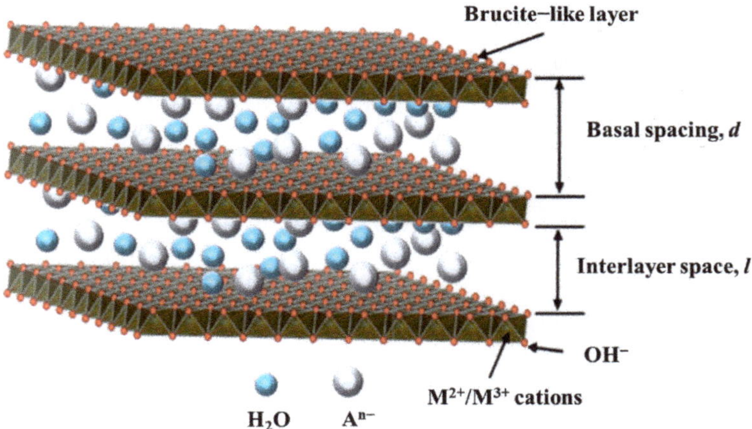

Figure 2. Schematic diagram of LDH structure [60].

3.2. Advantages of MgAl-LDH for CO_2 Photoreduction

As 2D materials with unique layer structure, LDHs have advantages for CO_2 photoreduction derived from physicochemical properties, morphology characteristics, and layer structure. The advantages of MgAl-LDH for CO_2 photoreduction are summarized in this section, while some advantages are common for LDHs and some advantages are unique for MgAl-LDH.

3.2.1. Tunable Composition for Demand

The metal species, element ratio and anion species for the construction of LDHs can be adjusted for demand in the catalytic reaction, which immediately leads to a change in optical absorption and chemical properties. The introduction of specific metal species into LDHs can regulate the bandgap to expand the light absorption range and reduce the energy barrier of CO_2 activation. Wang et al. [61] explored the mechanism of d-bands classification in LDHs for CO_2 photoreduction by changing metal cations M (Mg^{2+}, Ni^{2+}, Cu^{2+}, Zn^{2+}) in MAl LDHs and NiAl-LDH exhibited the highest CO yield. The upward shift of d-band center position could induce anti-bonding state above the Fermi level and strengthen CO_2 adsorption and activation. Different metal species in LDHs also have an influence on the product selectivity due to the change of reaction paths over different active sites. Xiong et al. [62] adjusted the metal cations in Zn-based LDHs for CO_2 photoreduction. The results indicated that the main product of LDHs constructed by Ti^{4+}, Ga^{3+} and Al^{3+} was CO or CH_4, and the main product of LDHs constructed by Fe^{3+} and Co^{3+} was H_2. The d-band center of specific metal (Ti^{4+}, Ga^{3+} and Al^{3+}) was close to Fermi level, resulting in the strong CO_2 adsorption and the low energy barrier of the conversion from CO_2 to CO and CH_4.

The metal element ratio of LDHs can be adjusted easily by changing the amount of metal precursors, and the chemical property and charge density of layers also change in this process. When the molar ratio of Mg^{2+}/Al^{3+} is at the range of 2.0–4.0, the octahedral structure is formed in MgAl-LDH and the positive charge of the layers is disordered. The charge density of layers and the electrostatic force decrease with the increase of the molar ratio of Mg^{2+}/Al^{3+}. $Mg(OH)_2$ and $Al(OH)_3$ maybe appear accompanied with the formation of MgAl-LDH if the molar ratio of Mg^{2+}/Al^{3+} is beyond the range of 2–4. The molar ratio of metal elements in LDHs influences the bandgap and CO_2 adsorption directly. Tao et al. [63] adjusted the molar ratio of Al^{3+}/Ce^{3+} in NiAlCe-LDH from 4 to 0 and found that the bandgap decreased from 2.07 to 1.81 eV. Yang et al. [28] found that the intensity of CO_2 adsorption over LDHs was strongest when the molar ratio of metal elements was in the range of 2–3. Mori et al. realized the isolated single-atom Ru bound on MgAl-LDH and found that the CO_2 adsorption capacity could be adjusted near Ru sites by changing the molar ratio of Mg^{2+}/Al^{3+}. When the molar ratio of Mg^{2+}/Al^{3+} was 5, the CO_2 adsorption capacity reached the maximum due to the variation of basicity [64].

The anion species in the interlayer can be adjusted by precursor selecting or ion exchange, resulting in the change of chemical property, stability and the interaction intensity between anions and layers. Generally, the volume, amount, valance state of anions and binding intensity between anions and hydroxyl determine the interlayer distance and space. Hirata et al. [65] synthesized ZnCr-LDH with different interlayer anions ($CO_3{}^{2-}$, Cl^-, $SO_4{}^{2-}$, $NO_3{}^-$) for O_2 evolution by photocatalytic H_2O decomposition and found that Cl^- intercalated ZnCr-LDH exhibited the highest photocatalytic activity. Previous studies have shown that $CO_3{}^{2-}$ intercalated LDHs had more stable structure and stronger interactions between anions and layers than $NO_3{}^-$ intercalated LDHs [66]. The larger interlayer distance of $NO_3{}^-$ intercalated LDHs makes the exfoliation of LDHs easier to implement. Xu et al. [67] obtained the exfoliated MgAl-LDH by the mechanical vibration after ion exchange from $CO_3{}^{2-}$ to $NO_3{}^-$. However, previous studies indicated that the existence of $NO_3{}^-$ was not beneficial to CO_2 adsorption due to the occupancy of active sites over MgAl-LDH [68].

3.2.2. Outstanding CO_2 Adsorption Performance

The basicity of LDHs can promote the adsorption and activation of CO_2 molecules on the surface of catalysts. Meanwhile, the basicity and acidity of LDHs can be adjusted by changing chemical compositions [69]. Generally, the acidity of LDHs is closely related to interlayer anions. When the common anions ($CO_3{}^{2-}$, $NO_3{}^-$) occupy the interlayer space, the acidity of LDHs is always quite weak. The basicity of LDHs is always determined by the metal species and metal element ratio [66]. Therefore, LDHs always exhibit outstanding CO_2 adsorption performance. In the photocatalytic reaction over LDHs, the adsorbed CO_2 may exist in three forms: (1) dissociative or captured $CO_3{}^{2-}$ in the interlayer of LDHs, (2) dissolved CO_2 in water, (3) adsorbed CO_2 on the surface of LDHs in the form of linear or nonlinear molecules [70–72]. Compared with the surrounding free CO_2 or $CO_3{}^{2-}$, the interlayer $CO_3{}^{2-}$ is easier to be reduced to target products in the photocatalytic reaction, which will be supplemented subsequently by surrounding carbon species [73]. The nonlinear CO_2 molecules adsorbed on the surface of LDHs are unstable and easy to be activated under illumination [74]. Extending the interlayer distance of LDHs can increase the CO_2 adsorption capacity, promote the diffusion of CO_2 and react with hydroxyl groups to form bicarbonate intermediates, resulting in the decrease of energy barrier in photocatalytic reduction [75,76]. Therein, MgAl-LDH shows stronger CO_2 adsorption capacity than other LDHs due to the stronger basicity of Mg^{2+} compared with Zn^{2+}, Ni^{2+} and Cr^{3+}. Hong et al. [73] synthesized MgAl-LDH modified g-C_3N_4 for photocatalytic CO_2 reduction and found that MgAl-LDH greatly enhanced the CO_2 adsorption capacity. MgAl-LDH exhibited much higher CO_2 adsorption capacity than NiAl-LDH, ZnAl-LDH and ZnCr-LDH at room temperature. Meanwhile, MgAl-LDH could realize the targeting activation of CO_2 in the photothermal reaction [77].

3.2.3. Layer Structure for Construction of Composite Catalysts

The layer structure of LDHs has unique advantages for the construction of composite catalysts. On the one hand, the hierarchical heterostructures containing 2D nanosheets have multiple intrinsic advantages in heterogeneous photocatalysis, including the increased light harvesting capacity, the accelerated charge separation and transfer, and the promotion of surface redox reactions [78]. On the other hand, the intercalation of LDHs can assemble the layer structure and specific guest anions, and the guest anions can overcome the force between the layers of LDHs to insert into the interlayers. Therefore, specific heterogeneous catalysts can be custom-made based on the characteristic of LDHs. Dou et al. [79] fabricated TiO_2@CoAl-LDH nanospheres with core–shell structure by in situ growth of LDHs on the surface of TiO_2 hollow nanospheres. The interaction between the core–shell structure and the matched band structure are favorable for photogenerated charge separation and transfer. Zhao et al. [80] utilized the confined interlayer space of LDHs to enhance the dispersion of metal active sites and inserted a series of metalloporphyrin molecules (TCPP-M, M = Zn^{2+}, Co^{2+}, Ni^{2+}, Fe^{2+}) into interlayer of CuAl-LDH by ion exchange, and the heterogeneous catalysts with highly dispersed metal sites were obtained after calcination. The framework of metalloporphyrin molecules and the confined interlayer space of LDHs effectively improved the dispersion of active sites on the surface of catalysts. Nayak et al. [81] synthesized g-C_3N_4/NiFe-LDH catalyst by an impregnation method and the exfoliated g-C_3N_4 sheets were dispersed on the layer of LDHs. The strong interaction promoted the charge transport and enhanced the lifetime of carriers, and the composite catalyst exhibited high activity in the photocatalytic H_2 production from water decomposition.

3.3. Drawbacks of MgAl-LDH

Although MgAl-LDH has many advantages in photocatalytic CO_2 reduction, some drawbacks still greatly limit the practical application of MgAl-LDH. The drawbacks related to the properties of MgAl-LDH are summarized below.

3.3.1. Wide Bandgap of MgAl-LDH

Visible light accounts for 45% in the solar spectrum, while ultraviolet light only accounts for 5%. MgAl-LDH has a wide bandgap and can only absorb the ultraviolet light in photocatalytic reaction. As shown in Figure 3, Xu et al. [57] calculated the bandgap and potentials of Mg-, Zn-, Co-, and Ni based LDHs by DFT calculation. The results indicated that the bandgaps of Co- and Ni-based LDHs were more narrowed than 3.1 eV and could absorb the visible light, but the bandgaps of Mg- and Zn-based LDHs were wider than 3.1 eV and could only absorb ultraviolet light. Therein, the bandgap of MgAl-LDH was 4.63 eV and the visible light could not be utilized in photocatalytic reaction. As the most widely used LDH material, the wide bandgap of MgAl-LDH seriously limits its potential application. In recent reports, the bandgaps of MgAl-LDH synthesized by Nayak et al. [82], Gao et al. [83], Chen et al. [84] and Ning et al. [85] are 3.20, 3.57, 3.66 and 4.18 eV, respectively. The bandgaps of MgAl-LDH in various reports are inconsistent, but none of them can absorb the visible light. The poor visible light absorption performance is ascribed to the lack of electrons in the d orbital. As shown in Figure 4, the lack of electrons in the d orbital leads to the absence of intermediate band and the electron transfer is quite difficult between the metal atoms. Therefore, the electron transition requires high energy and the utilization efficiency of visible light is quite unsatisfied.

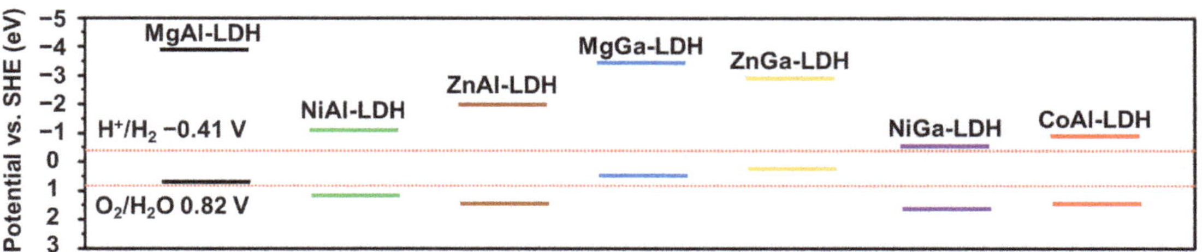

Figure 3. Band potentials of Mg, Zn, Co, Ni based LDHs [57].

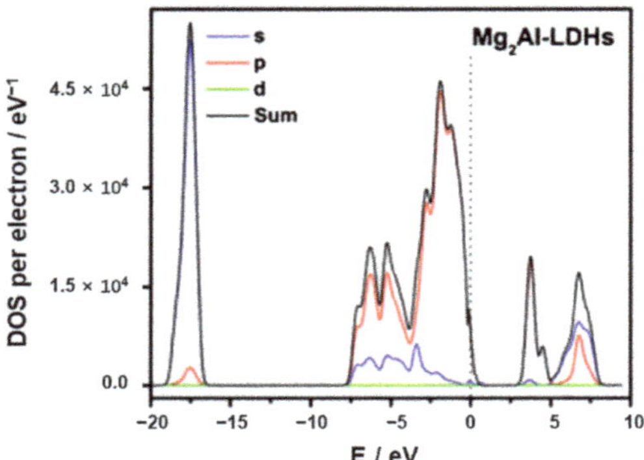

Figure 4. Densities of state for MgAl-LDH [86].

3.3.2. High Recombination Rate of Electron–Hole Pairs

Single LDH materials always have a high recombination rate of electron–hole pairs and the performance of photogenerated charge transfer is poor. As shown in Figure 5, the different steps and time scales in the photocatalytic reaction based on semiconductor catalysts are illustrated. The catalyst will generate high-energy electrons and holes under the excitation of photons (step I). Due to the existence of Coulomb force, some electrons and holes will recombine in the bulk of catalyst and release energy in the form of light or heat (step II). Meanwhile, the uncombined electrons and holes will migrate from the bulk to the surface of catalyst (step III), and some electrons and holes will also recombine on the surface of catalyst (step IV). The remaining electrons and holes finally reach the active sites on the surface and participate in the catalytic reaction (step V). Generally, the generation of electrons and holes is completed at the time scale of femtosecond (fs) and the migration of electrons and holes from bulk to surface is completed within hundreds of picoseconds (ps), while the catalytic reaction between charge carriers and absorbed reactants is completed within a few nanoseconds (ns) to a few microseconds (μs) [87]. In contrast, the recombination of electrons and holes in the bulk of catalyst is always completed within a few picoseconds (ps) while the recombination of electrons and holes on the surface is always completed within ten nanoseconds (ns) [88]. The process of charge recombination is much faster than the process of charge transfer and surface reaction. With the assistance of DFT calculation, Liu et al. [86] demonstrated that the bandgap of ZnCr-LDH is 2–3 eV and has satisfactory visible light absorption performance, but the rapid recombination of charge carriers leads to low separation efficiency of electron–hole pairs and limited photocatalytic activity. Therefore, most of the electrons and holes will be annihilated due to the high recombination rate and only a few electrons and holes can participate in the final surface redox reaction for single MgAl-LDH catalyst.

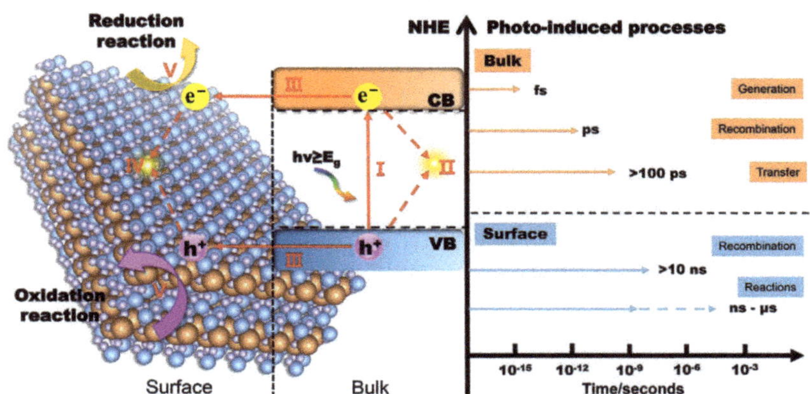

Figure 5. Different steps and time scales in the photocatalytic reaction based on semiconductor catalysts [89].

4. CO_2 Photocatalytic Reduction by MgAl-LDH-Based Materials

Due to the poor visible light absorption and fast charge recombination of MgAl-LDH, its application in photocatalysis is limited and studies on the application of MgAl LDH in photocatalytic CO_2 reduction are relatively few. In order to overcome the drawbacks of MgAl-LDH in visible light absorption and charge separation, many efforts have been devoted to developing different modification methods to improve the photocatalytic performance. Several strategies, such as defect engineering, morphology control, component regulation and material coupling, have been reported for the modification of MgAl-LDH. The specific modification methods are summarized in this section and the performance summary of MgAl-LDH based photocatalysts for CO_2 photoreduction is listed in Table 3.

4.1. The Introduction of Active Components into MgAl-LDH

Single MgAl-LDH always exhibits poor photocatalytic activity due to its own drawbacks. Iguchi et al. [74] compared the catalytic activity of different LDHs in the photocatalytic CO_2 reduction and found that NiAl-LDH exhibited the highest photocatalytic activity in the presence of Cl^- as the hole sacrificial agent. In contrast, MgAl-LDH only exhibited the poor CO evolution rate of 0.6 $\mu mol \cdot g^{-1} \cdot h^{-1}$ under UV light. Due to the ordered dispersion of metal cations in the layer structure of LDHs and flexible adjustment property, active components can be introduced into the layer structure by element doping to improve the visible light absorption and charge separation of MgAl-LDH.

Table 3. Summary of MgAl-LDH based photocatalysts for CO_2 photoreduction.

Catalyst	Synthesized Method	Reaction System	Light Source	Type of Reactor	Reaction Condition	Product Yield ($\mu mol \cdot g^{-1} \cdot h^{-1}$)	Selectivity	Ref.
MgAl-LDH	Coprecipitation	$Cl^- + H_2O$	400 W Hg lamp (UV light)	220 mL closed circulating system	7.7 mmol CO_2 + 350 mL H_2O	CO = 0.6 CH_4 = 0.3 H_2 = 1.2	CO: 28.6% CH_4: 14.3% H_2: 57.1%	[74]
MgAl-LDH	Microwave-hydrothermal	-	150 W Xe lamp (Simulated sunlight)	200 mL stainless cylindrical reactor	-	CO = 1.3 CH_4 = 0.1	CO: 92.9% CH_4: 7.1%	[68]
Fluorinated MgAl-LDH	Coprecipitation	$Cl^- + H_2O$	400 W Xe lamp (UV light)	Quartz inner-irradiation reactor	7.6 mmol CO_2 + 350 mL H_2O	CO = 4.3 H_2 = 10.5	CO: 29.1% H_2: 70.9%	[70]
MgAlTi-LDH	Coprecipitation	H_2O	400 W UV lamp (200–1000 nm)	Continuous-flow quartz tube reactor	CO_2 + 2.3 vol.% H_2O + 150 °C	CO = 10 (per gram of TiO_2)	-	[90]
CoMgAl-LDH	Coprecipitation	$[Ru(bpy)_3]Cl_2 \cdot 6H_2O$ + TEOA + MeCN + H_2O	300 W Xe lamp (>400 nm)	50 mL stainless reactor	1.8 bar + 30 °C	CO = 7700 H_2 = 5808.8	CO: 57.0% H_2: 43.0%	[85]
CrMgAl-LDH	Hydrothermal	$[Ru(bpy)_3]Cl_2 \cdot 6H_2O$ + TEOA + H_2O	300 W Xe lamp (420–780 nm)	250 mL quartz reactor	0.1 MPa CO_2 + 20 °C	CO = 3.91 H_2 = 2.05	CO: 65.6% H_2: 34.4%	[48]
Ultrathin MgAl-LDH	Titration	$[Ru(bpy)_3]Cl_2 \cdot 6H_2O$ + TEOA + MeCN + H_2O	300 W Xe lamp (400–800 nm)	50 mL stainless reactor	1.8 bar	CO = 700 H_2 = 1700	CO: 29.2% H_2: 70.8%	[91]
Ru/ultrathin MgAl-LDH	Impregnation	H_2	300 W Xe lamp	Flow photoreactor	5 $mL \cdot min^{-1}$ CO_2 + 20.5 $mL \cdot min^{-1}$ H_2 + ~350 °C	$CH_4 = 2.77 \times 10^5$	CH_4: ~100%	[77]
Monolayer MgAl-LDH	SNAS	$[Ru(bpy)_3]Cl_2 \cdot 6H_2O$ + TEOA + MeCN + H_2O	300 W Xe lamp (>400 nm)	50 mL stainless reactor	1.8 bar + 30 °C	CO = 170 H_2 = 210	CO: 44.7% H_2: 55.3%	[92]
Pt/MgAl-LDH	Electrostatic attraction	H_2O	300 W Xe lamp (UV light)	250 mL quartz reactor	0.1 MPa CO_2 + 20 °C	CO = 2.64 CH_4 = 0.2	CO: 93.0% CH_4: 7.0%	[67]
Ag/Ga_2O_3/MgAl-LDH	Coprecipitation	$NaHCO_3$	400 W Hg lamp (UV light)	Flowing batch system	30 $mL \cdot min^{-1}$ CO_2 + Room temperature	CO = 211.7 H_2 = 131.4	CO: 61.7% H_2: 38.3%	[93]
CeO_2/MgAl-LDH	Hydrothermal	$[Ru(bpy)_3]Cl_2 \cdot 6H_2O$ + TEOA + MeCN + H_2O	300 W Xe lamp (Visible light)		1.8 bar + 30 °C	CO = 85 H_2 = 110.5	CO: 42.1% H_2: 57.9%	[94]
Fe_3O_4/MgAl-LDH	Coprecipitation	NaOH + MeCN + H_2O	8 W UV light	300 mL quartz reactor	0.1 MPa CO_2 + 20 °C	CO = 442.2 CH_4 = 223.9	CO: 68.9% CH_4: 31.1%	[83]
Co-TCPP@MgAl-LDH	Esterification	H_2O	300 W Xe lamp (420–780 nm)	250 mL quartz reactor	-	CO = 0.13 CH_4 = 0.08	CO: 62.0% H_2: 38.0%	[95]
Pd/C_3N_4/MgAl-LDH	Self-assembly	H_2O	500 W Hg-Xe lamp (UV light)		0.03 MPa CO_2 + Room temperature	CH_4 = 0.7	-	[73]
Pt/MgAl-LDO/TiO_2	Coprecipitation	H_2O	300 W Xe lamp (260–400 nm)	Closed circulating system	0.1 MPa CO_2 + 40 mL H_2O + 20 °C	CO = 1.5 CH_4 = 2.3	CO: 39.5% CH_4: 60.5%	[96]
MgAl-LDO/N_v-CN	In situ deposition	H_2O	300 W Xe arc lamp (>400 nm)	280 mL double-necked reactor	-	CO = 20.47	-	[97]
Fe/MgAl-LDO	Thermal reduction	H_2	300 W Xe lamp (200–800 nm)	50 cm^3 stainless reaction chamber	1.8 bar + $CO_2/H_2/Ar$ = 15/60/25 + ~275 °C	CO_2 conversion = 50.1%	CO: 16.0% CH_4: 49.5% C_{2-4}: 31.2% C_5: 3.3%	[98]
MgAl-LDO/TiO_2	Coprecipitation	H_2O	100 W Hg lamp (<390 nm)	Continuous-flow quartz tube reactor	4.5 $mL \cdot min^{-1}$ He with H_2O + 200 °C	CO = 99 $\mu mol \cdot g^{-1}$	-	[99]
MgAl-LDO/TiO_2	Hydrothermal	H_2O	100 W Hg lamp (<390 nm)	Continuous-flow quartz tube reactor	4.5 $mL \cdot min^{-1}$ CO_2 with 2.3 vol.% H_2O + 150 °C	CO = 4.3	-	[100]
TiMgAl-LDH/GO	Coprecipitation	H_2O	300 W Xe lamp (>420 nm)	50 cm^3 stainless reactor	0.08 MPa	CO = 4.6 CH_4 = 3.8	CO: 54.8% CH_4: 45.2%	[101]

Doping suitable elements into MgAl-LDH to construct ternary LDHs can adjust the band structure, introduce the surface defects, improve the separation of electron–hole pairs and strength the surface reaction [102]. The introduction of Co, Ni, Ti and other metals by element doping is always applied to improve the visible light absorption of LDHs with wide bandgaps, which can enhance d-band electrons and introduce impurity levels in the forbidden band. Meanwhile, the introduced metal atoms are embedded into the layer structure of LDHs, which provide abundant active sites and create defect sites. The active sites and defect sites promote the surface reduction reaction of CO_2 and improve the charge separation performance. As shown in Figure 6, Ning et al. [85] synthesized Co-doped MgAl-LDH by a hydrothermal method for CO_2 photoreduction and the bandgap was reduced from 4.18 eV to 1.80 eV after the introduction of Co sites. The introduction of Co into MgAl-LDH led to the broad absorption peaks at 520 nm and 610 nm due to the d–d transitions of Co^{2+}. Hydroxyl vacancies (V_{OH}) were created after Co doping and the charge separation was greatly improved. Under the light irradiation above 400 nm, the TOF value of CO production for CoMgAl-LDH came to 11.57 h^{-1}, which was three times that of CoAl-LDH. Although the wavelength was up to 650 nm, CoMgAl-LDH still exhibited high activity with CO evolution rate of 0.35 $\mu mol \cdot h^{-1}$. It was noteworthy that the formation energy of hydroxyl vacancy (V_{OH}) in CoMgAl-LDH was lower than that of CoAl-LDH, indicating that the high content of Co made it harder to form unsaturated sites. Xu et al. [48] introduced five kinds of transition metal (Co, Ni, Cu, Fe, Cr) into the layer structure of MgAl-LDH and found that that the introduction of transition metal could effectively reduce bandgap and improve visible light absorption performance. Therein, Cr-doped MgAl-LDH exhibited the highest TOF value of 0.0053 h^{-1} for CO production under visible light irradiation (420–780 nm). The bandgap of MgAl-LDH was 5.42 eV and the introduction of Co, Ni, Cu, Fe, Cr reduced the bandgap to 2.13, 2.72, 3.94, 2.01 and 2.37 eV, respectively. Except Cu-doped MgAl-LDH, other metal-doped MgAl-LDH all exhibited visible light response ability. As shown in Figure 7, The introduction of metal elements introduced the intermediate bands, which led to the decreased OH vacancy formation energy and the increased defect density. Cr doping into MgAl-LDH provided the ability to continuously generate defects and the high surface defect density, which led to the highest charge transfer efficiency. Meanwhile, Cr-doped MgAl-LDH exhibited competitiveness in the release of CO during the surface reaction, which was confirmed by DFT calculation.

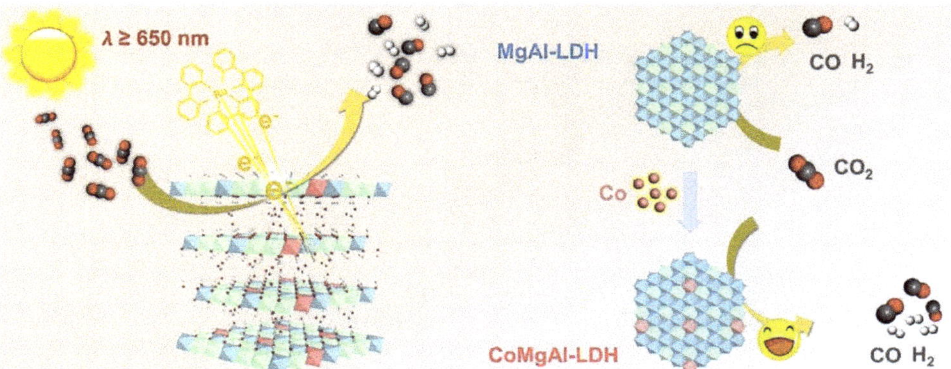

Figure 6. Schematic diagram for MgAl-LDH and CoMgAl-LDH for CO_2 photoreduction to generate CO and H_2 under $\lambda \geq 650$ nm combining with a Ru complex [85].

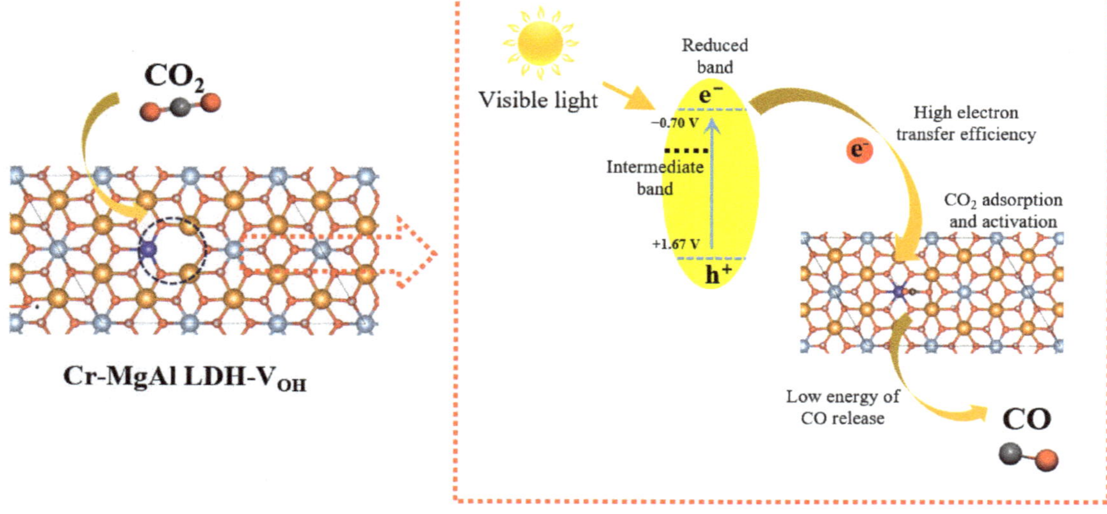

Figure 7. Mechanism diagram of Cr doping into MgAl-LDH in CO_2 photoreduction [48].

In addition to the metal doping into MgAl-LDH, the metal compounds can also be introduced into the layer structure to achieve the substitution of hydroxyl groups. Iguchi et al. [70] prepared a MgAl-LDH photocatalyst containing AlF_6^{3-} units by a coprecipitation method, which achieved the substitution of hydroxyl with fluorine by changing the precursor species. The fluorination of MgAl-LDH obviously promoted the evolution of CO in the photocatalytic CO_2 reduction with Cl^- as a hole scavenger and CO yield was twice that of pure MgAl-LDH. Meanwhile, the product selectivity of CO was also increased after the fluorination of MgAl-LDH. UV-vis spectrum indicated that the absorption edge of MgAl-LDH was not influenced by the fluorination of MgAl-LDH. The increased activity was due to the enhanced CO_2 adsorption on the surface and the high electronegativity of fluorine units greatly influenced the basicity of MgAl-LDH. In summary, the introduction of active components into MgAl-LDH always obtains the highly dispersed active sites, creates the unsaturated defect sites and changes the surface chemical property, resulting the increased photocatalytic activity.

4.2. Morphology Control of MgAl-LDH

The bulk LDHs always show limited catalytic activity in CO_2 photoreduction due to the limited specific surface area, fast recombination of electron–hole pairs and insufficient charge transfer [103]. Morphology control can improve the electronic structure and expose more active sites, which can further improve catalytic activity of bulk MgAl-LDH [104]. Ultrathin or monolayer MgAl-LDH nanosheets can be prepared by a bottom-up or top-down method [16,105], which always leads to the structural distortion and the formation of vacancies. The electronic structure and charge transfer will be modified due to the existence of vacancies, which directly influences the photocatalytic performance.

The ultrathin LDHs always show the increased specific surface area and more active sites can be exposed. The surface defects in the ultrathin nanosheets are considered to be important adsorption sites for CO_2 molecules and lead to the enhanced conductivity of LDHs [104]. Xu et al. [67] prepared the ultrathin MgAl-LDH by the mechanical exfoliation in formamide solution and obtained the nanosheets with the thickness of 1–2 nm. The exfoliated MgAl-LDH showed the increased specific surface area compared with bulk

MgAl-LDH due to the expansion of layer structure. Ren et al. [77] prepared Ru-loaded ultrathin MgAl-LDH (Ru@FL-LDHs) by ultrasonicated exfoliation and impregnation for photothermal CO_2 methanation. The thickness of the ultrathin MgAl-LDH was around 8 Å, which is consistent with the single basal spacing of MgAl-LDH. As shown in Figure 8, Ru@FL-LDHs showed the fastest initiation of the reaction within 30 min and the highest CO_2 conversion (96.3% after 60 min irradiation). Meanwhile, the highest selectivity of CH_4 was also achieved in photothermal CO_2 reduction over Ru@FL-LDHs. Compared with Ru-loaded bulk MgAl-LDH (Ru@LDHs), Ru@FL-LDHs exhibited the higher CO_2 adsorption capacity due to the stronger basicity of the exfoliated LDHs, while CO_2 molecules could be converted into $CO_3{}^{2-}$ that was easily activated. The exfoliated MgAl-LDH with abundant basic sites provided the ability to activate CO_2 molecules effectively. The excellent photothermal conversion of CO_2 was ascribed to the targeting activation of CO_2 and H_2 over ultrathin MgAl-LDH and Ru nanoparticles. Bai et al. [91] developed the strategy of manipulating d-electron configuration in ultrathin LDHs by changing the selection of divalent metal cations to regulate photocatalytic activity for CO_2 reduction. DFT calculation indicated that the bandgap of ultrathin MgAl-LDH was reduced to 3.33 eV effectively after the morphology control.

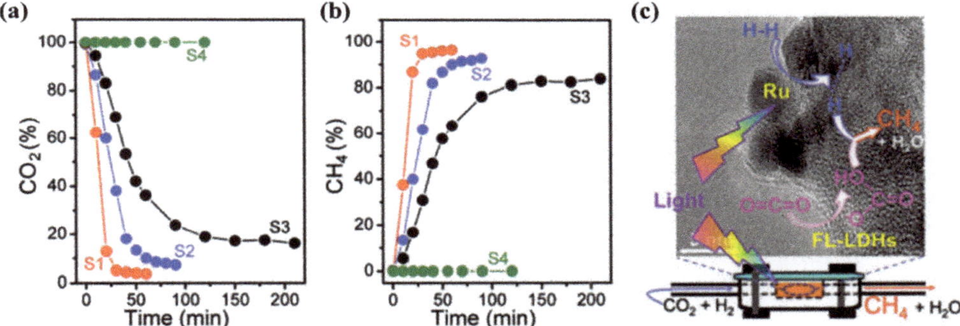

Figure 8. Concentration change of (**a**) CO_2 reactant and (**b**) CH_4 product in photothermal CO_2 methanation over different catalysts (S1: Ru@FL-LDHs; S2: Ru@LDHs; S3: Ru@SiO_2; S4: FL-LDHs). Reaction condition: 150 mg catalysts, H_2: 20.5 mL·min^{-1}, CO_2: 5.0 mL·min^{-1}, 300 W xenon lamp. (**c**) Schematic diagram of targeting activation of CO_2 and H_2 over Ru@FL-LDHs [77].

Ulteriorly, the monolayer LDHs have better advantages in charge separation efficiency than ultrathin LDHs benefiting from the thickness at the atomic level. Bai et al. [92] developed a facile method to synthesize monolayer LDHs by adding layer growth inhibitor (formamide) into a colloid mill reactor and obtained uniform monolayer MgAl-LDH with the thickness of ~1 nm. EXAFS results confirmed that lots of vacancies (oxygen and metal vacancies) existed on the surface of monolayer LDHs. The monolayer LDHs exhibited higher photocurrent density than multilayer LDHs, indicating that the vacancies in monolayer LDHs improved the electron transport performance. Meanwhile, as shown in Figure 9, different metal species in monolayer LDHs had an influence on the selectivity and evolution rate of products in CO_2 photoreduction. The synthetic method can be applied to prepare different monolayer LDHs and achieve the great increasement of preparation rate, which can be used for large-scale industrial production. Lai et al. [106] prepared monolayer MgAl-LDH by a similar method in colloid mill reactor. As shown in Figure 10, the TEM image and AFM image clearly indicate the atomic thickness of monolayer MgAl-LDH (~1 nm). N_2 adsorption–desorption isotherms were used to evaluate the specific surface areas of monolayer and multilayer LDHs. The specific surface areas of monolayer and multilayer MgAl-LDH were 294.7 m^2/g and 8.3 m^2/g, respectively, indicating that monolayer MgAl-LDH could provide much more active sites.

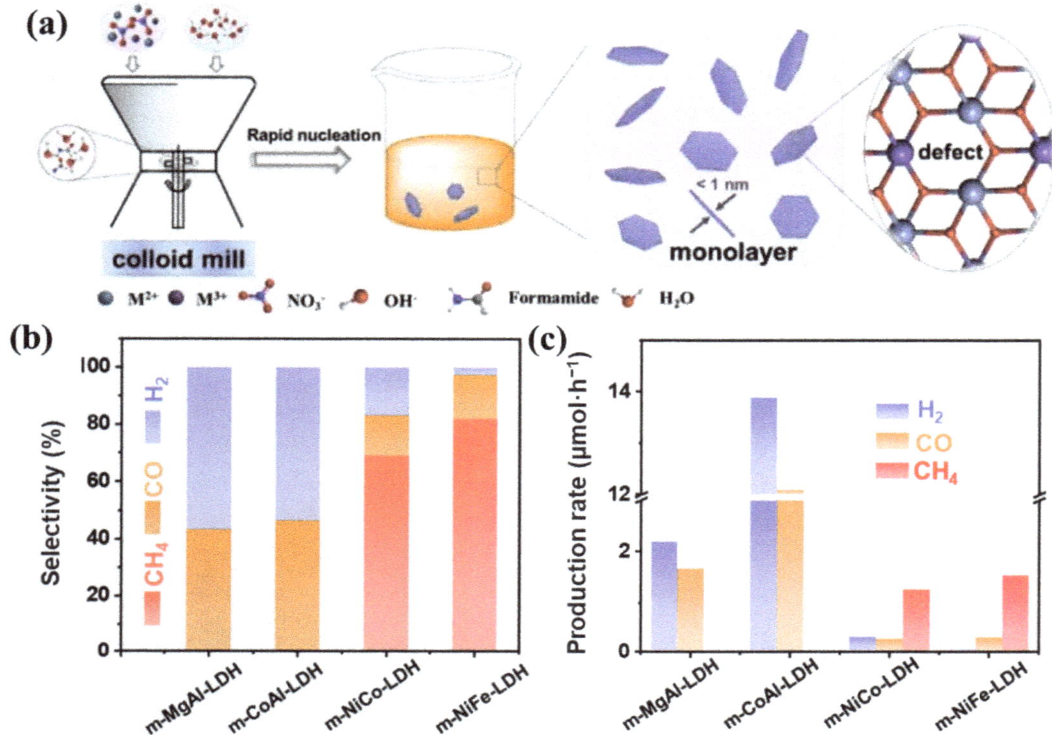

Figure 9. (**a**) Schematic illustration of synthetic process of monolayer LDHs; (**b**) selectivity and (**c**) production rate of CH_4, CO, and H_2 in photocatalytic CO_2 reduction over different monolayer LDHs under $\lambda \geq 400$ nm [92].

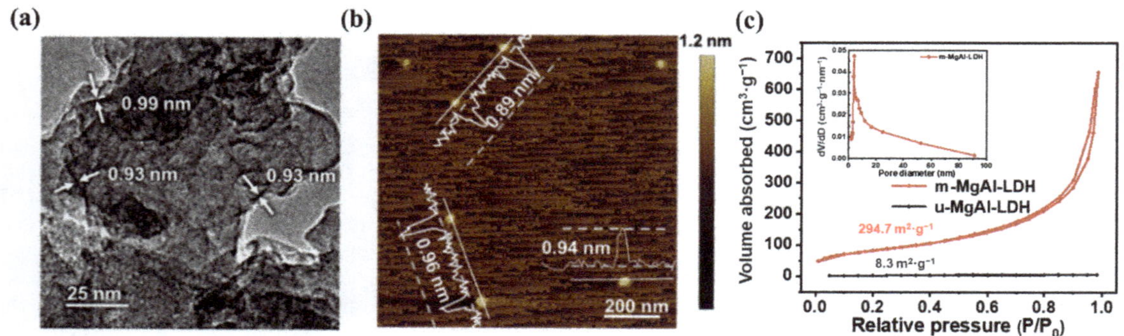

Figure 10. (**a**) HRTEM image and (**b**) AFM image of monolayer MgAl-LDH; (**c**) N_2 adsorption–desorption isotherms for monolayer MgAl-LDH and multilayer MgAl-LDH and pore size distribution of monolayer MgAl-LDH (inset) [106].

4.3. Supported Active Components over MgAl-LDH

The supporting of active components on the surface of MgAl-LDH is a potential method to improve the photocatalytic activity of catalysts, which provides the highly dispersed active sites and constructs the interaction between the two components [107]. Generally, multiple active components, such as metals, metal compounds and nonmetallic materials, can be loaded on the surface of LDHs to improve the photocatalytic activity. Depending on the different characteristics of various active components, the interface engineering can be designed minutely [108]. As a result, the light response range of composite catalysts can be widened and the transfer efficiency of photogenerated carriers can be improved.

The surface hydroxyl groups of MgAl-LDH have a strong immobilization effect for the loading of noble metal, so it is suitable to be the support to form stable noble-metal-based catalysts [109,110]. For example, Zhu et al. [111] anchored Ru atoms on the surface of MgAl-LDH by an impregnation method and a strong metal-support interaction between Ru and LDHs was constructed through the coordinated Ru species with one OH and three oxygen atoms, which have an important influence on the photocatalytic activity. Generally, the supported noble metal can act as the active sites to capture photogenerated electrons and promote the separation of electron–hole pairs. Xu et al. [67] achieved high dispersion of Pt nanoparticles on the surface of MgAl-LDH by electrostatic attraction for CO_2 photocatalytic reduction. The loading of Pt nanoparticles over MgAl-LDH greatly improved the charge transfer efficiency of catalysts, further increasing the photocatalytic activity. 0.1 wt.% Pt-loaded MgAl-LDH exhibited the highest activity with CO evolution rate of 2.64 $\mu mol \cdot g^{-1} \cdot h^{-1}$ under UV light, which is about 8.52 times that of 1 wt.% Pt-loaded MgAl-LDH. The high content of Pt species did not form the electron traps but promoted the formation of combination centers, which increased the combination rate of electron–hole pairs. This method provided a strategy that maintained high photocatalytic activity while reducing the dosage of noble metal. Iguchi et al. [93] prepared Ag-loaded MgAl-LDH/Ga_2O_3 by an impregnation method for CO_2 photocatalytic reduction under UV light. 0.25 wt.% Ag-loaded MgAl-LDH/Ga_2O_3 exhibited the CO evolution rate of 211.7 $\mu mol \cdot h^{-1}$ and CO selectivity of 61.7%. The loading of MgAl-LDH improved the CO_2 adsorption capacity and Ag nanoparticles acted as co-catalyst, which improved the product selectivity. Photogenerated electrons transferred from Ga_2O_3 to Ag nanoparticles via MgAl-LDH, which promoted the conversion of CO_2 to CO. In addition, the SPR effect of noble metals also improves the visible light response performance, which has been reported over LDHs [112].

The limitation of noble metals in CO_2 photoreduction is the high cost, and transition metal components have been reported to replace noble metals. The coupling of MgAl-LDH with suitable transition metal components can effectively improve the visible light absorption and charge separation. Tan et al. [94] prepared CeO_2 decorated MgAl-LDH for photocatalytic CO_2 reduction and found that the selectivity of syngas (CO/H_2) was closely related to the loading amount of CeO_2. $Mg_6Al_{0.85}Ce_{0.15}$-LDH (Ce-0.15) exhibited the CO productivity of 0.85 μmol, which was 4.7 and 9.4 times that of MgAl-LDH and CeO_2. As shown in Figure 11, the selectivity and productivity trend of CO achieved the highest point at Ce-0.15 whit the evolution rate of 85 $\mu mol \cdot g^{-1} \cdot h^{-1}$ and selectivity of 42.1%. UV-visible absorption indicated that Ce-0.15 exhibited the best light absorption performance in the visible region, which was due to the interaction between CeO_2 and MgAl-LDH. EIS spectra proved that the resistance of Ce-0.15 was smaller than that of MgAl-LDH, indicating that the loading of CeO_2 effectively promoted the separation and transfer of charge carriers. Gao et al. [83] prepared Fe_3O_4-loaded ultrathin MgAl-LDH by a coprecipitation method for photocatalytic CO_2 reduction, and it exhibited high catalytic activity with CO and CH_4 yields of 442.2 $\mu mol \cdot g^{-1} \cdot h^{-1}$ and 223.9 $\mu mol \cdot g^{-1} \cdot h^{-1}$ in the presence of NaOH and acetonitrile under UV light, which were 1.8 and 1.7 times than those of MgAl-LDH. Fe_3O_4 with the narrow bandgap and high conductivity, strikingly broadened the absorption edge to ~700 nm and promoted the transfer of f-photogenerated electrons. Ultrathin MgAl-LDH reduced the transfer resistance of charge carriers and provided abundant active sites. Fe_3O_4 could act as electron traps, which prolonged the lifetime of photogenerated electrons. The direct loading of transition metal components always obtains the weak interfacial interaction and it is difficult to ensure the dispersion of transition metal sites. Xu et al. [95] prepared Co-porphyrin-loaded flower-like MgAl-LDH for photocatalytic CO_2

reduction under visible light. The excellent light absorption of metalloporphyrin effectively broadened the absorption edge to visible region. The framework of metalloporphyrin could ensure the high dispersion of transition metal sites and the esterification effect could obtain tight intramolecular interfaces, which further promoted the separation of electrons and holes.

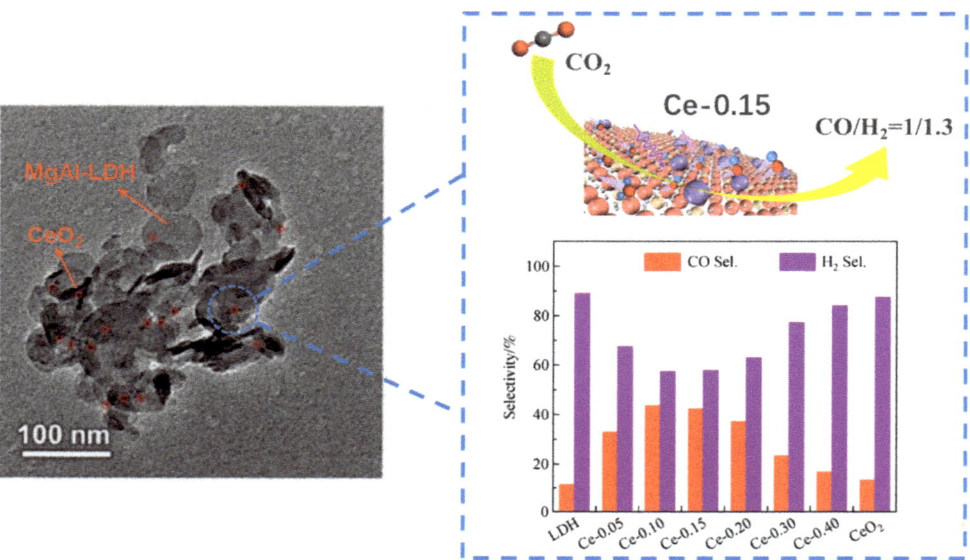

Figure 11. Schematic diagram of the tunable selectivity of syngas from photocatalytic CO_2 reduction by CeO_2/LDH [94].

In addition to the loading of metal components, the nonmetal materials are also reported to improve the photocatalytic activity of MgAl-LDH. Hong et al. [73] reported the self-assembling catalyst constructed by carbon nitride (C_3N_4) and MgAl-LDH for photocatalytic CO_2 reduction under UV light. In the presence of Pd cocatalyst, the CH_4 yield of C_3N_4/MgAl-LDH was 0.15 $\mu mol \cdot h^{-1}$, but the catalytic activity mainly came from C_3N_4, and the role of MgAl LDH is mainly to strengthen CO_2 adsorption and activation. As shown in Figure 12, the enriched CO_2 in the form of CO_3^{2-} in the interlayer could be reduced easily to CH_4 by photogenerated electrons from C_3N_4 at Pd active sites. The supported active components over MgAl-LDH always provide the visible light absorption ability and promote the separation of charge carriers, while MgAl-LDH always provide the considerable CO_2 adsorption performance.

4.4. MgAl-LDH as Precursor for Catalyst Preparation

LDO (layer double oxide), which is derived from LDHs precursor, is the mixed metal oxide with high dispersibility of metal sites and high surface area. The component, morphology and surface sites can be adjusted expediently by changing the metal cations of LDHs [113]. As the product after calcination, MgAl-LDO retains the highly dispersed metal sites and provides abundant surface basic or acid sites (OH^-, Mg^{2+}-O^{2-} pairs, Al^{3+}-O^{2-} pairs) derived from the diffusion of Al^{3+} into MgO lattice [114]. The Lewis basic sites are beneficial to adsorb CO_2 and form carbonates and bicarbonates, while the Lewis acid sites are beneficial to dissociate H_2O and provide abundant protons [100,115]. Chong et al. [96] prepared MgAl-LDO/TiO_2 by an in situ deposition method and then Pt cocatalyst was loaded on the surface of catalysts by an in situ photo deposition method for photocatalytic CO_2 reduction in the present of water vapor. CO and CH_4 evolution rates of Pt/MgAl-

LDO/TiO_2 were 0.030 and 0.046 μmol·h^{-1} under UV light, which were 2 and 11 times than those of Pt/TiO_2, respectively. The deposition of MgAl-LDO had a negligible effect on the light absorption of TiO_2 but changed CO_2 adsorption states and the adsorbed species. FTIR spectra indicated that the Lewis basic sites and Lewis acid sites enhanced CO_2 adsorption and H_2O dissociation. The recombination of electrons and holes was inhibited due to the formation of oxygen vacancy derived from the interaction between MgAl-LDO and TiO_2, and Pt cocatalyst effectively promoted the transfer of photogenerated electrons. Song et al. [97] prepared the composite catalyst between MgAl-LDO and nitrogen-deficient g-C_3N_4 (MgAl-LDO/N_v-CN) by an in situ deposition and calcination transformation method for photoreduction of carbon dioxide. 10% MgAl-LDO/N_v-CN exhibited the highest CO evolution rate (20.47 μmol·g^{-1}·h^{-1}) under visible light, which was 6 times that of CN (3.38 μmol·g^{-1}·h^{-1}) and 3 times that of N_v-CN (5.71 μmol·g^{-1}·h^{-1}). The introduction of MgAl-LDO on N_v-CN led to a slight blue-shift of absorption edge and the bandgap increased from 2.38 eV to 2.42 eV. As shown in Figure 13, Lewis base/acid sites provided the targeting activation for CO_2 and H_2O molecules. The introduction of nitrogen defects and the interfacial interaction between MgAl-LDO and N_v-CN facilitated the charge transfer and inhibited the recombination of photogenerated electrons and holes.

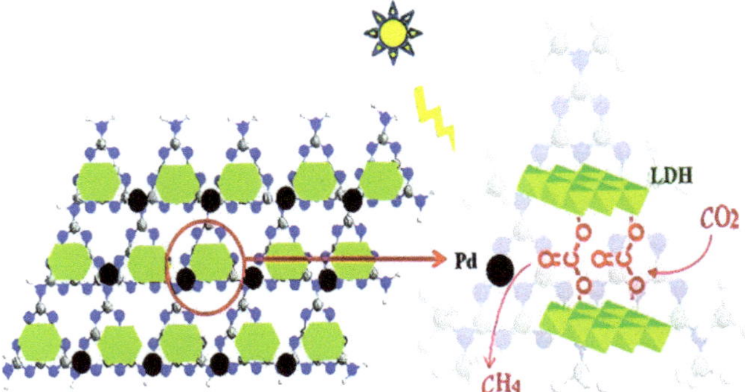

Figure 12. Schematic diagram of the self-assembly of carbon nitride (C_3N_4) and MgAl-LDH for CH_4 production with Pd as cocatalyst [73].

The introduction of active metal components into LDH precursors to prepare the supported metal catalysts is an efficient method to improve the photocatalytic activity. Due to the uniform and ordered distribution of metal cations in the layer structure of LDHs, active metal components will be highly dispersed on the surface of catalysts. The strong interaction between active metal components and supports will be constructed and the confined effect will prevent the sintering and aggregation of active metal components. The memory effect of LDHs can be used for the loading of active metal components with LDHs as the precursor. Zhao et al. [90] prepared Ti-embedded MgAl-LDH for CO_2 photoreduction by redispersing calcined MgAlTi-LDH in water to achieve the structural reconstruction. The sample calcined at 400 °C exhibited the increased photocatalytic activity under UV light compared with initial MgAl-LDH and the catalytic activity was closely related to the crystallinity and specific surface areas. The reduction treatment of LDH precursors can be used to prepare the supported metal catalysts. Li et al. [98] prepared Fe-based catalysts by reducing MgFeAl-LDH in the H_2/Ar atmosphere for direct photohydrogenation of CO_2. The catalyst reduced at 500 °C (Fe-500) exhibited high CO_2 conversion (50.1%) and significant C2+ product selectivity (52.9%) in the photohydrogenation reaction. Fe-500 consisted of Fe^0 nanoparticles and FeO_x supported on MgO-Al_2O_3. The Fe species suppressed the hydrogenation of -CH_2 and -CH_3 intermediates and promoted the coupling of C-C bonds, while MgO strengthened the adsorption of CO_2 molecules. The supported metal catalysts over MgAl-LDO can be used for the integration of CO_2 capture and photocatalytic conversion. Liu et al. [99] prepared MgAl-LDO/TiO_2 for CO_2 capture and photocatalytic conversion simultaneously. As shown in Figure 14, MgAl-LDO/TiO_2 adsorbed CO_2 at flue gas temperature, while then the desorption of gas-phase CO_2 and the conversion of adsorbed species to CO under UV light took place concurrently at the range of 100–200 °C. The hybrid material can be regenerated automatically for the next cycle. The incorporation of MgAl-LDO and TiO_2 enhanced CO_2 capture capacity and the sample containing Ti atoms of 43% (MgAl/Ti_{43}) exhibited the capacity of 0.648 mmol/g after 2 h, which was 24 times that of TiO_2 and 1.4 times that of MgAl-LDO. There is a trade-off between CO_2 capture and CO_2 conversion capability, while MgAl/Ti_{43} exhibited good CO_2 capture capacity and conversion efficiency (15.3%) as well.

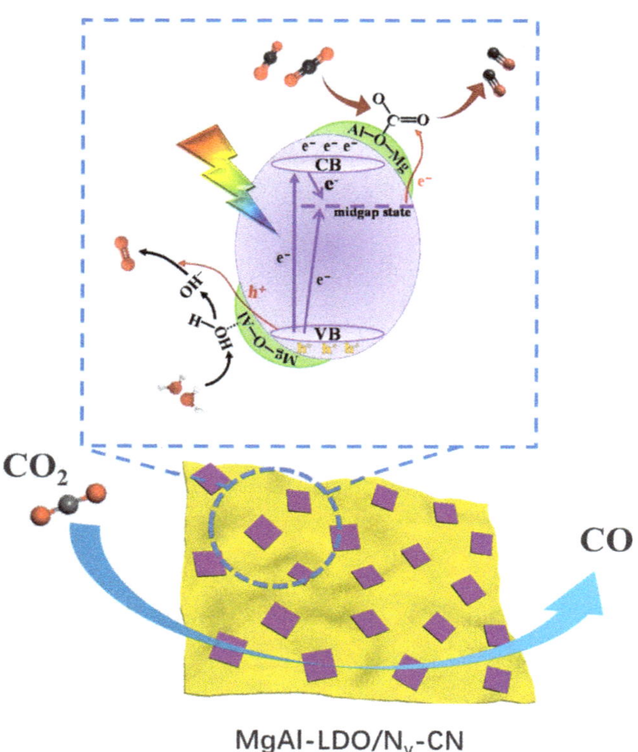

Figure 13. Schematic diagram of surface Lewis base/acid and nitrogen defect in MgAl-LDO/N_v-CN for CO_2 photoreduction [97].

In addition, the morphology of LDO can be designed by inheriting layer structure of LDHs precursor, which can be applied to develop the catalysts with special morphology [116,117]. Zhao et al. [100] prepared MgAl-LDO-grafted TiO_2 cuboids by hydrothermal and coprecipitation methods. As shown in Figure 15, the SEM images showed that the morphology of samples composed of MgAl-LDO grafted on TiO_2 cuboids and the platelet shape of MgAl-LDO was almost the same as MgAl-LDH precursor. The graft of MgAl-LDO on TiO_2 cuboids did not significantly improve the photocatalytic activity compared with bare TiO_2 due to the weak CO_2 adsorption of MgAl-LDO at low temperature. In the activity test at 150 °C under UV light, 10% MgAl-LDO/TiO_2 exhibited the increased activity with CO production of 4.3 $\mu mol \cdot g^{-1} \cdot h^{-1}$, which was 6.1 times that of bare TiO_2. The photoinduced electrons on TiO_2 transferred to the CO_2 adsorption sites at the interfaces and promoted the reduction reaction. The high loading of MgAl-LDO on TiO_2 cuboids did not improve the photocatalytic activity at 150 °C, which was ascribed to the weak contact between TiO_2 and adsorbed CO_2 and the limited light absorption due to the complete covering of TiO_2 by MgAl-LDO. In summary, the properties of LDHs endow the derived catalysts the adjustable composition and morphology, excellent CO_2 adsorption capacity, highly dispersed metal sites and large surface area, which ensure the good catalytic activity.

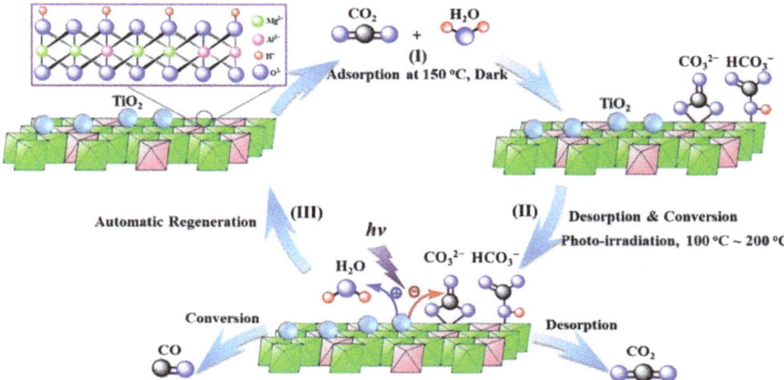

Figure 14. Schematic diagram of the integrated CO_2 capture and photocatalytic conversion over MgAl-LDO/TiO_2 [99].

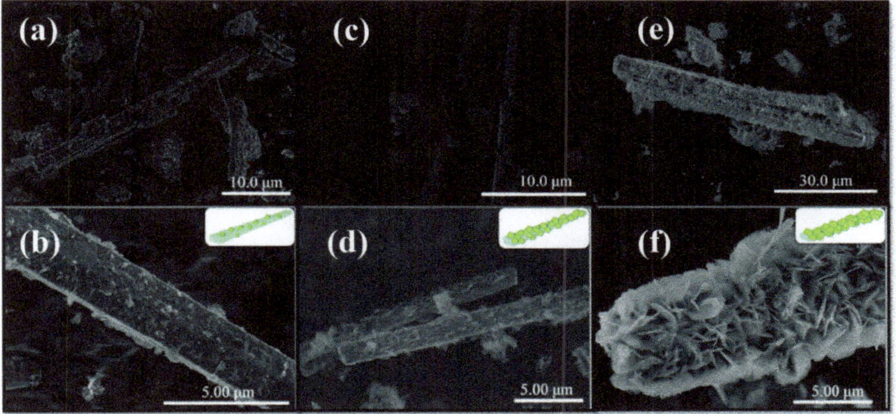

Figure 15. SEM images of MgAl-LDO/TiO_2 catalysts: (**a**,**b**) 8% MgAl-LDO/TiO_2, (**c**,**d**) 10% MgAl-LDO/TiO_2, and (**e**,**f**) 12% MgAl-LDO/TiO_2 [100].

4.5. Construction of Heterojunction

The separation efficiency of photogenerated charge can be effectively improved and the light absorption can be widened by constructing heterojunctions between LDHs and other semiconductors with matched band structures [118]. Studies on LDH-based heterojunctions, including Type-II [119], Z-scheme [120] and S-scheme [121] heterojunctions, have been widely reported for CO_2 photoreduction. However, the heterojunctions constructed by MgAl-LDH are rarely reported for CO_2 photoreduction due to the wide bandgap and it is difficult to form a stagger band structure. Yang et al. [122] prepared $ZnIn_2S_4$/MgAl-LDH heterojunction for photocatalytic Cr (VI) reduction and hydrogen evolution. MgAl-LDH promoted the expose of high-active (001) facets on $ZnIn_2S_4$ and acted as a hole storage layer, which realized the spatial separation of electrons and holes. As shown in Figure 16, MgAl-LDH cannot be excited under visible light and the holes in the VB of $ZnIn_2S_4$ transferred to the VB of MgAl-LDH to realize the separation of electrons and holes. Due to the wide band of MgAl-LDH, the other semiconductor should possess more positive VB or more negative CB to form stagger band structure, which is strict for the selectivity of semiconductors. Therefore, it is important to improve the wide bandgap of MgAl LDH and further construct efficient heterojunctions. Wang et al. [101] prepared ultrathin GO/TiMgAl-LDH heterojunction by the electrostatic self-assembly method for CO_2 photoreduction. When the proportion of GO was 5%, LDH/5GO exhibited the highest photocatalytic activity under visible light with CH_4 and CO evolution rate of 3.8 $\mu mol \cdot g^{-1} \cdot h^{-1}$ and 4.6 $\mu mol \cdot g^{-1} \cdot h^{-1}$, respectively. The ultrathin TiMgAl-LDH and GO created the unsaturated coordination, and introduced Ti^{3+}-V_o and electron-rich carbon defects. The presence of Ti^{3+}-V_o and GO expanded absorption edge to the visible light region, while the electron-rich carbon defects promoted the adsorption/activation of CO_2. As shown in Figure 17, electrons in the CB of the ultrathin TiMgAl-LDH would transfer to carbon defects on GO and react with adsorbed CO_2 to form superoxide radicals (CO_2^-). Simultaneously, the photoinduced holes would concentrate in the VB of TiMgAl-LDH, which achieved the spatial separation of electrons and holes.

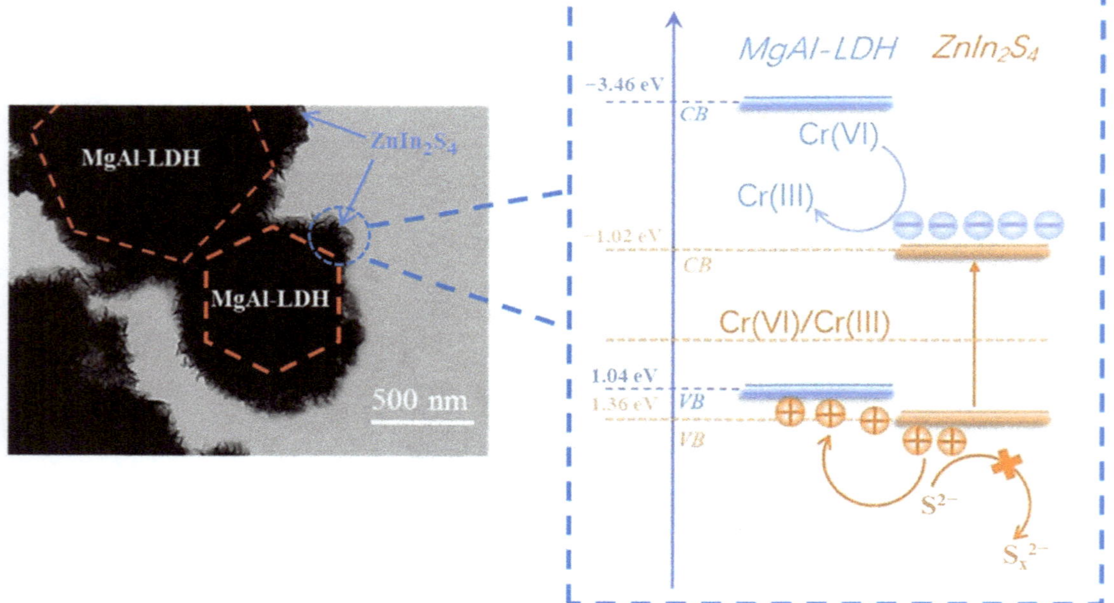

Figure 16. Mechanism schematic diagram of improving catalytic efficiency for $ZnIn_2S_4$/MgAl-LDH heterojunction [122].

In summary, construction of heterojunctions has been widely considered as an efficient method to promote the spatial separation of electrons and holes in LDHs, but MgAl-LDH is rarely selected to construct heterojunctions with other semiconductors due to the wide bandgap. The reduction of bandgap is of significance to improve visible light absorption and construct heterojunctions for MgAl-LDH.

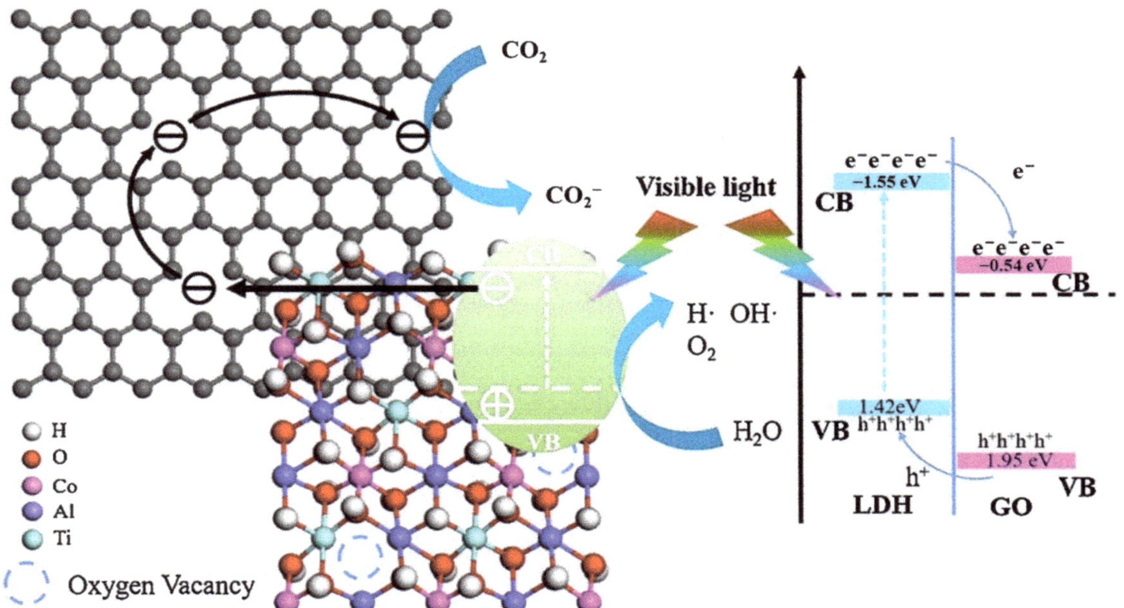

Figure 17. Schematic diagram of the proposed charge transfer and separation processes in GO/TiMgAl-LDH heterojunction for CO_2 photoreduction [101].

5. Conclusions and Perspectives

LDHs and LDH-based catalysts have been widely applied in the field of CO_2 photoreduction due to the excellent structural and chemical advantages. As the most widely used LDH material, MgAl-LDH has the advantages in CO_2 adsorption and low cost, but the poor visible light absorption and fast charge recombination of MgAl-LDH limit its application in CO_2 photoreduction. Due to the inherent drawbacks of MgAl-LDH, relatively few studies have focused on the improvement of photocatalytic activity for MgAl-LDH compared with other LDHs. More attention should be given to the modification of MgAl-LDH to promote the practical application of LDH materials in CO_2 photoreduction. Although some efforts have been devoted to overcome the drawbacks of MgAl-LDH in visible light absorption and charge separation, there are still some scientific issues that need further exploration. The development perspectives of MgAl-LDH in CO_2 photoreduction are as follows.

(1) Monolayer MgAl-LDH is beneficial to improve photocatalytic activity in CO_2 photoreduction by exposing as many defect sites and metal sites as possible. Bai et al. [92] developed the separate nucleation and aging steps (SNAS) method for scale-up synthesis of monolayer LDHs and more novel methods with the superior adjustment, convenience and feasibility for large-scale production should be further explored. The structure stability of monolayer LDHs in the process of preservation and reaction for practical application should be considered.

(2) A few studies focused on improving the dispersion of active components over MgAl-LDH by utilizing the confinement effect. For example, the anchoring effect of layer structure and the limited space of interlayer can obtain the extremely dispersed active components, such as single-atom catalysts, which may effectively improve the catalytic activity in CO_2 photoreduction. More efforts should be devoted to obtaining the highest catalytic efficiency of active components over MgAl-LDH.

(3) The methods to reduce the bandgap of MgAl-LDH are still limited, and basically rely on element doping. More methods should be developed to regulate the bandgap of

MgAl-LDH that further improve visible light absorption and make it suitable to construct heterojunction catalysts.

(4) The heterojunction catalysts based on MgAl-LDH for CO_2 photoreduction are still scarce and more efforts should be devoted to exploring the suitable semiconductor materials to construct heterojunction with MgAl-LDH. In the process, experiment and theoretical calculation should be combined to explore the properties of different materials and judge the matching degree of band structure. This can provide guidance for the design and development of heterojunction catalysts based on MgAl-LDH.

(5) In the condition without H_2 or at low temperature, it is quite difficult to generate C2 product over LDHs in CO_2 photoreduction. Developing new methods to regulate the conversion of CO_2 to C2 product is important to produce more valuable industrial products.

In summary, MgAl-LDH possesses many advantages in CO_2 photoreduction and has the potential to be widely applied. The application of MgAl-LDH in the field of photocatalysis can be further popularized through solving the above problems in the future.

Funding: This work was supported financially by the National Natural Science Foundation of China (No. 52276112), Key R&D Project (Social Development) in Xuzhou (No. KC21290), Joint Funds of the National Natural Science Foundation of China (No. U20A20302) and Innovative group projects in Hebei Province (No. E2021202006).

Data Availability Statement: No data was reported in the study.

Conflicts of Interest: The authors declare no conflict of interest.

References

1. Hoegh-Guldberg, O.; Mumby, P.J.; Hooten, A.J.; Steneck, R.S.; Greenfield, P.; Gomez, E.; Harvell, C.D.; Sale, P.F.; Edwards, A.J.; Caldeira, K.; et al. Coral Reefs under Rapid Climate Change and Ocean Acidification. *Science* **2007**, *318*, 1737–1742. [CrossRef] [PubMed]
2. Albright, R.; Caldeira, L.; Hosfelt, J.; Kwiatkowski, L.; Maclaren, J.K.; Mason, B.M.; Nebuchina, Y.; Ninokawa, A.; Pongratz, J.; Ricke, K.L.; et al. Reversal of ocean acidification enhances net coral reef calcification. *Nature* **2016**, *531*, 362–365. [CrossRef]
3. Yang, Y.; Cao, J.; Hu, Y.; Sun, J.; Yao, S.; Li, Q.; Li, Z.; Zhou, S.; Liu, W. Eutectic doped Li4SiO4 adsorbents using the optimal dopants for highly efficient CO_2 removal. *J. Mater. Chem. A* **2021**, *9*, 14309–14318. [CrossRef]
4. Pacala, S.; Socolow, R. Stabilization Wedges: Solving the Climate Problem for the Next 50 Years with Current Technologies. *Science* **2004**, *305*, 968–972. [CrossRef]
5. Din, I.U.; Shaharun, M.S.; Alotaibi, M.A.; Alharthi, A.I.; Naeem, A. Recent developments on heterogeneous catalytic CO_2 reduction to methanol. *J. CO2 Util.* **2019**, *34*, 20–33. [CrossRef]
6. Domínguez-Espíndola, R.B.; Arias, D.M.; Rodríguez-González, C.; Sebastian, P.J. A critical review on advances in TiO_2-based photocatalytic systems for CO_2 reduction. *Appl. Therm. Eng.* **2022**, *216*, 119009. [CrossRef]
7. Sun, C.; Yang, J.; Xu, M.; Cui, Y.; Ren, W.; Zhang, J.; Zhao, H.; Liang, B. Recent intensification strategies of SnO_2-based photocatalysts: A review. *Chem. Eng. J.* **2022**, *427*, 131564. [CrossRef]
8. Lee, Y.Y.; Jung, H.S.; Kang, Y.T. A review: Effect of nanostructures on photocatalytic CO_2 conversion over metal oxides and compound semiconductors. *J. CO2 Util.* **2017**, *20*, 163–177. [CrossRef]
9. Xiang, Q.; Cheng, B.; Yu, J. Graphene-Based Photocatalysts for Solar-Fuel Generation. *Angew. Chem. Int. Ed.* **2015**, *54*, 11350–11366. [CrossRef]
10. Khan, A.A.; Tahir, M. Well-designed 2D/2D Ti3C2TA/R MXene coupled g-C_3N_4 heterojunction with in-situ growth of anatase/rutile TiO_2 nucleates to boost photocatalytic dry-reforming of methane (DRM) for syngas production under visible light. *Appl. Catal. B Environ.* **2021**, *285*, 119777. [CrossRef]
11. Tang, J.-Y.; Guo, R.-T.; Zhou, W.-G.; Huang, C.-Y.; Pan, W.-G. Ball-flower like NiO/g-C_3N_4 heterojunction for efficient visible light photocatalytic CO2 reduction. *Appl. Catal. B Environ.* **2018**, *237*, 802–810. [CrossRef]
12. Low, J.; Cheng, B.; Yu, J. Surface modification and enhanced photocatalytic CO_2 reduction performance of TiO_2: A review. *Appl. Surf. Sci.* **2017**, *392*, 658–686. [CrossRef]
13. Wang, M.; Shen, M.; Jin, X.; Tian, J.; Shao, Y.; Zhang, L.; Li, Y.; Shi, J. Exploring the enhancement effects of hetero-metal doping in CeO2 on CO2 photocatalytic reduction performance. *Chem. Eng. J.* **2022**, *427*, 130987. [CrossRef]
14. Hailili, R.; Jacobs, D.L.; Zang, L.; Wang, C. Morphology controlled synthesis of CeTiO4 using molten salts and enhanced photocatalytic activity for CO_2 reduction. *Appl. Surf. Sci.* **2018**, *456*, 360–368. [CrossRef]
15. Bie, C.; Zhu, B.; Xu, F.; Zhang, L.; Yu, J. In Situ Grown Monolayer N-Doped Graphene on CdS Hollow Spheres with Seamless Contact for Photocatalytic CO_2 Reduction. *Adv. Mater.* **2019**, *31*, 1902868. [CrossRef]
16. Xu, M.; Wei, M. Layered Double Hydroxide-Based Catalysts: Recent Advances in Preparation, Structure, and Applications. *Adv. Funct. Mater.* **2018**, *28*, 1802943. [CrossRef]

17. Bi, Z.-X.; Guo, R.-T.; Hu, X.; Wang, J.; Chen, X.; Pan, W.-G. Research progress on photocatalytic reduction of CO_2 based on LDH materials. *Nanoscale* **2022**, *14*, 3367–3386. [CrossRef]
18. Cavani, F.; Trifirò, F.; Vaccari, A. Hydrotalcite-type anionic clays: Preparation, properties and applications. *Catal. Today* **1991**, *11*, 173–301. [CrossRef]
19. Oh, J.-M.; Hwang, S.-H.; Choy, J.-H. The effect of synthetic conditions on tailoring the size of hydrotalcite particles. *Solid State Ion.* **2002**, *151*, 285–291. [CrossRef]
20. Fang, D.; Huang, L.; Fan, J.; Xiao, H.; Wu, G.; Wang, Y.; Zeng, Z.; Shen, F.; Deng, S.; Ji, F. New insights into the arrangement pattern of layered double hydroxide nanosheets and their ion-exchange behavior with phosphate. *Chem. Eng. J.* **2022**, *441*, 136057. [CrossRef]
21. Szabados, M.; Kónya, Z.; Kukovecz, Á.; Sipos, P.; Pálinkó, I. Structural reconstruction of mechanochemically disordered CaFe-layered double hydroxide. *Appl. Clay Sci.* **2019**, *174*, 138–145. [CrossRef]
22. Valeikiene, L.; Paitian, R.; Grigoraviciute-Puroniene, I.; Ishikawa, K.; Kareiva, A. Transition metal substitution effects in sol-gel derived Mg3-xMx/Al1 (M = Mn, Co, Ni, Cu, Zn) layered double hydroxides. *Mater. Chem. Phys.* **2019**, *237*, 121863. [CrossRef]
23. Lv, S.; Kong, X.; Wang, L.; Zhang, F.; Lei, X. Flame-retardant and smoke-suppressing wood obtained by the in situ growth of a hydrotalcite-like compound on the inner surfaces of vessels. *New J. Chem.* **2019**, *43*, 16359–16366. [CrossRef]
24. Zhang, H.-M.; Zhang, S.-H.; Stewart, P.; Zhu, C.-H.; Liu, W.-J.; Hexemer, A.; Schaible, E.; Wang, C. Thermal stability and thermal aging of poly(vinyl chloride)/MgAl layered double hydroxides composites. *Chin. J. Polym. Sci.* **2016**, *34*, 542–551. [CrossRef]
25. Li, T.; Hao, X.; Bai, S.; Zhao, Y.; Song, Y.-F. Controllable synthesis and scale-up production prospect of monolayer layered double hydroxide nanosheets. *Acta Phys. Chim. Sin.* **2020**, *36*, 1912005–1912021.
26. Evans, D.G.; Duan, X. Preparation of layered double hydroxides and their applications as additives in polymers, as precursors to magnetic materials and in biology and medicine. *Chem. Commun.* **2006**, *5*, 485–496. [CrossRef]
27. Zhao, Y.; Li, F.; Zhang, R.; Evans, D.G.; Duan, X. Preparation of Layered Double-Hydroxide Nanomaterials with a Uniform Crystallite Size Using a New Method Involving Separate Nucleation and Aging Steps. *Chem. Mater.* **2002**, *14*, 4286–4291. [CrossRef]
28. Yang, Z.-Z.; Wei, J.-J.; Zeng, G.-M.; Zhang, H.-Q.; Tan, X.-F.; Ma, C.; Li, X.-C.; Li, Z.-H.; Zhang, C. A review on strategies to LDH-based materials to improve adsorption capacity and photoreduction efficiency for CO_2. *Coord. Chem. Rev.* **2019**, *386*, 154–182. [CrossRef]
29. Jerome, M.P.; Alahmad, F.A.; Salem, M.T.; Tahir, M. Layered double hydroxide (LDH) nanomaterials with engineering aspects for photocatalytic CO_2 conversion to energy efficient fuels: Fundamentals, recent advances, and challenges. *J. Environ. Chem. Eng.* **2022**, *10*, 108151. [CrossRef]
30. Bian, X.; Zhang, S.; Zhao, Y.; Shi, R.; Zhang, T. Layered double hydroxide-based photocatalytic materials toward renewable solar fuels production. *InfoMat* **2021**, *3*, 719–738. [CrossRef]
31. Dewangan, N.; Hui, W.M.; Jayaprakash, S.; Bawah, A.-R.; Poerjoto, A.J.; Jie, T.; Jangam, A.; Hidajat, K.; Kawi, S. Recent progress on layered double hydroxide (LDH) derived metal-based catalysts for CO_2 conversion to valuable chemicals. *Catal. Today* **2020**, *356*, 490–513. [CrossRef]
32. Chang, X.; Wang, T.; Gong, J. CO_2 photo-reduction: Insights into CO_2 activation and reaction on surfaces of photocatalysts. *Energy Environ. Sci.* **2016**, *9*, 2177–2196. [CrossRef]
33. Jiang, Q.; Chen, Z.; Tong, J.; Yang, M.; Jiang, Z.; Li, C. Direct thermolysis of CO_2 into CO and O_2. *Chem. Commun.* **2017**, *53*, 1188–1191. [CrossRef]
34. Zhang, K.; Harvey, A.P. CO_2 decomposition to CO in the presence of up to 50% O_2 using a non-thermal plasma at atmospheric temperature and pressure. *Chem. Eng. J.* **2021**, *405*, 126625. [CrossRef]
35. Kattel, S.; Ramírez, P.J.; Chen, J.G.; Rodriguez, J.A.; Liu, P. Active sites for CO_2 hydrogenation to methanol on Cu/ZnO catalysts. *Science* **2017**, *355*, 1296–1299. [CrossRef]
36. Xu, M.; Yu, D.; Yao, H.; Liu, X.; Qiao, Y. Coal combustion-generated aerosols: Formation and properties. *Proc. Combust. Inst.* **2011**, *33*, 1681–1697. [CrossRef]
37. Yu, D.; Xu, M.; Yao, H.; Sui, J.; Liu, X.; Yu, Y.; Cao, Q. Use of elemental size distributions in identifying particle formation modes. *Proc. Combust. Inst.* **2007**, *31*, 1921–1928. [CrossRef]
38. Guo, Z.; Yu, F.; Yang, Y.; Leung, C.-F.; Ng, S.-M.; Ko, C.-C.; Cometto, C.; Lau, T.-C.; Robert, M. Photocatalytic Conversion of CO_2 to CO by a Copper(II) Quaterpyridine Complex. *ChemSusChem* **2017**, *10*, 4009–4013. [CrossRef]
39. Jiang, D.; Zhou, Y.; Zhang, Q.; Song, Q.; Zhou, C.; Shi, X.; Li, D. Synergistic Integration of AuCu Co-Catalyst with Oxygen Vacancies on TiO_2 for Efficient Photocatalytic Conversion of CO_2 to CH4. *ACS Appl. Mater. Interfaces* **2021**, *13*, 46772–46782. [CrossRef]
40. Kumar, S.; Yadav, R.K.; Ram, K.; Aguiar, A.; Koh, J.; Sobral, A.J.F.N. Graphene oxide modified cobalt metallated porphyrin photocatalyst for conversion of formic acid from carbon dioxide. *J. CO2 Util.* **2018**, *27*, 107–114. [CrossRef]
41. Nakata, K.; Ozaki, T.; Terashima, C.; Fujishima, A.; Einaga, Y. High-Yield Electrochemical Production of Formaldehyde from CO_2 and Seawater. *Angew. Chem. Int. Ed.* **2014**, *53*, 871–874. [CrossRef] [PubMed]
42. Yousaf, M.; Ahmad, M.; Zhao, Z.-P. Rapid and highly selective conversion of CO_2 to methanol by heterometallic porous ZIF-8. *J. CO2 Util.* **2022**, *64*, 102172. [CrossRef]

43. Inoue, T.; Fujishima, A.; Konishi, S.; Honda, K. Photoelectrocatalytic reduction of carbon dioxide in aqueous suspensions of semiconductor powders. *Nature* **1979**, *277*, 637–638. [CrossRef]
44. Shehzad, N.; Tahir, M.; Johari, K.; Murugesan, T.; Hussain, M. A critical review on TiO_2 based photocatalytic CO_2 reduction system: Strategies to improve efficiency. *J. CO2 Util.* **2018**, *26*, 98–122. [CrossRef]
45. Hou, W.; Hung, W.H.; Pavaskar, P.; Goeppert, A.; Aykol, M.; Cronin, S.B. Photocatalytic Conversion of CO_2 to Hydrocarbon Fuels via Plasmon-Enhanced Absorption and Metallic Interband Transitions. *ACS Catal.* **2011**, *1*, 929–936. [CrossRef]
46. Zhang, K.-L.; Liu, C.-M.; Huang, F.-Q.; Zheng, C.; Wang, W.-D. Study of the electronic structure and photocatalytic activity of the BiOCl photocatalyst. *Appl. Catal. B Environ.* **2006**, *68*, 125–129. [CrossRef]
47. Tu, W.; Zhou, Y.; Zou, Z. Photocatalytic Conversion of CO_2 into Renewable Hydrocarbon Fuels: State-of-the-Art Accomplishment, Challenges, and Prospects. *Adv. Mater.* **2014**, *26*, 4607–4626. [CrossRef]
48. Xu, J.; Liu, X.; Zhou, Z.; Deng, L.; Liu, L.; Xu, M. Surface defects introduced by metal doping into layered double hydroxide for CO_2 photoreduction: The effect of metal species in light absorption, charge transfer and CO_2 reduction. *Chem. Eng. J.* **2022**, *442*, 136148. [CrossRef]
49. Zhou, Y.; Zhang, Q.; Shi, X.; Song, Q.; Zhou, C.; Jiang, D. Photocatalytic reduction of CO_2 into CH4 over Ru-doped TiO_2: Synergy of Ru and oxygen vacancies. *J. Colloid Interface Sci.* **2022**, *608*, 2809–2819. [CrossRef]
50. Han, B.; Ou, X.; Deng, Z.; Song, Y.; Tian, C.; Deng, H.; Xu, Y.-J.; Lin, Z. Nickel Metal–Organic Framework Monolayers for Photoreduction of Diluted CO_2: Metal-Node-Dependent Activity and Selectivity. *Angew. Chem. Int. Ed.* **2018**, *57*, 16811–16815. [CrossRef]
51. Izumi, Y. Recent advances in the photocatalytic conversion of carbon dioxide to fuels with water and/or hydrogen using solar energy and beyond. *Coord. Chem. Rev.* **2013**, *257*, 171–186. [CrossRef]
52. Morikawa, M.; Ogura, Y.; Ahmed, N.; Kawamura, S.; Mikami, G.; Okamoto, S.; Izumi, Y. Photocatalytic conversion of carbon dioxide into methanol in reverse fuel cells with tungsten oxide and layered double hydroxide photocatalysts for solar fuel generation. *Catal. Sci. Technol.* **2014**, *4*, 1644–1651. [CrossRef]
53. Miao, Y.-F.; Guo, R.-T.; Gu, J.-W.; Liu, Y.-Z.; Wu, G.-L.; Duan, C.-P.; Pan, W.-G. Z-Scheme Bi/$Bi_2O_2CO_3$/Layered Double-Hydroxide Nanosheet Heterojunctions for Photocatalytic CO_2 Reduction under Visible Light. *ACS Appl. Nano Mater.* **2021**, *4*, 4902–4911. [CrossRef]
54. Khan, A.I.; O'Hare, D. Intercalation chemistry of layered double hydroxides: Recent developments and applications. *J. Mater. Chem.* **2002**, *12*, 3191–3198. [CrossRef]
55. Deng, L.; Liu, X.; Xu, J.; Zhou, Z.; Feng, S.; Wang, Z.; Xu, M. Transfer hydrogenation of CO_2 into formaldehyde from aqueous glycerol heterogeneously catalyzed by Ru bound to LDH. *Chem. Commun.* **2021**, *57*, 5167–5170. [CrossRef]
56. Iglesias, A.H.; Ferreira, O.P.; Gouveia, D.X.; Souza Filho, A.G.; de Paiva, J.A.C.; Mendes Filho, J.; Alves, O.L. Structural and thermal properties of Co–Cu–Fe hydrotalcite-like compounds. *J. Solid State Chem.* **2005**, *178*, 142–152. [CrossRef]
57. Xu, S.-M.; Pan, T.; Dou, Y.-B.; Yan, H.; Zhang, S.-T.; Ning, F.-Y.; Shi, W.-Y.; Wei, M. Theoretical and Experimental Study on MIIMIII-Layered Double Hydroxides as Efficient Photocatalysts toward Oxygen Evolution from Water. *J. Phys. Chem. C* **2015**, *119*, 18823–18834. [CrossRef]
58. Miyata, S. Anion-Exchange Properties of Hydrotalcite-Like Compounds. *Clays Clay Miner.* **1983**, *31*, 305–311. [CrossRef]
59. Weiyang, L.; Ji'an, S.; Yuyuan, Y.; Miao, D.; Qiang, Z. Morphology control of layered double hydroxide and its application in water remediation. *Prog. Chem.* **2021**, *32*, 2049.
60. Boumeriame, H.; Da Silva, E.S.; Cherevan, A.S.; Chafik, T.; Faria, J.L.; Eder, D. Layered double hydroxide (LDH)-based materials: A mini-review on strategies to improve the performance for photocatalytic water splitting. *J. Energy Chem.* **2022**, *64*, 406–431. [CrossRef]
61. Wang, R.; Qiu, Z.; Wan, S.; Wang, Y.; Liu, Q.; Ding, J.; Zhong, Q. Insight into mechanism of divalent metal cations with different d-bands classification in layered double hydroxides for light-driven CO_2 reduction. *Chem. Eng. J.* **2022**, *427*, 130863. [CrossRef]
62. Xiong, X.; Zhao, Y.; Shi, R.; Yin, W.; Zhao, Y.; Waterhouse, G.I.N.; Zhang, T. Selective photocatalytic CO_2 reduction over Zn-based layered double hydroxides containing tri or tetravalent metals. *Sci. Bull.* **2020**, *65*, 987–994. [CrossRef]
63. Tao, X.; Han, Y.; Sun, C.; Huang, L.; Xu, D. Plasma modification of NiAlCe–LDH as improved photocatalyst for organic dye wastewater degradation. *Appl. Clay Sci.* **2019**, *172*, 75–79. [CrossRef]
64. Mori, K.; Taga, T.; Yamashita, H. Isolated Single-Atomic Ru Catalyst Bound on a Layered Double Hydroxide for Hydrogenation of CO_2 to Formic Acid. *ACS Catal.* **2017**, *7*, 3147–3151. [CrossRef]
65. Hirata, N.; Tadanaga, K.; Tatsumisago, M. Photocatalytic O_2 evolution from water over Zn–Cr layered double hydroxides intercalated with inorganic anions. *Mater. Res. Bull.* **2015**, *62*, 1–4. [CrossRef]
66. Yan, K.; Liu, Y.; Lu, Y.; Chai, J.; Sun, L. Catalytic application of layered double hydroxide-derived catalysts for the conversion of biomass-derived molecules. *Catal. Sci. Technol.* **2017**, *7*, 1622–1645. [CrossRef]
67. Xu, J.; Liu, X.; Zhou, Z.; Deng, L.; Liu, L.; Xu, M. Platinum Nanoparticles with Low Content and High Dispersion over Exfoliated Layered Double Hydroxide for Photocatalytic CO_2 Reduction. *Energy Fuels* **2021**, *35*, 10820–10831. [CrossRef]
68. Flores-Flores, M.; Luévano-Hipólito, E.; Martínez, L.M.T.; Morales-Mendoza, G.; Gómez, R. Photocatalytic CO_2 conversion by MgAl layered double hydroxides: Effect of Mg2+ precursor and microwave irradiation time. *J. Photochem. Photobiol. A Chem.* **2018**, *363*, 68–73. [CrossRef]

69. Bravo-Suárez, J.J.; Páez-Mozo, E.A.; Ted Oyama, S. Microtextural properties of layered double hydroxides: A theoretical and structural model. *Microporous Mesoporous Mater.* **2004**, *67*, 1–17. [CrossRef]
70. Iguchi, S.; Teramura, K.; Hosokawa, S.; Tanaka, T. Photocatalytic conversion of CO_2 in water using fluorinated layered double hydroxides as photocatalysts. *Appl. Catal. A Gen.* **2016**, *521*, 160–167. [CrossRef]
71. Teramura, K.; Iguchi, S.; Mizuno, Y.; Shishido, T.; Tanaka, T. Photocatalytic Conversion of CO_2 in Water over Layered Double Hydroxides. *Angew. Chem. Int. Ed.* **2012**, *51*, 8008–8011. [CrossRef]
72. Wang, K.; Zhang, L.; Su, Y.; Shao, D.; Zeng, S.; Wang, W. Photoreduction of carbon dioxide of atmospheric concentration to methane with water over CoAl-layered double hydroxide nanosheets. *J. Mater. Chem. A* **2018**, *6*, 8366–8373. [CrossRef]
73. Hong, J.; Zhang, W.; Wang, Y.; Zhou, T.; Xu, R. Photocatalytic Reduction of Carbon Dioxide over Self-Assembled Carbon Nitride and Layered Double Hydroxide: The Role of Carbon Dioxide Enrichment. *ChemCatChem* **2014**, *6*, 2315–2321. [CrossRef]
74. Iguchi, S.; Teramura, K.; Hosokawa, S.; Tanaka, T. Photocatalytic conversion of CO2 in an aqueous solution using various kinds of layered double hydroxides. *Catal. Today* **2015**, *251*, 140–144. [CrossRef]
75. Saliba, D.; Ezzeddine, A.; Sougrat, R.; Khashab, N.M.; Hmadeh, M.; Al-Ghoul, M. Cadmium–Aluminum Layered Double Hydroxide Microspheres for Photocatalytic CO_2 Reduction. *ChemSusChem* **2016**, *9*, 800–805. [CrossRef]
76. Ahmed, N.; Morikawa, M.; Izumi, Y. Photocatalytic conversion of carbon dioxide into methanol using optimized layered double hydroxide catalysts. *Catal. Today* **2012**, *185*, 263–269. [CrossRef]
77. Ren, J.; Ouyang, S.; Xu, H.; Meng, X.; Wang, T.; Wang, D.; Ye, J. Targeting Activation of CO_2 and H_2 over Ru-Loaded Ultrathin Layered Double Hydroxides to Achieve Efficient Photothermal CO_2 Methanation in Flow-Type System. *Adv. Energy Mater.* **2017**, *7*, 1601657. [CrossRef]
78. Wang, S.; Wang, Y.; Zang, S.-Q.; Lou, X.W. Hierarchical Hollow Heterostructures for Photocatalytic CO_2 Reduction and Water Splitting. *Small Methods* **2020**, *4*, 1900586. [CrossRef]
79. Dou, Y.; Zhang, S.; Pan, T.; Xu, S.; Zhou, A.; Pu, M.; Yan, H.; Han, J.; Wei, M.; Evans, D.G.; et al. TiO_2@Layered Double Hydroxide Core–Shell Nanospheres with Largely Enhanced Photocatalytic Activity Toward O_2 Generation. *Adv. Funct. Mater.* **2015**, *25*, 2243–2249. [CrossRef]
80. Zhao, F.; Zhan, G.; Zhou, S.-F. Intercalation of laminar Cu–Al LDHs with molecular TCPP(M) (M = Zn, Co, Ni, and Fe) towards high-performance CO2 hydrogenation catalysts. *Nanoscale* **2020**, *12*, 13145–13156. [CrossRef]
81. Nayak, S.; Mohapatra, L.; Parida, K. Visible light-driven novel g-C3N4/NiFe-LDH composite photocatalyst with enhanced photocatalytic activity towards water oxidation and reduction reaction. *J. Mater. Chem. A* **2015**, *3*, 18622–18635. [CrossRef]
82. Nayak, S.; Parida, K.M. Nanostructured CeO_2/MgAl-LDH composite for visible light induced water reduction reaction. *Int. J. Hydrogen Energy* **2016**, *41*, 21166–21180. [CrossRef]
83. Gao, G.; Zhu, Z.; Zheng, J.; Liu, Z.; Wang, Q.; Yan, Y. Ultrathin magnetic Mg-Al LDH photocatalyst for enhanced CO_2 reduction: Fabrication and mechanism. *J. Colloid Interface Sci.* **2019**, *555*, 1–10. [CrossRef] [PubMed]
84. Chen, J.; Wang, C.; Zhang, Y.; Guo, Z.; Luo, Y.; Mao, C.-J. Engineering ultrafine NiS cocatalysts as active sites to boost photocatalytic hydrogen production of MgAl layered double hydroxide. *Appl. Surf. Sci.* **2020**, *506*, 144999. [CrossRef]
85. Ning, C.; Wang, Z.; Bai, S.; Tan, L.; Dong, H.; Xu, Y.; Hao, X.; Shen, T.; Zhao, J.; Zhao, P.; et al. 650 nm-driven syngas evolution from photocatalytic CO2 reduction over Co-containing ternary layered double hydroxide nanosheets. *Chem. Eng. J.* **2021**, *412*, 128362. [CrossRef]
86. Liu, X.; Zhao, X.; Zhu, Y.; Zhang, F. Experimental and theoretical investigation into the elimination of organic pollutants from solution by layered double hydroxides. *Appl. Catal. B Environ.* **2013**, *140*, 241–248. [CrossRef]
87. Kubacka, A.; Fernández-García, M.; Colón, G. Advanced Nanoarchitectures for Solar Photocatalytic Applications. *Chem. Rev.* **2012**, *112*, 1555–1614. [CrossRef]
88. Schneider, J.; Matsuoka, M.; Takeuchi, M.; Zhang, J.; Horiuchi, Y.; Anpo, M.; Bahnemann, D.W. Understanding TiO_2 Photocatalysis: Mechanisms and Materials. *Chem. Rev.* **2014**, *114*, 9919–9986. [CrossRef]
89. Chen, F.; Ma, T.; Zhang, T.; Zhang, Y.; Huang, H. Atomic-Level Charge Separation Strategies in Semiconductor-Based Photocatalysts. *Adv. Mater.* **2021**, *33*, 2005256. [CrossRef]
90. Zhao, H.; Xu, J.; Liu, L.; Rao, G.; Zhao, C.; Li, Y. CO2 photoreduction with water vapor by Ti-embedded MgAl layered double hydroxides. *J. CO2 Util.* **2016**, *15*, 15–23. [CrossRef]
91. Bai, S.; Wang, Z.; Tan, L.; Waterhouse, G.I.N.; Zhao, Y.; Song, Y.-F. 600 nm Irradiation-Induced Efficient Photocatalytic CO_2 Reduction by Ultrathin Layered Double Hydroxide Nanosheets. *Ind. Eng. Chem. Res.* **2020**, *59*, 5848–5857. [CrossRef]
92. Bai, S.; Li, T.; Wang, H.; Tan, L.; Zhao, Y.; Song, Y.-F. Scale-up synthesis of monolayer layered double hydroxide nanosheets via separate nucleation and aging steps method for efficient CO_2 photoreduction. *Chem. Eng. J.* **2021**, *419*, 129390. [CrossRef]
93. Iguchi, S.; Hasegawa, Y.; Teramura, K.; Kidera, S.; Kikkawa, S.; Hosokawa, S.; Asakura, H.; Tanaka, T. Drastic improvement in the photocatalytic activity of Ga_2O_3 modified with Mg–Al layered double hydroxide for the conversion of CO_2 in water. *Sustain. Energy Fuels* **2017**, *1*, 1740–1747. [CrossRef]
94. Tan, L.; Peter, K.; Ren, J.; Du, B.; Hao, X.; Zhao, Y.; Song, Y.-F. Photocatalytic syngas synthesis from CO_2 and H_2O using ultrafine CeO_2-decorated layered double hydroxide nanosheets under visible-light up to 600 nm. *Front. Chem. Sci. Eng.* **2021**, *15*, 99–108. [CrossRef]
95. Xu, J.; Liu, X.; Zhou, Z.; Deng, L.; Liu, L.; Xu, M. Visible Light-Driven CO_2 Photocatalytic Reduction by Co-porphyrin-Coupled MgAl Layered Double-Hydroxide Composite. *Energy Fuels* **2021**, *35*, 16134–16143. [CrossRef]

96. Chong, R.; Su, C.; Du, Y.; Fan, Y.; Ling, Z.; Chang, Z.; Li, D. Insights into the role of MgAl layered double oxides interlayer in Pt/TiO_2 toward photocatalytic CO_2 reduction. *J. Catal.* **2018**, *363*, 92–101. [CrossRef]
97. Song, Q.; Zhou, Y.; Hu, J.; Zhou, C.; Shi, X.; Li, D.; Jiang, D. Synergistic effects of surface Lewis Base/Acid and nitrogen defect in MgAl layered double Oxides/Carbon nitride heterojunction for efficient photoreduction of carbon dioxide. *Appl. Surf. Sci.* **2021**, *563*, 150369. [CrossRef]
98. Li, Z.; Liu, J.; Shi, R.; Waterhouse, G.I.N.; Wen, X.-D.; Zhang, T. Fe-Based Catalysts for the Direct Photohydrogenation of CO_2 to Value-Added Hydrocarbons. *Adv. Energy Mater.* **2021**, *11*, 2002783. [CrossRef]
99. Liu, L.; Zhao, C.; Xu, J.; Li, Y. Integrated CO_2 capture and photocatalytic conversion by a hybrid adsorbent/photocatalyst material. *Appl. Catal. B Environ.* **2015**, *179*, 489–499. [CrossRef]
100. Zhao, C.; Liu, L.; Rao, G.; Zhao, H.; Wang, L.; Xu, J.; Li, Y. Synthesis of novel MgAl layered double oxide grafted TiO2 cuboids and their photocatalytic activity on CO_2 reduction with water vapor. *Catal. Sci. Technol.* **2015**, *5*, 3288–3295. [CrossRef]
101. Wang, K.; Miao, C.; Liu, Y.; Cai, L.; Jones, W.; Fan, J.; Li, D.; Feng, J. Vacancy enriched ultrathin TiMgAl-layered double hydroxide/graphene oxides composites as highly efficient visible-light catalysts for CO_2 reduction. *Appl. Catal. B Environ.* **2020**, *270*, 118878. [CrossRef]
102. Hao, X.; Tan, L.; Xu, Y.; Wang, Z.; Wang, X.; Bai, S.; Ning, C.; Zhao, J.; Zhao, Y.; Song, Y.-F. Engineering Active Ni Sites in Ternary Layered Double Hydroxide Nanosheets for a Highly Selective Photoreduction of CO_2 to CH_4 under Irradiation above 500 nm. *Ind. Eng. Chem. Res.* **2020**, *59*, 3008–3015. [CrossRef]
103. Zhao, Y.; Jia, X.; Waterhouse, G.I.N.; Wu, L.-Z.; Tung, C.-H.; O'Hare, D.; Zhang, T. Layered Double Hydroxide Nanostructured Photocatalysts for Renewable Energy Production. *Adv. Energy Mater.* **2016**, *6*, 1501974. [CrossRef]
104. Tan, L.; Xu, S.-M.; Wang, Z.; Xu, Y.; Wang, X.; Hao, X.; Bai, S.; Ning, C.; Wang, Y.; Zhang, W.; et al. Highly Selective Photoreduction of CO_2 with Suppressing H_2 Evolution over Monolayer Layered Double Hydroxide under Irradiation above 600 nm. *Angew. Chem. Int. Ed.* **2019**, *58*, 11860–11867. [CrossRef] [PubMed]
105. Wang, C.J.; Wu, Y.A.; Jacobs, R.M.J.; Warner, J.H.; Williams, G.R.; O'Hare, D. Reverse Micelle Synthesis of Co—Al LDHs: Control of Particle Size and Magnetic Properties. *Chem. Mater.* **2011**, *23*, 171–180. [CrossRef]
106. Lai, T.; Wang, J.; Xiong, W.; Wang, H.; Yang, M.; Li, T.; Kong, X.; Zou, X.; Zhao, Y.; O'Hare, D.; et al. Photocatalytic CO_2 reduction and environmental remediation using mineralization of toxic metal cations products. *Chem. Eng. Sci.* **2022**, *257*, 117704. [CrossRef]
107. Chen, W.; Han, B.; Xie, Y.; Liang, S.; Deng, H.; Lin, Z. Ultrathin Co-Co LDHs nanosheets assembled vertically on MXene: 3D nanoarrays for boosted visible-light-driven CO_2 reduction. *Chem. Eng. J.* **2020**, *391*, 123519. [CrossRef]
108. Xu, J.; Zhang, A.; Zhou, Z.; Wang, C.; Deng, L.; Liu, L.; Xia, H.; Xu, M. Elemental Mercury Removal from Flue Gas over Silver-Loaded CuS-Wrapped Fe_3O_4 Sorbent. *Energy Fuels* **2021**, *35*, 13975–13983. [CrossRef]
109. Miao, C.; Hui, T.; Liu, Y.; Feng, J.; Li, D. Pd/MgAl-LDH nanocatalyst with vacancy-rich sandwich structure: Insight into interfacial effect for selective hydrogenation. *J. Catal.* **2019**, *370*, 107–117. [CrossRef]
110. Wang, Z.; Song, Y.; Zou, J.; Li, L.; Yu, Y.; Wu, L. The cooperation effect in the Au–Pd/LDH for promoting photocatalytic selective oxidation of benzyl alcohol. *Catal. Sci. Technol.* **2018**, *8*, 268–275. [CrossRef]
111. Zhu, P.; Gao, M.; Zhang, J.; Wu, Z.; Wang, R.; Wang, Y.; Waclawik, E.R.; Zheng, Z. Synergistic interaction between Ru and MgAl-LDH support for efficient hydrogen transfer reduction of carbonyl compounds under visible light. *Appl. Catal. B Environ.* **2021**, *283*, 119640. [CrossRef]
112. Kawamura, S.; Puscasu, M.C.; Yoshida, Y.; Izumi, Y.; Carja, G. Tailoring assemblies of plasmonic silver/gold and zinc–gallium layered double hydroxides for photocatalytic conversion of carbon dioxide using UV–visible light. *Appl. Catal. A Gen.* **2015**, *504*, 238–247. [CrossRef]
113. Fan, G.; Li, F.; Evans, D.G.; Duan, X. Catalytic applications of layered double hydroxides: Recent advances and perspectives. *Chem. Soc. Rev.* **2014**, *43*, 7040–7066. [CrossRef]
114. Wang, Z.; Xu, S.-M.; Tan, L.; Liu, G.; Shen, T.; Yu, C.; Wang, H.; Tao, Y.; Cao, X.; Zhao, Y.; et al. 600 nm-driven photoreduction of CO2 through the topological transformation of layered double hydroxides nanosheets. *Appl. Catal. B Environ.* **2020**, *270*, 118884. [CrossRef]
115. Li, P.; Yu, Y.; Huang, P.-P.; Liu, H.; Cao, C.-Y.; Song, W.-G. Core–shell structured MgAl-LDO@Al-MS hexagonal nanocomposite: An all inorganic acid–base bifunctional nanoreactor for one-pot cascade reactions. *J. Mater. Chem. A* **2014**, *2*, 339–344. [CrossRef]
116. He, S.; An, Z.; Wei, M.; Evans, D.G.; Duan, X. Layered double hydroxide-based catalysts: Nanostructure design and catalytic performance. *Chem. Commun.* **2013**, *49*, 5912–5920. [CrossRef]
117. Li, C.; Wei, M.; Evans, D.G.; Duan, X. Layered Double Hydroxide-based Nanomaterials as Highly Efficient Catalysts and Adsorbents. *Small* **2014**, *10*, 4469–4486. [CrossRef]
118. Wu, M.J.; Wu, J.Z.; Zhang, J.; Chen, H.; Zhou, J.Z.; Qian, G.R.; Xu, Z.P.; Du, Z.; Rao, Q.L. A review on fabricating heterostructures from layered double hydroxides for enhanced photocatalytic activities. *Catal. Sci. Technol.* **2018**, *8*, 1207–1228. [CrossRef]
119. Miao, Y.-F.; Guo, R.-T.; Gu, J.-W.; Liu, Y.-Z.; Wu, G.-L.; Duan, C.-P.; Zhang, X.-D.; Pan, W.-G. Fabrication of β-In2S3/NiAl-LDH heterojunction photocatalyst with enhanced separation of charge carriers for efficient CO_2 photocatalytic reduction. *Appl. Surf. Sci.* **2020**, *527*, 146792. [CrossRef]
120. Yang, Y.; Wu, J.; Xiao, T.; Tang, Z.; Shen, J.; Li, H.; Zhou, Y.; Zou, Z. Urchin-like hierarchical CoZnAl-LDH/RGO/g-C3N4 hybrid as a Z-scheme photocatalyst for efficient and selective CO_2 reduction. *Appl. Catal. B Environ.* **2019**, *255*, 117771. [CrossRef]

121. Han, X.; Lu, B.; Huang, X.; Liu, C.; Chen, S.; Chen, J.; Zeng, Z.; Deng, S.; Wang, J. Novel p- and n-type S-scheme heterojunction photocatalyst for boosted CO_2 photoreduction activity. *Appl. Catal. B Environ.* **2022**, *316*, 121587. [CrossRef]
122. Yang, Z.; Li, S.; Xia, X.; Liu, Y. Hexagonal MgAl-LDH simultaneously facilitated active facet exposure and holes storage over ZnIn2S4/MgAl-LDH heterojunction for boosting photocatalytic activities and anti-photocorrosion. *Sep. Purif. Technol.* **2022**, *300*, 121819. [CrossRef]

Article

Generation and Emission Characteristics of Fine Particles Generated by Power Plant Circulating Fluidized Bed Boiler

Heming Dong [1], Yu Zhang [1,*], Qian Du [1,*], Jianmin Gao [1], Qi Shang [1], Dongdong Feng [1] and Yudong Huang [2]

[1] School of Energy Science and Engineering, Harbin Institute of Technology, Harbin 150001, China
[2] School of Chemical Engineering and Technology, Harbin Institute of Technology, Harbin 150001, China
* Correspondence: zhang.y@hit.edu.cn (Y.Z.); duqian@hit.edu.cn (Q.D.)

Abstract: The generation and emission characteristics of fine particulates ($PM_{2.5}$) from three 300 MW power plant circulating fluidized bed boilers were investigated. One boiler had an external bed and used an electrostatic precipitator, the other two used an electrostatic filter precipitator and fabric filter, respectively. The particle size distribution of fine particles was performed by an electrical low-pressure impactor. $PM_{2.5}$ samplers were used at the same time to collect fine particles for subsequent laboratory analysis. The results show that the number size distributions of fine particles presented one single peak, but there was no peak in mass size distributions. The mass concentrations of three CFB boilers were similar, but the number concentration of the external bed CFB boiler was much higher than that of the general CFB boiler. The minimum removal efficiencies of the precipitator appeared between 0.1~1 μm, but the locations of the minimum point were different. The morphology of fine particles was mostly irregular. The highest content of fine particles was insoluble oxides and the content of S element was also high. Different precipitators have different removal effects on Si, Al, Ca, S and Fe in fine particles, but they all have poor removal effects on Na and K as well as OC and EC.

Keywords: fine particles; CFB boiler; PSDs; dust removal; physicochemical properties

Citation: Dong, H.; Zhang, Y.; Du, Q.; Gao, J.; Shang, Q.; Feng, D.; Huang, Y. Generation and Emission Characteristics of Fine Particles Generated by Power Plant Circulating Fluidized Bed Boiler. *Energies* **2022**, *15*, 6892. https://doi.org/10.3390/en15196892

Academic Editor: Andres Siirde

Received: 2 August 2022
Accepted: 7 September 2022
Published: 21 September 2022

1. Introduction

With the massive emission of CO_2 and fine particles, environmental problems such as the global greenhouse effect and smog are becoming more serious. Fine particles have always been a main environmental atmospheric pollutant and a research hotspot in China. Fine particles whose aerodynamic diameter are less than 2.5 μm are also called $PM_{2.5}$. Because of their low sedimentation velocity, caused by small aerodynamic diameter, fine particles can be suspended for several months in the air and transmitted thousands of miles away. Studies have shown that between city and suburb, the PM_{10} concentrations are different, but $PM_{2.5}$ concentrations are much closer [1]. $PM_{2.5}$ can enter the alveolus region of the lungs and then enter the blood circulatory system [2,3]. Due to their small volume, the surface area of fine particles is very large, and lots of organic matter (e.g., polycyclic aromatic hydrocarbons), bacteria and other harmful substances in the atmosphere are absorbed on them [4]. Fine particles emitted from stationary combustion sources is often rich in toxic trace elements that are harmful to humans [5].

In the Twentieth Century for 80 years, PM had concerned many foreign scholars. At that time, the classical theory that particles emitted from coal combustion in pulverized coal boilers (PCB) had bimodal distribution [6] has been widely accepted. Markowski and Ensor [7] measured a pulverized coal utility boiler in 1980 and they found a sharp peak of PM number and mass distribution in the submicron region appeared both upstream and the downstream of the dust remover. The mode (fine-mode) and chemical composition of these submicron aerosols were obviously different from a large fly ash aerosol which was formed by char fragmentation and surface ash aggregation. It appeared to result from a gasification–coagulation mechanism. However, in 2000, Linak etc. [8] burned three different coals in a small fire-tube boiler, a laboratory-scale refractory-lined combustor and a pulverized coal combustor. Their experimental results showed that, in addition to the fine mode and coarse mode, coal combustion also generated a central mode which showed a peak at 0.8~2.0 μm (aerodynamic diameter).

In China, emissions from power generation with coal as fuel are a main source of fine particles in the environmental atmosphere [9]. PCBs are the most widely used coal-fired boilers in power plants, and most studies of fine particles have been focused on PCBs, with little research on circulating fluidized bed (CFB) boilers. Recently though, the application of CFB boilers in power generation has become more extensive. This is because CFB boilers have an excellent fuel adaptability, so can burn almost all solid fuel, especially low-grade fuel. Therefore, recently, some researchers had studied mineral transformation and ash deposition during the combustion of brown coal and other fuels at laboratory scale CFB [10,11]. In addition, the SO_2 and NO_x emissions from CFB boilers are very low, so there are needless desulphurization facilities and denitration facilities.

The gas–solid flow pattern of CFB boilers was between fixed bed (grate-fired boilers, GFB) and pneumatic conveying state (PCBs) and the feed size of coal was usually several millimeters in diameter, which was much larger than pulverized coal. In addition, coal only constituted a few percent in the dense suspension of CFB boilers and its residence time was much longer than that in pulverized coal boilers [12]. When coal was burning in CFB boiler furnaces, the temperature was lower and heat transfer was more efficient. Because of the above reasons, the generation characteristics and physical/chemical properties of fine particles emissions from coal-fired CFB boilers was quite different with PCBs. Some Finnish scholars have conducted research on small-scale CFB boilers with biomass or waste combustion [13,14]. Ruan et al. [15] compared the particulate matter characteristics discharged by an industrial CFB (150 t/h) and an industrial GFB (100 t/h), and found that the $PM_{2.5}$ generation of CFBs was higher than that of BFBs, but the removal effect of a fabric filter on $PM_{2.5}$ of CFBs was better. Liu et al. [16] studied the generation and removal of $PM_{2.5}$ from two 135 MW CFBs, and discussed the effect of adding limestone into the furnace on the generation characteristics of $PM_{2.5}$. However, studies on fine particles of large coal-fired CFB boilers in power plants are relatively rare to date.

Three 300 MW power plant CFB boilers were studied in this work, which were equipped with an electrostatic filter precipitator (EFP), fabric filter (FF) and electrostatic precipitator (ESP), respectively. We determined fine particle generation and emission characteristics, including particle size distributions, morphology, and chemical constituents as well as removal characteristics of fine particles for three kinds of dust removal device. The experimental data are very helpful to understand the fine particle generation and emission characteristics of CFB power stations in China.

2. Methods

2.1. Boilers and Measuring Points

We conducted studies on three 300 MW CFB boilers in power plants, with dust removal methods including EFP, FF and ESP. The method of desulfurization was feeding limestone as a sorbent into the furnace, and there was no denitration facility. The basic situation of each boiler is shown in Table 1. During the test period, same kind of fuel was fed into the furnace and the boiler operated under stable load. Proximate analyses and ultimate analyses of coal feed into each CFB boiler are in Table 2.

Table 1. Basic situation of three CFB boilers.

No.	Location	Boiler Type	Coal	Load	Pollution Control Technology		
					Denitration	Dusting	Desulfurization
1	Shanxi	CFB boiler	Lignite mixed with peat	300 MW	—	EFP	Adding limestone sorbent in the furnace
2	Inner Mongolia	CFB boiler	Bituminous coal	300 MW	—	FF	Adding limestone sorbent in the furnace
3	Shandong	CFB boiler with external bed	Bituminous coal	300 MW	—	ESP	Adding limestone sorbent in the furnace

Table 2. Proximate analyses and ultimate analyses of coal feed in the furnace.

No.	Proximate Analysis (wt%)				Ultimate Analysis (wt%)				Calorific Value
	M_{ad} [a]	Vad [b]	A_{ad} [c]	FC_{ad} [d]	C_{ad}	H_{ad}	N_{ad}	S_{ad}	Q_{ad} MJ/kg
1	1.83	22.57	42.44	33.16	40.98	3.58	0.54	1.52	14.97
2	2.16	26.79	31.01	40.04	50.75	3.28	0.79	0.86	19.95
3	5.36	20.53	33.63	40.48	46.53	3.53	1.21	1.56	22.38

[a] M_{ad} means Moisture, air dry; [b] V_{ad} means Volatile, air dry; [c] A_{ad} means ash, air dry; [d] FC_{ad} means Fixed carbon, air dry.

Figure 1 is the schematic of the CFB boiler and the arrangement of measuring points. The measuring point before the dust collector was used for measuring fine particle generation characteristics of the CFB boiler. The measuring point after the dust collector was used for measuring fine particle emission characteristics of the CFB boiler and researching the effects of different dust collectors for fine particulate emissions. Because fine particles have good following characteristics and are approximately evenly distributed in the flue gas, the position of the sampling probe was fixed at every measuring point during the experiment. The compositions and temperatures of fuel gas measured by TH-880f flue gas parallel sampler (Wuhan Tianhong intelligent instrument factory) are shown in Table 3. The measurement error of the flue gas temperature is ±3 °C, the measurement error of O_2 concentration is ±2.5%, and the error of other gas concentrations is ±3%.

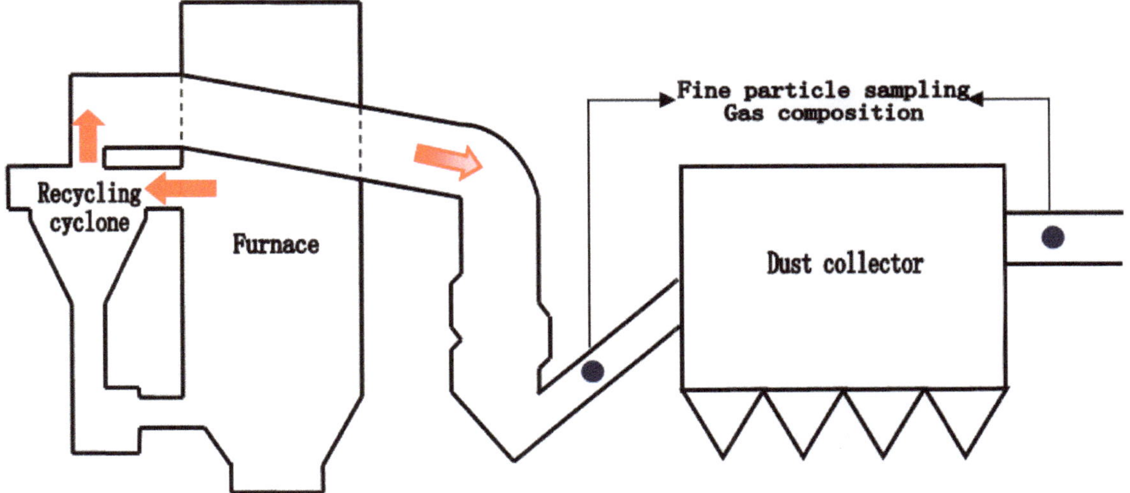

Figure 1. Schematic of CFB boiler and the measuring points arrangement.

Table 3. Flue gas analysis results.

Location	SO_2 mg/m^3	NO mg/m^3	CO mg/m^3	O_2 %	NO_2 mg/m^3	Temperature °C
Boiler 1 before dust removal	178.05	174.13	58.00	5.63	0.00	139.54
Boiler 1 after dust removal	209.16	156.84	61.00	6.09	0.00	115.00
Boiler 2 before dust removal	220.00	283.00	52.00	5.24	1.00	167.80
Boiler 2 after dust removal	371.00	263.00	49.00	5.41	0.00	142.70
Boiler 3 before dust removal	159.20	47.80	92.20	8.32	0.00	179.40
Boiler 3 after dust removal	168.67	69.50	60.00	6.97	0.00	148.80

2.2. Fine Particle Sampling and Measurement

Figure 2 is the diagram of dilution measurement and sampling system used during this work. The flue gas was successively passed by an isokinetic sampling probe, a cyclone for PM_{10} and two dilution devices that diluted using filtered air. When the pressure in the filtered air supply system was stable at 0.2 MPa, throughout the sampling process, the dilution ratio (about 9-fold) was stable, so the entire dilution system could keep a stable dilution ratio of about 80 times. The first stage diluter needed to be heated and injected with heated air to avoid water and acid condensation.

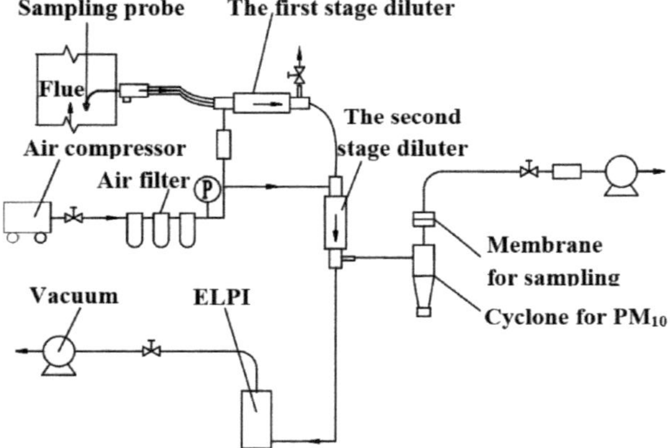

Figure 2. Schematic of fine particles sampling and measurement.

This work used an electrical low-pressure impactor (ELPI) (Dekati, Finland) to measure fine particle number and mass size distributions. The response time of ELPI was generally less than 5 s, so it could measure instantaneous concentration of particulate matter. The measuring range of ELPI was 0.007~10 μm that was divided into 12 levels by different impactors. For fine particles research, only level 1 to level 10 were needed. Average particle diameters and measurement ranges of all 10 impactors are given in Table 4; the 11th stage impactor has a cutting diameter of 2.48 μm.

During fine particle sample collection, two dedicated fine particle samplers (Wuhan Tianhong Instruments Co., Ltd., Wuhan, China) were connected to the tail of a primary or secondary diluter. As different analytical methods have certain requirements for the filter, this work selected a Teflon membrane and quartz fiber for analysis of the chemical composition of fine particles, and polycarbonate membranes for micrographs of single fine particles.

Table 4. Average particle diameters and measurement Ranges of all 10 impactor.

Stage	Di [μm]	Cutting Diameter [μm]	Mass Min [μg/m^3]	Mass Max [mg/m^3]	Number Min [1/cm^3]	Number Max [1/cm^3]
10	1.9	1.59	6.3	630	0.36	4×10^4
9	1.2	0.943	3.5	350	0.8	8×10^4
8	0.76	0.609	2	200	1.6	2×10^5
7	0.48	0.38	1	90	3	3×10^5
6	0.31	0.26	0.4	40	5	5×10^5
5	0.20	0.154	0.17	17	9	9×10^5
4	0.12	0.093	0.078	7.8	15	2×10^6
3	0.073	0.057	0.035	3.5	26	3×10^6
2	0.041	0.029	0.015	1.5	50	5×10^6
1	0.021	0.007	0.005	0.5	90	9×10^6

2.3. Analytical Methods for Physical and Chemical Properties of Fine Particles

2.3.1. Micrographs

Thermal scanning electron microscopy (SEM) produced by German Zeiss company (Model: EVOMA10) was used to analyze morphology of single fine particles. The sampling time had to be controlled within 5~10 s, to make sure that enough fine particles were collected by the polycarbonate membrane, and particle accumulation did not occur. Because fine particle samples, whose main constituent was aluminosilicate-based non-metallic compounds, have poor electrical and thermal conductivity, the samples were sprayed with about 25 nm Au before testing.

2.3.2. Elemental Analysis

Two kinds of laboratory instruments were used for elemental analysis of fine particles in this testing: EVOMA10 consists of SEM and energy-dispersive X-ray spectroscopy (EDX), which means that it can also analyze the elemental composition of the tested sample. However, due to the lower resolution of EDX results, we also used an X-ray fluorescence (XRF) analyzer to determine fine particle sample on a Teflon membrane. Compared to EDX, XRF gave a more accurate proportional relationship between various elements, but the test result was smaller due to the insufficient thickness of fine particle sample. For this reason, the final result of the elemental analysis was the combination of EDX and XRF analysis.

2.3.3. Organic Carbon and Elemental Carbon

Organic carbon (OC) and elemental carbon (EC) in the study of particulate matter composition are not strictly defined substances, and they represent two kinds of extremely complex substances with common physicochemical properties. OC represents aromatic compounds, aliphatic compounds, organic acids and other organic compounds; EC contains soot and some oxygen-containing functional groups (such as alcohol, phenol, acyl and carboxyl). In this work, a carbon analyzer developed by the Desert Research Institute of the United States, which adopts the method of thermo optical reflection (TOR), was used to test OC and EC in fine particle samples.

3. Results and Discussion

3.1. Concentration and Size Distributions of Fine Particle

In this research, concentrations of fine particles emitted by three CFB boilers were characterized before and after dust removal devices. Table 5 shows the number/mass concentration of fine particles and the ratio of the concentration of different particle size ranges. Fine particle number concentrations of the CFB boiler before dust removal were millions per cubic centimeter, much lower than in earlier PCB testing (millions to more than ten million/cm^3); however, the mass concentrations are about 2 g/m^3, much larger than PCB (hundreds mg/m^3) [17]. The number concentration of boiler 3 was much higher than the other two, but mass concentrations of the three boilers were same. Before the precipitator, the concentration of fine particles mainly depended on particles in the size range of 0.38~2.5 μm ($PM_{0.38\sim2.5}$), except for boiler 3. This this means that fine particles generated by the CFB boiler are mostly residual ash. As a result of highly efficient dust removal devices, fine particle concentrations of the CFB boiler decreased significantly after dust removal. After dust removal, the mass concentration of fine particles still depended on $PM_{0.38\sim2.5}$, but fine particle number concentration was mainly dependent on $PM_{0.38}$. This is due to the dust collection efficiency of small particles being lower than for large particles, which causes the ratio of $PM_{0.38}/PM_{2.5}$ to rise.

Table 5. Number/mass concentration of fine particle from three CFB boilers and the ratio of the concentration of different particle size ranges.

No.	Type	Location	Fine Particles ($PM_{2.5}$) Concentration	$PM_{2.5}/PM_{10}$ (%)	$PM_{1.0}/PM_{2.5}$ (%)	$PM_{0.38}/PM_{2.5}$ (%)
Boiler 1	Number (/cm^3)	Before dust removal	2,899,674.8	96.54	75.01	27.86
		After dust removal	13,180.1	99.31	97.04	86.24
	Mass (mg/m^3)	Before dust removal	1629.0	31.46	12.36	0.37
		After dust removal	1.1	46.73	21.27	3.23
Boiler 2	Number (/cm^3)	Before dust removal	3,084,830.0	94.32	72.20	23.30
		After dust removal	2316.9	98.34	91.36	65.62
	Mass (mg/m^3)	Before dust removal	2026.5	21.70	12.36	0.31
		After dust removal	0.5	24.21	18.39	1.65
Boiler 3	Number (/cm^3)	Before dust removal	7,625,854.6	97.68	89.65	72.63
		After dust removal	23,984.2	99.74	96.77	81.46
	Mass (mg/m^3)	Before dust removal	2043.9	30.60	11.12	1.28
		After dust removal	2.0	48.01	29.74	5.78

The number/mass concentration distribution of CFB boiler fine particles before and after dust removal is shown in Figure 3. The number size distribution was unimodal distribution that was significantly different with bimodal distribution of PCBs [17]. Before dust removal, the peak of boiler 1 and boiler 2 was at 0.76 μm, which was similar to central modal among three modal distributions [8]. However, for boiler 3, the peak was at 0.12 μm. This peak was more likely to be fine-modal [7]. The possible reason of the number size distribution difference between boiler 3 and the others was that boiler 3 was equipped with a series of external heat changers which were equivalent to a series of bubbling fluidized beds (BFBs) and the concentration of ultrafine particles generated by the BFB boiler was much higher than the CFB boiler [18]. After dust removal, the concentration of each level of ELPI decreased significantly and the peak of number concentration distribution had moved to the direction of small particles. As shown in Figure 3, the mass size distribution of fine particles of the CFB boiler had no peak, and the mass concentration of particles in each level increased with particle diameter increasing.

From the foregoing analysis, fine particles generated by power plant CFB boilers were mainly concentrated in the range of 0.38~2.5 μm, and the broken-coalescence mechanism was their main generation mechanism. Because the ash content of coal burned in the CFB boiler was extremely high and the severe abrasion caused by strong horizontal and vertical mixing promoted coal crushing, residual ash particles in the fuel gas of CFB boilers were much more than in PCBs. Coal crushing in CFBs is mainly mechanical crushing (mechanical abrasion) caused by violent collision between coal particles (0–13 mm) and bed material or heating surface and thermal crushing. Particles generated by mechanical abrasion are smaller than for thermal crushing, and the process of producing ultrafine particles by mechanical abrasion is called ultrafine abrasion. Some some researchers [19] believe that ultrafine abrasion is the main generation mechanism of sub-micron particles in CFB boilers. Since the temperature in CFB boiler furnaces was much lower (850~950 °C) than that of PCBs (1250~1300 °C), the volatilization of minerals in coal was suppressed and ultrafine particles ($PM_{0.38}$) formed by vaporization–condensation mechanisms were rare. The flue gas discharged from CFB boiler furnaces had to pass through several cyclone separators to collect the unburned coal char particles. These separators had a large cut diameter and their direct capture effect of fine particles was poor. However, they increased the residence time and collision of particles, which increased the trapping of large particles to ultrafine particles and caused further reduction of $PM_{0.38}$. Some articles claim that adding limestone in the furnace as a desulfurization sorbent can reduce $PM_{1.0}$ emissions of a CFB boiler: On one hand, CaO generated by limestone when it is heated could adsorb volatile substances steam or react with it [20]; on the other hand, the large surface area of limestone not only made the absorption of minerals steam more efficient but also provided attachment sites for nanoscale particles in flue gas [21]. However, limestone with a high degree of fragmentation was also a source of fly ash of the CFB boiler. Due to the $CaCO_3$/CaO content and particle size distribution of limestone used in power plants every day varying greatly, this paper cannot clarify the effect of adding limestone in the furnace on the generation characteristics of fine particles of CFB boilers.

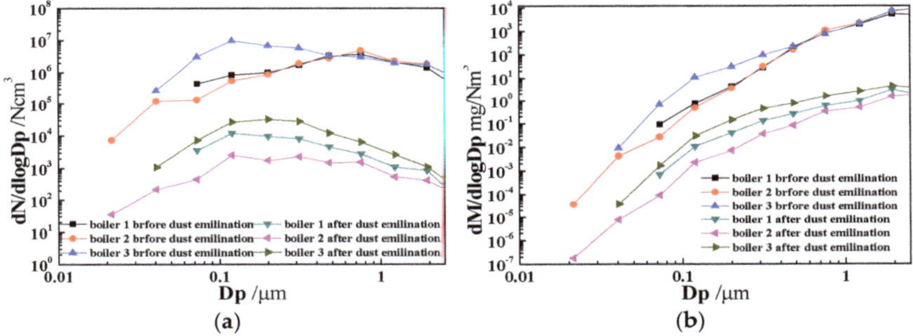

Figure 3. Concentration size distributions of fine particle at upstream and downstream of CFB boiler dust remover. (**a**) number concentration; (**b**) mass concentration.

3.2. The Influence of Dust Removal Device to Fine Particle

The key parameters of the dust collectors of three boilers are shown in Table 6. The total removal efficiency of number/mass of fine particles emitted by three CFB boilers is shown in Table 7. For all three kinds of dust removal device, the mass removal efficiency was higher than the number removal efficiency, and the removal effect of FF on fine particles (boiler 2) was better than that of ESP (boiler 3). Size-classified removal efficiency of fine particles by different dust removal devices was shown in Figure 4. The size-classified removal efficiency of EFP, FF and ESP on fine particles all had a minimum point in 0.1~1.0 μm, but the locations of the minimum points were different. After the minimum point, the removal efficiency of ESP and FF was higher and rose faster than for ESP. The EFP of boiler 1 in Figure 4 did not show the advantages of a composite dust removal device in particulate matter removal, especially for the low efficiency of submicron particle removal, which was caused by the use of needle felt with a poor filtering effect as filter material. Since an electrostatic precipitator was arranged in front of EFP, and fabric filter was arranged at the rear, the classified dust removal efficiency curve of EFP for fine particles is similar to that of FF.

Table 6. Key parameters of the dust collectors of three boilers.

Boiler No.	1	2	3
Type	Electrostatic-fabric filter integrated precipitator (air blowback)	Fabric filter (flue gas blowback)	Electrostatic precipitator (5 electric field)
Chamber × field	2 × (2-field + 2-baghouse)	2 × 4	2 × 5
Full load flue gas flow(hot), Nm^3/s	597.23	541.73	577.57
Flue gas temperature, °C	139.54	167.80	179.40
Sectional area, m^2	576		660.96
Flow velocity, m/s	1.04		0.874
Specific collection area, $m^2/m^3/s$	35.7		109.86
Field length of Single electric field, m	4.0		3.84
Total filter area, m^2	36,363	34,696	
Filter velocity, m/min	1.2	0.8	
Filter material	PTFE + PPS	85%PPS + 15%P84	

Table 7. Total removal efficiency of number/mass of fine particles emitted by CFB boiler.

Removal Efficiency (%)	Boiler 1	Boiler 2	Boiler 3
Removal efficiency of number	99.545	99.925	99.685
Removal efficiency of mass	99.936	99.975	99.902

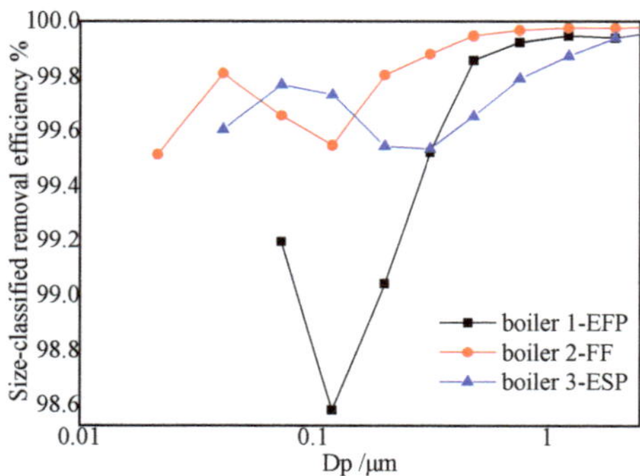

Figure 4. Size-classified removal efficiency of fine particle emitted by CFB boiler through different dust removal devices.

The removal process of particles through fabric filter was mainly influenced by the interception of a filter layer and ash layer, inertial impaction and diffusion collision. For larger particles, inertial impaction was the main trapping mechanism, which means the particles deviate from the streamline due to inertia and collide with interceptors. However, the capture of ultrafine particles was largely dependent on the diffusion effect. The Brownian motion free path of ultrafine particles was less than the filter layer clearance, so ultrafine particles would collide with the filter layer or ash layer and be adsorbed when it made a random motion [22]. The minimum point (0.12 μm) of size-classified removal efficiency was in the transition zone between these two mechanisms.

There was also a minimum value of size-classified removal efficiency of ESP in 0.1~1 μm [23,24]. Static electricity was the main force that influenced particle removal in ESP. For particles less than 0.1 μm, their charge mainly depended on collision with ionized gas during irregular Brownian motion; and charge of particles larger than 1 μm was mainly dependent on the electric field. However, 0.1~1 μm was the transitional stage of these two charging mechanisms. Charge of particles in this transitional stage was less effective, and thus the collection efficiency was relatively low. Some studies have shown that due to the high resistance of limestone in fly ash, ESP may not be well-suited for CFB boiler dust removal [25,26]. Since the collection efficiency of FF is not affected by fly ash resistivity, it is especially suitable for use in CFB boiler dust removal.

3.3. Morphological Characteristics of Fine Particle Generated by CFB Boiler

The micrographs of fine particles formed by power plant CFB boiler are shown in Figure 5. Figure 5a is a micrograph of a great quantity of fine particles. It shows that fine particles generated by CFB were mostly irregular in shape, which means their specific surface area was much bigger than fine particles with the same particle diameter but generated by PCB. This was because the temperature in the CFB boiler furnace was low, so the majority of minerals in coal would not completely melt and reduce the gasification amount. Residual ash particles were mostly keeping the original morphology in coal, and ultrafine particles formed via nucleation of vaporized minerals and grown via coagulation and heterogeneous condensation were very small, so irregular particles were the major component of fine particles generated by the CFB boiler. Figure 5b–d are the main morphology of irregular fine particles formed by the CFB boiler. Ultrafine particles formed via nucleation of vaporized minerals were rare, and since most minerals did not melt, the adhesion between particles was smaller than for PCB, so the surface of large particles rarely adhered to ultrafine particles. However, there was occurrence of gasification–condensation and melting in the CFB boiler, the surface of particles in Figure 5e adhered to some circular ultrafine particles, and particles in Figure 5f had partially melted.

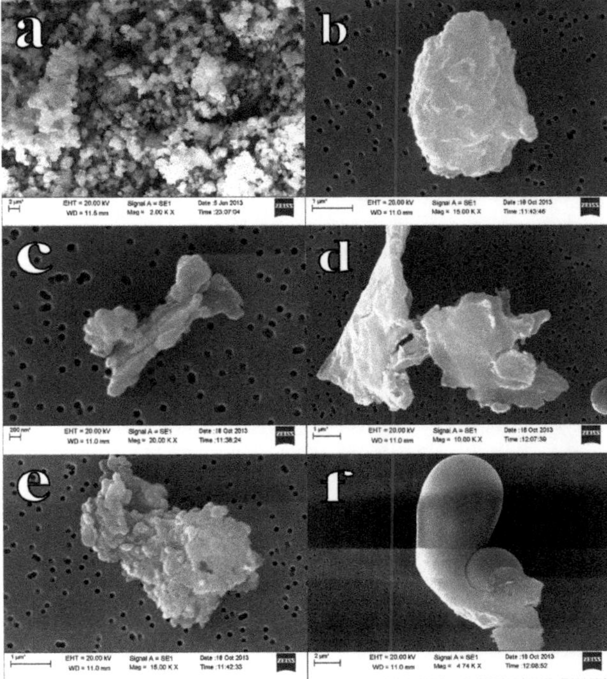

Figure 5. Micrographs of fine particle formed by power plant CFB boiler. (**a**) micrograph of a great quantity of fine particles; (**b**–**d**) main morphology of irregular fine particles; (**e**) particles with ultrafine particles adsorbed on the surface; (**f**). partially melted particles.

3.4. Study of Chemical Constitutes of Fine Particle Emitted by CFB Boiler

The content of different elements in fine particles varies greatly, wherein Al, Si, S, Ca, Fe, etc. accounts for the vast majority. Therefore, the test results of these elements were credible and representative, and in this section the influence of fluidized bed combustion and dust removal device on fine particle generation and emission were determined by studying the content of these elements. According to fine particle formation mechanisms and research methods of the past researchers [27], the high content or representative elements were broadly divided into four kinds. (These elements are usually present in fine particles as salts or oxides, but in this section they are expressed in terms of oxides.):

(1) Difficult melting oxides, including Al_2O_3, SiO_2, CaO, Fe_2O_3;
(2) Alkali metal oxides, including Na_2O and K_2O;
(3) SO_3, assuming S exists in the form of sulfur oxides;
(4) Other elements, including MgO, compounds of trace elements, OC and EC, etc.

According to the classification method described above, the percentage of the different components contained in fine particles collected before dust removal were calculated, as shown in Figure 6.

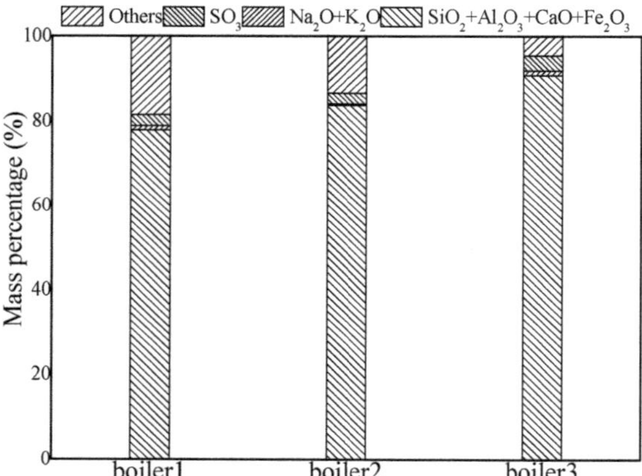

Figure 6. The percentage of representative elements contained in fine particles collected at the outlet of the furnace of a CFB boiler.

As shown in Figure 6, in fine particles generated by the CFB boiler, the content of difficult melting oxides was the highest and the content of alkali metal oxides was the lowest. Al, Si, Ca, Fe were also the highest mineral elements in coal. Compared with PCB, chemical composition of fine particles generated by the CFB boiler was more akin to the ash composition [28,29]. The content of S in fine particles was also high, which means that there were a lot of adsorbent particles in the CFB boiler fine particles. Terttaliisa Lind et al. [24] detected CFB boiler fly ash using CC-SEM and found that there were a lot of particles with Ca or Ca/S as the main material. Fly ash particles formed by these adsorbents may also be an important source of fine particles from a CFB boiler.

To analyze the effects of different types of precipitator on representative elements contained in fine particles of a CFB boiler, Figure 7 shows the contrast of representative elements content of fine particles before and after dust removal (OC and EC were also given separately in this report).

As it shown in Figure 7, after dust removal, the content of Na and K increased, because Na and K allow easy gasification and enrichment in $PM_{0.38}$, and the removal efficiency of each precipitator for $PM_{0.38}$ is lower than $PM_{0.38\sim2.5}$. The removal effect of the dust remover on different elements was different. After FF, the content of Ca, S and Fe decreased significantly but the content of Si and Al increased, which means that the removal effect of FF on fine particles rich in Ca, S and Fe was better than particles rich in Si and Al. This may be due to the different removal efficiency of FF on different-sized particles, which are enriched with different elements. Since Ca can greatly increase the fly ash resistivity [24,25], ESP found it difficult to remove Ca-rich particles, and after the dust removal device, the reducing of flue gas temperature can cause sulfuric acid generation and condensation on fine particle surfaces. Therefore, after ESP (boiler 3), the content of Ca and S increased. ESP was not as effective as FF in the removal of Fe-containing particles. Because the element-removal characteristics were controlled by two types of dust removal device, the changes in the content of elements other than Ca and S in fine particles after EFP are the same as FF. The content of Ca and S increased due to the influence of the electrostatic precipitator in EFP.

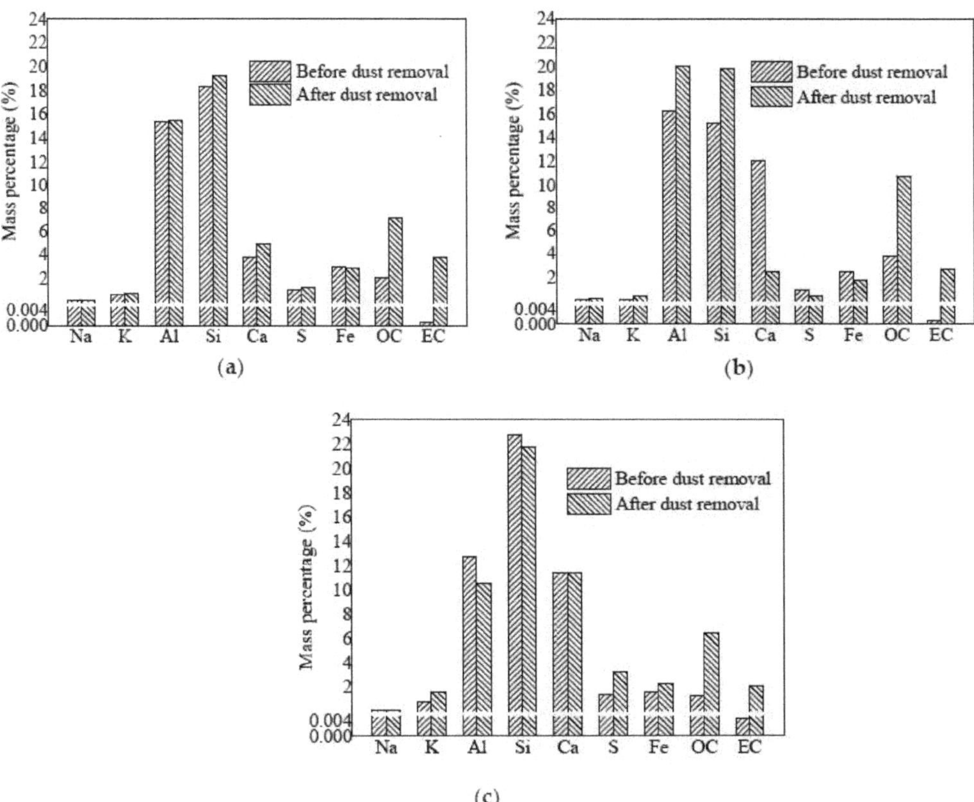

Figure 7. The contrast before and after dust removal of representative elements content of fine particle of CFB boiler. (**a**) Boiler 1; (**b**) Boiler 2; (**c**) Boiler 3.

After dedusting equipment, the content of OC and EC in fine particles increased. With flue gas temperature decreasing in the dust remover, some gaseous organics were transformed into particulate organics, which increased the OC content in fine particles. Soot particles, the main component of EC, had a great quantity of oxygen-containing functional groups on the surface (which affected the charging of soot particles [30,31]) and had a special chain or cluster structure (which affected the charging and aerodynamic characteristics of soot [32,33]). Obviously, the capture efficiency of soot aggregates by the two dust removal methods was lower than for other particles, which was an important reason for the increase of EC content in fine particles after dust removal.

4. Conclusions

In this work, fine particle characteristics of three 300 MW power plant CFB boilers were determined before and after a dust remover. Generation and emission characteristics were measured in situ with ELPI. Physical and chemical properties of fine particles were performed by SEM, EDX, XRF and TOR. This comprehensive analysis of particle concentration distribution, physical morphology and elemental composition was helpful to understand the formation mechanism of fine particles in the fluidized bed combustion process. It can also be compared with other or follow-up studies on the formation mechanism of fine particles in pulverized coal boilers and layer fired boilers. Conclusions are as follows:

(1) The broken-coalescence mechanism was the fine particle main generation mechanism in the CFB furnace. Ultrafine abrasion was the main generation mechanism of ultrafine particles. The number size distribution of fine particles of the CFB boiler was unimodal, the peak of the general CFB boiler was at 0.76 μm, but for the CFB boiler with external bed, the peak was at 0.12 μm.

(2) The size-classified removal efficiency of fine particles through three kinds of dust remover all had a minimum point in 0.1~1.0 μm, but the locations of minimum points are different. Due to the high resistance of fly ash caused by limestone, ESP may not be well suited for CFB boiler dust removal.

(3) Fine particles generated by CFB were mostly irregular in shape and had a large surface area. Because the temperature in the CFB boiler furnace was low, residual ash particles were mostly keeping original morphology in coal, and ultrafine particles formed via nucleation of vaporized minerals were very low. Therefore, irregular particles without ultrafine particles in the surface was the major component of fine particles generated by CFB.

(4) In fine particles from the CFB boiler, the content of difficult melting oxides was the highest, the content of SO_3 followed, and the content of alkali metal oxides was the lowest. After dust removal, the content of Na and K increased. The content of Al and Si increased after EFP and FF, but decreased after ESP. After EFP and ESP, the content of Ca and S increased. However, after FF, the content of these two elements decreased significantly. After three kinds of dust remover, the content of OC and EC in fine particles increased.

Author Contributions: Formal analysis, Y.H.; Investigation, D.F.; Resources, Q.D.; Supervision, J.G.; Validation, Q.S.; Writing—original draft, Y.Z.; Writing—review & editing, H.D. All authors have read and agreed to the published version of the manuscript.

Funding: This work was supported by Heilongjiang Province Postdoctoral Science Foundation "Formation mechanism of high-performance hollow carbon spheres prepared by pyrolysis of sodium salt and coal tar" (Funder: Heilongjiang Provincial Department of Human Resources and Social Security, Funding number: LBH-Z21134).

Conflicts of Interest: The authors declare no conflict of interest.

References

1. Wei, F.; Teng, E.; Wu, G.; Hu, W.; Wilson, W.; Chapman, R.; Pau, J.C.; Zhang, J. Ambient concentrations and elemental compositions of PM10 and PM2.5 in four Chinese cities. *Environ. Sci. Technol.* **1999**, *33*, 4188–4193. [CrossRef]
2. Morawska, L.; Zhang, J.J. Combustion sources of particles. 1. Health relevance and source signatures. *Chemosphere* **2002**, *49*, 1045–1058. [CrossRef]
3. Gibbs, A.R.; Pooley, F.D. Analysis and interpretation of inorganic mineral particles in "lung" tissues. *Thorax* **1996**, *51*, 327–334. [CrossRef]
4. Xu, M.; Yu, D.; Yao, H.; Liu, X.; Qiao, Y. Coal combustion-generated aerosols: Formation and properties. *Proc. Combust. Inst.* **2011**, *33*, 1681–1697. [CrossRef]
5. Senior, C.L.; Helble, J.J.; Sarofim, A.F. Emissions of mercury, trace elements, and fine particles from stationary combustion sources. *Fuel Process. Technol.* **2000**, *65*, 263–288. [CrossRef]
6. Flagan, R.; Friedlander, S. Particle formation in pulverized coal combustion—A review. *Recent Dev. Aerosol Sci.* **1978**, *2*, 25–59.
7. Markowski, G.; Ensor, D.; Hooper, R.; Carr, R. A submicron aerosol mode in flue gas from a pulverized coal utility boiler. *Environ. Sci. Technol.* **1980**, *14*, 1400–1402. [CrossRef]

8. Linak, W.P.; Miller, C.A.; Wendt, J.O. Comparison of particle size distributions and elemental partitioning from the combustion of pulverized coal and residual fuel oil. *J. Air Waste Manag. Assoc.* **2000**, *50*, 1532–1544. [CrossRef]
9. Yao, Q.; Li, S.-Q.; Xu, H.-W.; Zhuo, J.-K.; Song, Q. Reprint of: Studies on formation and control of combustion particulate matter in China: A review. *Energy* **2010**, *35*, 4480–4493. [CrossRef]
10. Liu, Z.; Li, J.; Wang, Q.; Lu, X.; Zhang, Y.; Zhu, M.; Zhang, Z.; Zhang, D. An experimental investigation into mineral transformation, particle agglomeration and ash deposition during combustion of Zhundong lignite in a laboratory-scale circulating fluidized bed. *Fuel* **2019**, *243*, 458–468. [CrossRef]
11. Liu, Y.; Cheng, L.; Ji, J.; Zhang, W. Ash deposition behavior in co-combusting high-alkali coal and bituminous coal in a circulating fluidized bed. *Appl. Therm. Eng.* **2019**, *149*, 520–527. [CrossRef]
12. Hernberg, R.; Stenberg, J.; Zethræus, B. Simultaneous in situ measurement of temperature and size of burning char particles in a fluidized bed furnace by means of fiberoptic pyrometry. *Combust. Flame* **1993**, *95*, 191–205. [CrossRef]
13. Valmari, T.; Kauppinen, E.I.; Kurkela, J.; Jokiniemi, J.K.; Sfiris, G.; Revitzer, H. Fly ash formation and deposition during fluidized bed combustion of willow. *J. Aerosol Sci.* **1998**, *29*, 445–459. [CrossRef]
14. Lind, T.; Hokkinen, J.; Jokiniemi, J.K. Fine particle and trace element emissions from waste combustion—Comparison of fluidized bed and grate firing. *Fuel Process. Technol.* **2007**, *88*, 737–746. [CrossRef]
15. Ruan, R.; Xu, X.; Tan, H.; Zhang, S.; Lu, X.; Zhang, P.; Han, R.; Xiong, X. Emission Characteristics of Particulate Matter from Two Ultralow-Emission Coal-Fired Industrial Boilers in Xi'an, China. *Energy Fuels* **2019**, *33*, 1944–1954. [CrossRef]
16. Liu, X.; Xu, Y.; Fan, B.; Lv, C.; Xu, M.; Pan, S.; Zhang, K.; Li, L.; Gao, X. Field measurements on the emission and removal of PM2.5 from coal-fired power stations: 2. Studies on two 135 MW circulating fluidized bed boilers respectively equipped with an electrostatic precipitator and a hybrid electrostatic filter precipitator. *Energy Fuels* **2016**, *30*, 5922–5929. [CrossRef]
17. Du, Q.; Dong, H.; Lv, D.; Su, L.; Gao, J.; Zhao, Z.; Wang, M. Field measurements on the generation and emission characteristics of PM2.5 generated by utility pulverized coal boiler. *J. Energy Inst.* **2018**, *91*, 1009–1020. [CrossRef]
18. Lind, T.; Kauppinen, E.I.; Maenhaut, W.; Shah, A.; Huggins, F. Ash vaporization in circulating fluidized bed coal combustion. *Aerosol Sci. Technol.* **1996**, *24*, 135–150. [CrossRef]
19. Brems, A.; Chan, C.W.; Seville, J.; Parker, D.; Baeyens, J. The transport disengagement height of fine and coarse particles. In Proceedings of the World Congress on Particle Technology (WCPT), Nuremberg, Germany, 26–29 April 2010.
20. Qu, C.-R.; Xu, B.; Wu, J.; Liu, J.-X.; Wang, X.-T. Effect of limestone addition on PM2.5 formation during fluidized bed coal combustion under O_2/CO_2 atmosphere. *J. Fuel Chem. Technol.* **2013**, *41*, 1020–1024.
21. Lu, J.Y.; Li, D.K. Experimental Study on PM10, PM2.5, PM1 Emission Features Influenced by Different Conditions in Pulverized Coal Combustion. *Proc. Chin. Soc. Electr. Eng.* **2006**, *26*, 103–107.
22. Wang, S.P.; Chen, P. On comparison of the static precipitator and the bag house. *Technol. Innov. Appl.* **2013**, *23*, 54–55.
23. Strand, M.; Pagels, J.; Szpila, A.; Gudmundsson, A.; Swietlicki, E.; Bohgard, M.; Sanati, M. Fly ash penetration through electrostatic precipitator and flue gas condenser in a 6 MW biomass fired boiler. *Energy Fuels* **2002**, *16*, 1499–1506. [CrossRef]
24. Lind, T. *Ash Formation in Circulating Fluidised Bed Combustion of Coal and Solid Biomass*; Technical Research Centre of Finland: Espoo, Finland, 1999.
25. Sotiropoulos, D.; Georgakopoulos, A.; Kolovos, N. Impact of free calcium oxide content of fly ash on dust and sulfur dioxide emissions in a lignite-fired power plant. *J. Air Waste Manag. Assoc.* **2005**, *55*, 1042–1049. [CrossRef]
26. Zhang, W.; Hu, M.; Liu, T.; Liu, Z.; Hu, Z. Effect of desulfurization in CFB on the electrostatic precipitator. *Ind. Saf. Dust Control* **2001**, *4*, 12–15.
27. Buhre, B.; Hinkley, J.; Gupta, R.; Nelson, P.; Wall, T. Fine ash formation during combustion of pulverised coal–coal property impacts. *Fuel* **2006**, *85*, 185–193. [CrossRef]
28. Yang, S.; Song, G.; Na, Y.; Song, W.; Qi, X.; Yang, Z. Transformation characteristics of Na and K in high alkali residual carbon during circulating fluidized bed combustion. *J. Energy Inst.* **2019**, *92*, 62–73. [CrossRef]
29. Yang, S.; Song, G.; Na, Y.; Yang, Z. Alkali metal transformation and ash deposition performance of high alkali content Zhundong coal and its gasification fly ash under circulating fluidized bed combustion. *Appl. Therm. Eng.* **2018**, *141*, 29–41. [CrossRef]
30. Shin, W.G.; Qi, C.; Wang, J.; Fissan, H.; Pui, D.Y. The effect of dielectric constant of materials on unipolar diffusion charging of nanoparticles. *J. Aerosol Sci.* **2009**, *40*, 463–468. [CrossRef]
31. Keller, A.; Fierz, M.; Siegmann, K.; Siegmann, H.; Filippov, A. Surface science with nanosized particles in a carrier gas. *J. Vac. Sci. Technol. A Vac. Surf. Films* **2001**, *19*, 1–8. [CrossRef]
32. Oh, H.; Park, H.; Kim, S. Effects of particle shape on the unipolar diffusion charging of nonspherical particles. *Aerosol Sci. Technol.* **2004**, *38*, 1045–1053. [CrossRef]
33. Rogak, S.N.; Flagan, R.C.; Nguyen, H.V. The mobility and structure of aerosol agglomerates. *Aerosol Sci. Technol.* **1993**, *18*, 25–47. [CrossRef]

Article

Effects of Solubilizer and Magnetic Field during Crystallization Induction of Ammonium Bicarbonate in New Ammonia-Based Carbon Capture Process

Linhan Dong [1], Dongdong Feng [1,*], Yu Zhang [1], Heming Dong [1], Zhiqi Zhao [1], Jianmin Gao [1], Feng Zhang [2], Yijun Zhao [1], Shaozeng Sun [1] and Yudong Huang [3]

1 School of Energy Science and Engineering, Harbin Institute of Technology, Harbin 150001, China
2 School of Mechatronics Engineering, Harbin Institute of Technology, Harbin 150001, China
3 School of Chemical Engineering and Technology, Harbin Institute of Technology, Harbin 150001, China
* Correspondence: 08031175@163.com

Abstract: As a chemical absorption method, the new ammonia carbon capture technology can capture CO_2. Adding ethanol to ammonia can reduce the escape of ammonia to a certain extent and increase the absorption rate of CO_2. The dissolution and crystallization of ethanol can realize the crystallization of ammonium bicarbonate and generate solid products. The induction of the crystallization process is influenced by many parameters, such as solution temperature, supersaturation, and solvating precipitant content. The basic nucleation theory is related to the critical size of nucleation. Accurate measurement of the induction period and investigating relevant factors can help to assess the nucleation kinetics. The effects of solubilizer content, temperature, and magnetic field on the induction period of the crystallization process of ammonium bicarbonate in the ethanol–H_2O binary solvent mixture and determining the growth mechanism of the crystal surface by solid–liquid surface tension and surface entropy factor are investigated. The results indicate that under the same conditions of mixed solution temperature, the crystallization induction period becomes significantly longer, the solid–liquid surface tension increases, and the nucleation barrier becomes more significant and less likely to form nuclei as the content of solvating precipitants in the components increases. At the same solubilizer content, there is an inverse relationship between the solution temperature and the induction period, and the solid–liquid surface tension decreases. The magnetic field can significantly reduce the induction period of the solvate crystallization process. This gap tends to decrease with an increase in supersaturation; the shortening reduces from 96.9% to 84.0%. This decreasing trend becomes more and more evident with the rise of solvent content in the solution. The variation of surface entropy factor under the present experimental conditions ranges from 0.752 to 1.499. The growth mode of ammonium bicarbonate in the ethanol–H_2O binary solvent mixture can be judged by the surface entropy factor as continuous growth.

Keywords: crystallization induction period; ammonium bicarbonate; binary blend solvent; crystal surface growth mechanism

Citation: Dong, L.; Feng, D.; Zhang, Y.; Dong, H.; Zhao, Z.; Gao, J.; Zhang, F.; Zhao, Y.; Sun, S.; Huang, Y. Effects of Solubilizer and Magnetic Field during Crystallization Induction of Ammonium Bicarbonate in New Ammonia-Based Carbon Capture Process. *Energies* **2022**, *15*, 6231. https://doi.org/10.3390/en15176231

Academic Editor: Fernando Rubiera González

Received: 5 August 2022
Accepted: 24 August 2022
Published: 26 August 2022

1. Introduction

As the primary source of greenhouse gas in recent years, CO_2 affects the living environment of humans. Carbon capture and storage (CCS) technology is considered the most economical and feasible way to reduce greenhouse gas emissions and slow down global warming on a large scale in a short period of the future. The new ammonia-based carbon capture technology using ammonia water as an absorbent is essentially a liquid-phase CO_2 capture technology of chemical absorption. It is one of the most potent ways to effectively achieve large-scale CO_2 emission reduction [1–4].

Since the new absorbent ammonia was proposed, many researchers have explored the reaction mechanism, absorption, and regeneration law of ammonia decarbonization with various reaction devices and processes [5–9]. The CCS technology route for industry contains pre-combustion decarbonization, oxygen-enriched combustion, chemical chain combustion, and post-combustion decarbonization [10]. For conventional coal-fired power plant units, the flue gas volume is about 4×10^5 to 5×10^5 Nm^3/h per 100 MW unit, with a CO_2 concentration of about 8% to 16% in the main composition and the rest containing 4% to 10% O_2, 70% to 75% N_2, 4% to 6% H_2O, and some micro/trace components. The post-combustion CO_2 capture technology is the most mature, complete, applicable to most industrial plants, and easy to retrofit. Chemical absorption, the most feasible post-combustion carbon capture technology, shows great promise for industrial applications. Currently, the primary industrial method is alcohol-amine solution ethanolamine (MEA) absorption. Still, in the practical application, there are many insurmountable problems such as high energy consumption for regeneration, severe equipment corrosion, and easy solvent volatilization and decomposition [11]. The ideal CO_2 chemical absorbent should have a high CO_2 absorption rate, low reaction heat, low corrosiveness, low viscosity, low degradation rate, low raw material price, and environmental friendliness. Therefore, ammonia decarbonization technology was probably one of the most likely technologies to be applied to industrial large-scale post-combustion carbon capture. Compared with MEA, the ammonia solution as a CO_2 absorber has many advantages [12,13]: CO_2 solid absorption capacity, low heat of absorption reaction, low degradation by O_2 in flue gas, low corrosiveness, and low raw material price, which helps to form an integrated system of energy gradient utilization and integrated removal of multiple pollutants. At the same time, its by-products also have some agricultural utilization value. In a typical post-combustion ammonia decarbonization system flow, the typical existing processes of this technology are as follows: the Alstom frozen ammonia process mainly uses ammonium carbonate and ammonium bicarbonate mixed slurry as recycled absorbent and can achieve an absorption capacity of up to 1.2 kg CO_2/kg NH_3 and its removal rate is as high as 90% with regeneration energy consumption of only 1.0 GJ/t CO_2 [2]. The Powerspan ECO_2 process does not require cooling like the CAP method, and the ECO_2 pilot plant has a power loss of only 16% and energy consumption of 1.1 GJ/t CO_2, which is only 27% of the energy consumption of the MEA method and can achieve a removal efficiency of more than 90% [14]. The CSIRO ammonia process in Australia has an absorption temperature of 15–30 °C; ammonia concentration below 6 wt.%; carbon burden of the lean liquor between 0.2 and 0.4; CO_2 removal efficiency of more than 85% [15]. The new ammonia carbon capture technology is promising, and the addition of ethanol can reduce the ammonia escape to a certain extent and increase the rate of CO_2 absorption, among other advantages. The application of this technology for industry is reflected in the mixture of CO_2 and ammonia to produce ammonium bicarbonate, which can provide a new way to produce ammonium fertilizer by solubilization and crystallization. It can also be coupled with sucrose production to synthesize nitrogen fertilizer. The new ammonia carbon capture technology can provide a new way to synthesize nitrogen fertilizer and other similar fertilizers, and can also be used for industrial applications in industries such as agricultural products and heating and power generation.

The induction period and metastable zone width are two essential parameters in the crystal nucleation process and play an important role in the crystallization process and the design of the crystallizer [16–18]. The core process of ammonium bicarbonate dissolution and crystallization in the new ammonia-based carbon capture system is the crystallization of ammonium bicarbonate. Nucleation has a significant impact on crystallization products. Improper operation easily causes problems such as small particle size, wide particle size distribution, and poor fluidity [19–21], which affect the application of the overall carbon capture process. Therefore, it is necessary to systematically study the induction period of the ammonium bicarbonate crystallization process.

The induction period can be generally divided into several parts. The relaxation time t_r requires for the system to reach the quasi-steady state of the molecular cluster. The time t_n is required to form a stable nucleus and the growth time t_g is required for the nucleus to grow to the size detected by the detection device [22–24]. The length of the induction period mainly depends on the supersaturation of the solution. The larger the supersaturation of the solution is, the shorter the induction time. When the magnitude of supersaturation tends to the middle line of the first metastable zone and the second metastable zone, the induction period is more significant. If the supersaturation is in the first metastable zone and no crystalline seed is added, no crystals are produced [23,25].

In the crystallization process, the degree of influence of the solution itself and external forces on the degree of supersaturation is usually explored. Then the degree of supersaturation is controlled in the appropriate metastable zone to shorten the induction period while obtaining crystalline products with larger and more uniform particle sizes [26–28]. Maheswata Lenka et al. [19] systematically studied the induction period and the width of the metastable zone of cooling crystallization of l-aspartic acid hydrate, the variation of the induction period at different temperatures and in the range of supersaturation, and the resulting calculation of parameters such as surface tension and critical nucleation. You et al. [29] measured the metastable zone and induction period of the crystallization process of aluminum ammonium sulfate in water by the aggregated laser reflectivity method. The effect of temperature and supersaturation on the induction period was systematically investigated. The mechanism of primary homogeneous phase nucleation with liquid surface energy was determined using Sangwal's classical three-dimensional nucleation method. Wang et al. [30] systematically measured the induction period and the width of the metastable zone of sodium vanadate crystallization in NaOH at different temperatures using laser scattering and analyzed the effects of NaOH concentration, supersaturation, stirring rate, cooling rate, and other influencing factors on the induction period, and determined the primary nucleation kinetics from the results of the induction period and metastable zone. Feng et al. [1] investigated the process of CO_2 mass transfer in an amine absorption reactor under the action of a static magnetic field. They elucidated that ammonia as an absorbent can lead to effective mass transfer in CO_2 absorption that the addition of an external influencing factor, the static magnetic field, can enhance the facilitation of the mass transfer process, and that the facilitation of the static magnetic field is more pronounced for low concentration CO_2 absorption mass transfer.

In summary, it can be seen that the solution temperature, ultrasound, solubilizer content, magnetic field, and other factors influence the supersaturation of the induction period of the crystallization process. The induction period can calculate the nucleation rate, and accurate measurement can provide essential parameters for assessing the crystallization kinetics. There is no report on the induction period of the crystallization process of ammonium bicarbonate in the ethanol–H_2O binary solvent mixture in the new ammonia carbon capture system. Therefore, this paper mainly studies the influence factors of the induction period of ammonium bicarbonate crystallization in the ethanol–H_2O binary solvent mixture, investigates the influence of the ethanol content of the solvating agent and the magnetic field on the induction period, and determines the growth mechanism of crystal surface by calculating the solid–liquid surface tension and surface entropy factor. The relevant characteristics of nucleation and the growth of the ammonium bicarbonate crystallization process in ethanol–H_2O binary mixed solvent, which has important guiding significance for the selection of temperature and solvent ratio in the core process of ammonium bicarbonate crystallization in the new ammonia carbon capture system, are investigated.

2. Materials and Methods

2.1. Experimental Chemicals

The drugs used in the experimental procedure to determine the crystallization induction period are shown in Table 1.

Table 1. List of experimental chemicals.

Name	Chemical Formula	Purity/Concentration
Ammonia	$NH_3 \cdot H_2O$	25~28 wt.%
Ammonium bicarbonate	NH_4HCO_3	Analysis of pure
Ammonium carbamate	NH_2COONH_4	Analysis of pure
Anhydrous ethanol	CH_3CH_2OH	Analysis of pure
Sulfuric acid	H_2SO_4	0.05 mol/L

2.2. Experimental System

The experimental setup used during the experiment is shown in Figure 1. The photoelectric transmitter emits laser light, the laser receiver receives laser light, and the recorder converts the received optical signal into an electrical signal and outputs the relevant information. The glass jacketed crystallizer serves as the core reactor for the reaction with an inner diameter of 10 cm and a height of 12.0 cm. A super thermostatic water bath with a temperature control accuracy of ± 0.1 °C is used to control the temperature of the induction phase process in the dissolver. A magnetic stirrer powers the magnetic stirring rotor and provides a specific stirring rate for the solution in the dissolver. A precision thermometer is used to record the temperature changes of the reaction process in the dissolver.

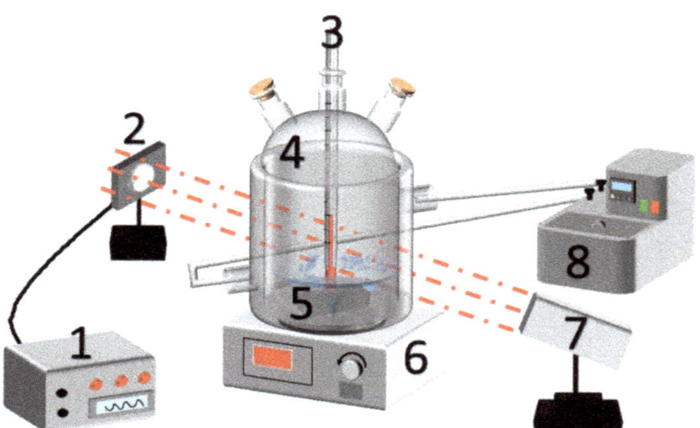

Figure 1. Schematic diagram of the induction period determination device (1—recorder; 2—laser receiver; 3—precision thermometer; 4—dissolver; 5—magnetic stirring rotor; 6—magnetic stirrer; 7—photoelectric transmitter; 8—super thermostat water bath).

2.3. Experimental Steps

The steps for the determination of crystallization induction of ammonium bicarbonate in ethanol–H_2O binary solvent are as follows: (a) Accurately weigh a certain mass of solute, configure a certain solvent ratio of mixed solvent, dissolve the weighed solute in the mixed solvent, and prepare the solution at a specific temperature. (b) Add the prepared solution into the crystallizer, turn on the super thermostat water bath, turn on the photoelectric emitter and the receiver, set a specific stirring rate, and preheat for 30 min. (c) Pour a certain amount of reaction solvent with the same set temperature into the crystallizer quickly, and start the recorder simultaneously. (d) Stop the recorder when there is a sudden change in the data displayed by the receiver. At this time, the data displayed by the recorder is the experimental data of the crystallization induction period. (e) Adjust the temperature and composition of the reaction solvent, adjust the operating conditions, and measure the induction period under each working condition. (f) Cycle the above experimental steps to obtain the induction periods under different supersaturation, temperature, and solvent composition.

2.4. Analytical Methods

The induction period can be expressed by Equation (1):

$$t_{ind} = t_r + t_n + t_g \tag{1}$$

According to the crystal nucleation theory, it is known that the solid–liquid surface tension mainly expresses the physical properties of crystals, and the crystal nucleation mechanism is primarily determined by the solid–liquid surface tension [31–33]. Meanwhile, the solid–liquid surface tension is an essential thermodynamic parameter for the optimization process of the solvation and precipitation crystallization process, as well as for describing the nucleation of primary crystals and the crystal growth process [34,35]. Homogeneous nucleation can be generated only by overcoming the nuclear energy barrier, and the free energy required in the crystal nucleation process can be calculated using the following equation [36–38]:

$$\Delta G = \frac{\left(K_A i^{\frac{2}{3}} v^{\frac{2}{3}} \gamma\right)}{K_V + iVkT \ln S} \tag{2}$$

When the change in free energy satisfies $\frac{d(\Delta G)}{di} = 0$, it is determined that a critical nucleus is formed [39]. The minimum number of molecules required to form a critical nucleus can be obtained and expressed as i_{cr}, and the minimum size to form a critical nucleus can be defined as r_{cr}.

$$r_{cr} = \frac{(2K_A v\gamma)}{(3K_V VkT \ln S)} \tag{3}$$

$$i_{cr} = \left[\frac{2K_A v^{\frac{2}{3}} \gamma}{3K_V^{\frac{2}{3}} VkT \ln S}\right]^3 \tag{4}$$

$$\Delta G = \frac{4K_a^3 v \gamma^3}{27K_V^2 V^2 k^2 T^2 \ln^2 S} \tag{5}$$

For spherical particles, the above equation can be simplified to [40]:

$$\Delta G_{cr} = \frac{16\pi v^2 \gamma^3}{3k^2 T^2 \ln^2 S} \tag{6}$$

Using the induction periods obtained for different supersaturation solution conditions, we can get the surface tension, which can be obtained by Equation (7) [41]:

$$\ln \tau = \ln B + \frac{\Delta G_{cr}}{kT} \tag{7}$$

In the equation, T is the solution temperature, ΔG_{cr} is the critical nucleation free energy change, B is a constant, and k is the Boltzmann constant. Bringing (6) into (7), the following equation is obtained:

$$\ln \tau = \ln B + \frac{16\pi \gamma^3 v^2}{3k^3 T^3 \ln^2 S} \tag{8}$$

$\ln\tau$ is linearly related to $1/\ln^2 S$ [42], and the solid–liquid surface tension can be obtained by calculating the slope of the straight line [25,43].

The surface entropy factor is defined as:

$$f = \frac{\varepsilon \Delta H_m}{RT} \tag{9}$$

In the equation, ΔH_m denotes the heat of melting, and ε denotes the surface anisotropy factor.

The smoothness of the crystal surface follows with the increase in the surface entropy factor. As the smoothness of the crystal surface increases with the rise of the surface entropy factor, the growth rate of the crystal decreases and the crystal growth at this time belongs to helical dislocation-type growth [44–46]. Conversely, when the surface entropy factor is low, the growth rate of the crystal surface is faster when the crystal exhibits reduced surface smoothness and when the growth satisfies the continuum-type growth mode [47,48]. However, if we follow Equation (9), calculating the surface entropy factor is difficult to achieve, so we need to simplify this equation. Barata et al. [49] proposed to predict the surface entropy factor in terms of molecular volume v, temperature, and solid–liquid surface tension γ, three variables.

$$f = \frac{4v^{\frac{2}{3}}\gamma}{kT} \tag{10}$$

3. Results and Discussion

3.1. Effect of Lysis Precipitant Content on Induction Period

The relationship between the induction period and the supersaturation of the solution can be used in the experimental process to select a suitable supersaturation solution for the solvation crystallization test and the solid–liquid surface tension under different working conditions calculated to help understand the crystallization process mechanistically. Figure 2a,c,e are the relationships between the induction periods and the corresponding supersaturation degrees at 15 °C, 20 °C, and 25 °C, for different solubilizer contents in the mixed solvents. It can be found that the nucleation induction period is significantly shortened with the increase in the supersaturation of the solution under the condition of certain solubilizer content, and the supersaturation is mainly concentrated between 1.1 and 2.4 during this experiment. The shortening of the induction period is about 93.3% at 0.2359 mol/mol, 87.2% at 0.1707 mol/mol, and 60.0% at 0.1169 mol/mol. The shortening of the induction period is evident with the increase in supersaturation at high solubilizer contents. Meanwhile, under the same temperature and the same supersaturation, the crystallization induction period is prolonged significantly with the increase in ethanol content of the solubilizer in the mixed solvent fraction, and the induction period time is as high as 220 s at the solution temperature of 15 °C and supersaturation of 1.25. Because the solubility of the mixed solvent decreases when the solubilizer content increases, the solute in the mixed solvent decreases under the same supersaturation condition, which reduces the chance of practical collision and is not conducive to the generation of spontaneous nucleation, resulting in a significantly more extended induction period.

Figure 2b,d,f are the graphs of supersaturation versus induction period at different solubilizer contents in the mixed solvents at temperatures of 15 °C, 20 °C, and 25 °C. If the crystallization is a spontaneous homogeneous nucleation process, $\ln t_{ind}$~$1/\ln^2 S$ should be a straight line according to Equation (8). From Figure 2b,d,f, it can be found that the logarithm of the induction period in the binary ethanol–H_2O solvent mixture at different temperatures in the experimental supersaturation range is linearly related to the inverse of the square of the logarithm of the supersaturation. The linearity of the two data sets indicates that the experimental process is homogeneous nucleation under these conditions. Since no crystalline species were added and the experiments were performed under dust-free conditions, no heterogeneous nucleation occurred. The homogeneous nucleation solid–liquid interfacial tension of the ethanol–H_2O binary solvent system can be calculated from the slope of the straight line. When the solution temperature is 15 °C, the straight line slope is 0.0896 when the solubilizer content is 0.1169 mol/mol, 0.1128 when the solubilizer content is 0.1707 mol/mol, and 0.3719 when the solubilizer content is 0.2359 mol/mol.

It can also be found that at the same temperature, with the increase in the molar fraction of ethanol in the ethanol–H_2O binary solvent mixture, the slope of the $\ln t_{ind}$~$1/\ln^2 S$ straight line increases and the induction period time is prolonged, indicating that the nucleation barrier becomes more significant and less likely to form nuclei.

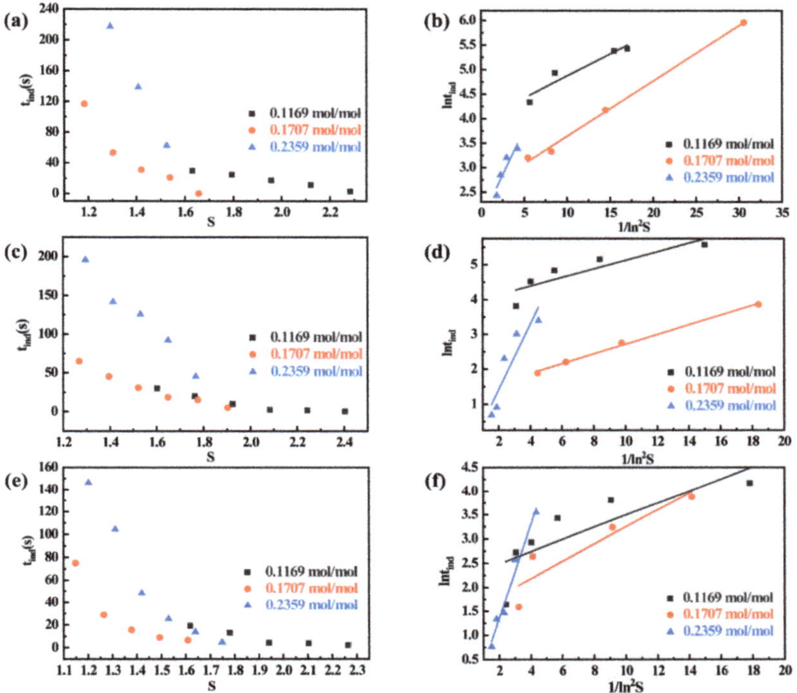

Figure 2. (**a**) Relationship between supersaturation and induction period at different solubilizer contents (t = 15 °C). (**b**) Relationship between $\ln t_{ind}$ and $1/\ln^2 S$ at different solubilizer contents (t = 15 °C). The lines are the linear fit curve of $\ln t_{ind}$ and $1/\ln^2 S$. (**c**) Relationship between supersaturation and induction period at different solubilizer contents (t = 20 °C). (**d**) Relationship between $\ln t_{ind}$ and $1/\ln^2 S$ at different solubilizer contents (t = 20 °C). The lines are the linear fit curve of $\ln t_{ind}$ and $1/\ln^2 S$. (**e**) Relationship between supersaturation and induction period at different solubilizer contents (t = 25 °C). (**f**) Relationship between $\ln t_{ind}$ and $1/\ln^2 S$ at different solubilizer contents (t = 25 °C). The lines are the linear fit curve of $\ln t_{ind}$ and $1/\ln^2 S$.

3.2. Effect of Temperature on Induction Period

Figure 3a,c,e are the relationships between the induction period and the corresponding supersaturation at different temperatures of the solutions when the molar fractions of the solubilizer ethanol are 0.1169, 0.1707, and 0.2359 mol/mol. The induction period of the solution temperature of 15 °C is more extended than that of the solution temperature of 20 °C and 25 °C for a solution content of 0.1169 mol/mol under the premise of equal supersaturation. The above trend is also satisfied for other solubilizer content conditions. The inverse relationship between the solution temperature and the induction period is observed at a certain solubilizer content. The higher the temperature, the shorter the induction period, which indicates that the high temperature could promote homogeneous nucleation to a certain extent. The influence of temperature on the spontaneous nucleation of crystals is mainly reflected in two aspects: on the one hand, the increase in temperature can increase the diffusion coefficient of solute and strengthen the diffusion process of the solute; on the other hand, the temperature rise can make the energy of solute molecules increase, so that the probability of practical collision of solute molecules increases, and thus spontaneous nucleation is more likely to occur.

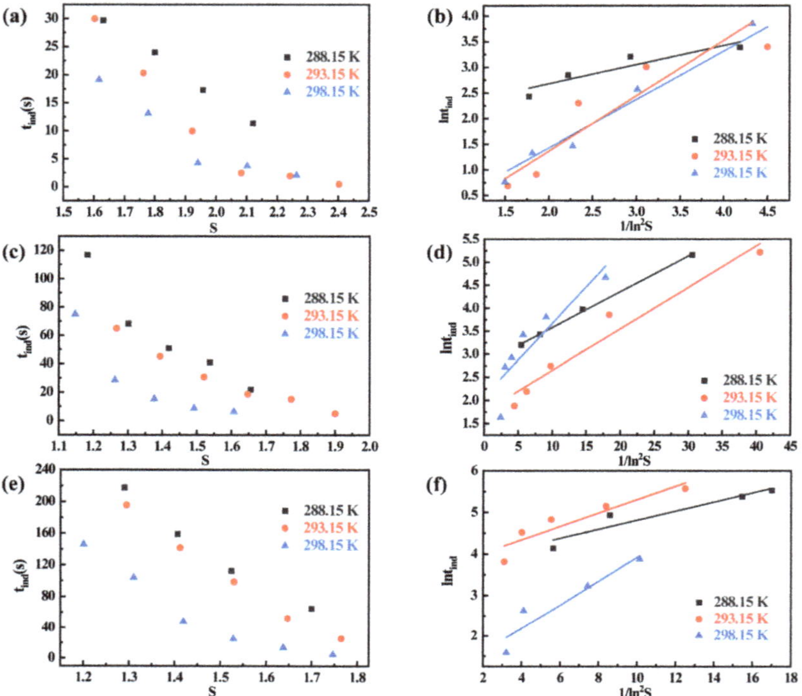

Figure 3. (**a**) Relationship between supersaturation and induction period at different temperatures (x_c = 0.1169). (**b**) Relationship between $\ln t_{ind}$ and $1/\ln^2 S$ at different temperatures (x_c = 0.1169). The lines are the linear fit curve of $\ln t_{ind}$ and $1/\ln^2 S$. (**c**) Relationship between supersaturation and induction period at different temperatures (x_c = 0.1707). (**d**) Relationship between $\ln t_{ind}$ and $1/\ln^2 S$ at different temperatures (x_c = 0.1707). The lines are the linear fit curve of $\ln t_{ind}$ and $1/\ln^2 S$. (**e**) Relationship between supersaturation and induction period at different temperatures (x_c = 0.2359). (**f**) Relationship between $\ln t_{ind}$ and $1/\ln^2 S$ at different temperatures (x_c = 0.2359). The lines are the linear fit curve of $\ln t_{ind}$ and $1/\ln^2 S$.

Figure 3b,d,f are the graphs of supersaturation versus induction period at different temperatures for the molar fractions of ethanol, and 0.1169, 0.1707, and 0.2359 mol/mol of a solubilizer. The slope of the straight line is 0.1092 when the solution temperature is 15 °C, 0.1608 when the solution temperature is 20 °C, and 0.2878 when the solution temperature is 25 °C under the condition of 0.2359 mol/mol solubilizer content. It is found that the slope of the straight line $\ln t_{ind} \sim 1/\ln^2 S$ decreases with the increase in temperature for the same solubilizer content. Moreover, it can be seen that the solid–liquid surface tension between the solution and crystal surfaces tends to decrease when the temperature increases, indicating that the rise in temperature helps to lower the nucleation barrier and promote the homogeneous nucleation process. This is verified in Section 3.4.

3.3. *Effect of Magnetic Field on Induction Period*

Figure 4a–c show the relationship between the induction periods and the corresponding supersaturation degrees at temperatures of 15 °C, 20 °C, and 25 °C with and without magnetic fields. It can be found that when the temperature and solubilizer content are the same, the induction period under the magnetic field is smaller than that without the magnetic field. The shortening of the induction period tends to decrease with the supersaturation increase. The induction period's shortening becomes more evident with the rise of the content of the solubilizer in the solution. Because a magnetic field changes the hydrogen bonding in water molecules, the polarity of the water is enhanced. The viscosity of the water is reduced [50,51], so the diffusion process of solute is accelerated under the effect of the magnetic field, the range of motion of solute molecules is more extensive and more accessible, there are more opportunities for solute molecules to collide, the practical collision is increased, and the nucleation process is promoted. On the other hand, the presence of a magnetic field decreases the surface tension of the solution, and the free energy of the liquid–solid transition during crystallization decreases, which reduces the critical nucleation radius [52,53]. In summary of the above two reasons, the presence of a magnetic field can shorten the induction period time.

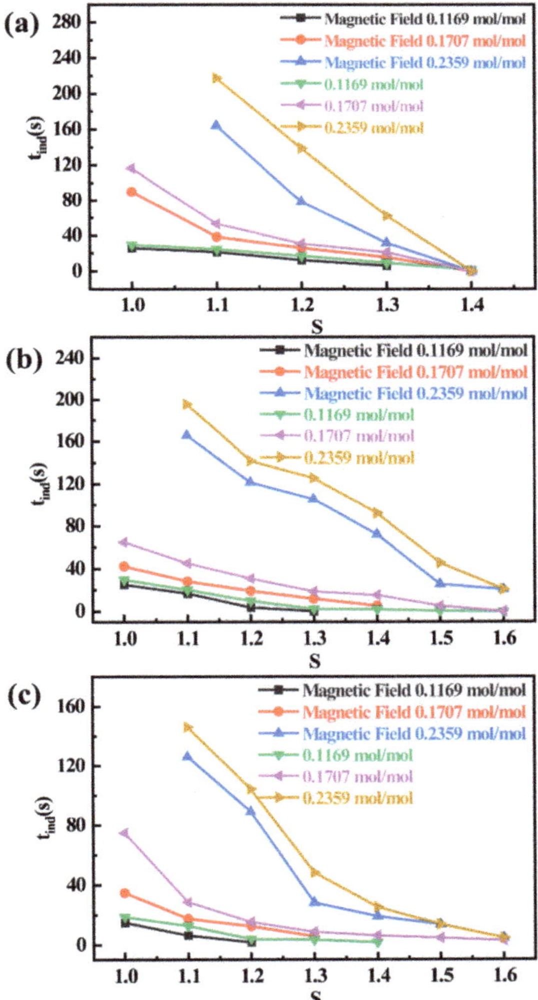

Figure 4. (**a**) Variation of induction period with supersaturation in the presence and absence of a magnetic field (t = 15 °C). (**b**) Variation of induction period with supersaturation in the presence and absence of a magnetic field (t = 20 °C). (**c**) Variation of induction period with supersaturation in the presence and absence of a magnetic field (t = 25 °C).

Comparing Figure 4a–c, it can be found that the effect of a magnetic field on shortening the induction period becomes weaker as the temperature increases. At a solubilizer content of 0.2359 mol/mol, the shortening of the induction period is about 96.9% at a solution temperature of 15 °C, 87.5% at a solution temperature of 20 °C, and 84.0% at a solution temperature of 25 °C. The above trends are also satisfied under the conditions of other solubilizer contents. This is mainly because when the temperature is lower, the average spacing between molecules is smaller, and the water molecules are bound together by hydrogen bonds [54]. When the temperature increases, the magnetization rate tends to decrease with the increased temperature, the intermolecular spacing increases, and the interaction between the magnetic moments of water molecules is weakened. At the same time, the thermal motion of molecules increases with temperature, and the orientation of molecular magnetic moments is disrupted, which eventually leads to the weakening of the magnetic field on the solution. The effect of a magnetic field shortens the induction period and decreases with temperature.

3.4. Crystal Surface Growth Mechanism

The line between $\ln t_{ind}$ and $1/\ln^2 S$ is straight, and the slope of the line can be obtained and combined with Equation (8), and the solid–liquid surface tension value can be found. The solid–liquid surface tension calculated using the relationship between supersaturation and induction period under this experimental condition is listed in Table 2. In studying crystal growth mechanisms, the surface entropy factor can theoretically be used to determine the crystal surface growth mechanism. In the experimental study of this paper, Equation (10) is used to calculate the surface entropy factor, and the calculation results are listed in Table 2.

Table 2. The calculated results of interfacial tension and surface entropy factor.

T/K	Solvent Composition/(mol·mol^{-1})	γ/(J·m^{-2})	*f*
283.15	0.1169	0.004187	0.752
283.15	0.1707	0.004683	0.8612
283.15	0.2359	0.008404	1.499
288.15	0.1169	0.00414	0.777
288.15	0.1707	0.00433	0.818
288.15	0.2359	0.008167	1.403
293.15	0.1169	0.003664	0.804
293.15	0.1707	0.003957	0.876
293.15	0.2359	0.00589	1.127

Comparing the data in the table, the solid–liquid surface tension and surface entropy factor of ammonium bicarbonate in the binary ethanol–H_2O solvent mixture show decreasing trends with increasing temperature. At the same temperature, the solid–liquid surface tension increases with the ethanol content of the solubilizer in the mixed solvent. In the range of experimental study conditions in this paper, the variation of surface entropy factor ranges from 0.752 to 1.499. The Monte Carlo method is used to simulate the crystal surface's growth. The range of entropy factor variation can be determined when the crystal growth mechanism changes [55–57], and the results are shown in Table 3. It is found that the solubilizer content and temperature do not significantly affect the growth mechanism of ammonium bicarbonate crystal surface within the scope of this experiment, and the growth mode of ammonium bicarbonate in the binary mixture of ethanol and water could be determined as continuous growth. It can also be seen from the above data that as the temperature decreases and the content of ethanol in the binary solvent mixture increases, the surface entropy factor gradually increases, the crystal surface becomes smoother, and the crystal growth energy barrier gradually increases. The nucleation and growth characteristics of ammonium bicarbonate crystallization in ethanol–H_2O binary solvents are essential for selecting the temperature and solubilizer ratio in the crystallization process.

Table 3. Relationship between surface entropy factor and crystal growth.

Surface Entropy Factor	Growing Position	Roughness of Crystal Surface	Growth Type
$f < 3$	A bumpy ride into the crystal plane	Rougher	Continuous Growth
$3 < f < 5$	Direct growth onto the crystal surface, and also diffusion to the crystal surface of the step	Smoother	Transmissive growth
$f > 5$	Dislocations through the surface into the lattice	Very smooth	Spiral dislocation growth

4. Conclusions

(1) Under the same mixed solution temperature, the crystallization induction period becomes significantly longer with the increase in ethanol content of the solubilizer in the components. The nucleation induction period becomes significantly shorter with the increase in the solution's supersaturation under certain solubilizer content, and the shortening is increased from 60.0% to 87.2% and 93.3%. The increase in supersaturation could significantly enhance the homogeneous nucleation process.

(2) Under the same solubilizer content, there is an inverse relationship between solution temperature and induction period. The increase in mixed solution temperature decreases the solid–liquid surface tension, and the induction period is significantly shortened. Temperature can promote homogeneous nucleation to a certain extent.

(3) At the same temperature of the mixed solution, the presence of a magnetic field can significantly reduce the induction period of the solvation crystallization process, and this gap tends to decrease with the increase in supersaturation. This decreasing trend becomes more and more evident with the rise of the solvating agent content in the solution. As the temperature increases, the magnetic field shortens the induction period less and less, and the shortening reduces from 96.9% to 84.0%.

(4) The solid–liquid surface tension and surface entropy factor gradually increase with the decrease in temperature and the increase in ethanol content of the solubilizer in the binary solvent mixture, and the variation of surface entropy factor under the present experimental conditions ranges from 0.752 to 1.499. The crystal surface becomes smoother and smoother, and the crystal growth energy barrier steadily increases. It can be judged that the growth mode of ammonium bicarbonate in the ethanol–H_2O binary solvent mixture is continuous growth.

(5) This paper investigates the efficiency of the mass transfer characteristics and the enhancement of the CO_2 absorption rate of the new carbon capture system from a quantitative point of view under the conditions of the new ammonia carbon capture system with the addition of the additional field, the solubilizer ethanol, and the addition of the static magnetic field. The new ammonia carbon capture technology can provide a new way to synthesize nitrogen fertilizer and other similar fertilizers and can also realize industrial applications for industries such as agricultural products and heating and power generation.

Author Contributions: Conceptualization, Y.Z. (Yu Zhang) and Y.H.; Data curation, L.D., H.D., Z.Z., Y.Z. (Yijun Zhao) and S.S.; Formal analysis, F.Z.; Funding acquisition, D.F., F.Z., Y.Z. (Yijun Zhao), S.S. and Y.H.; Investigation, Z.Z.; Methodology, J.G.; Project administration, H.D. and F.Z.; Resources, D.F. and J.G.; Software, Y.Z. (Yu Zhang); Validation, Y.Z. (Yu Zhang) and J.G.; Visualization, L.D.; Writing – original draft, L.D.; Writing – review & editing, D.F. All authors have read and agreed to the published version of the manuscript.

Funding: This work is supported by the National Natural Science Foundation of China (52006047), China Postdoctoral Science Foundation Funded Project (2020M670908), Heilongjiang Provincial Postdoctoral Science Foundation (LBH-Z19151), Foundation of State Key Laboratory of High-efficiency Utilization of Coal and Green Chemical Engineering (2021-K45), and Heilongjiang Higher Education Teaching Reform Project (SJGY20210322).

Conflicts of Interest: The authors declare no conflict of interest.

References

1. Feng, D.; Gao, J.; Zhang, Y.; Li, H.; Du, Q.; Wu, S. Mass transfer in ammonia-based CO_2 absorption in bubbling reactor under static magnetic field. *Chem. Eng. J.* **2018**, *338*, 450–456. [CrossRef]
2. Kozak, F.; Petig, A.; Morris, E.B.; Rhudy, R.; Thimsen, D. Chilled ammonia process for CO_2 capture. *Energy Procedia* **2009**, *1*, 1419–1426. [CrossRef]
3. Yang, Y.; Liew, R.K.; Tamothran, A.M.; Foong, S.Y.; Yek, P.N.Y.; Chia, P.W.; Van Tran, T.; Peng, W.; Lam, S.S. Gasification of refuse-derived fuel from municipal solid waste for energy production: A review. *Environ. Chem. Lett.* **2021**, *19*, 2127–2140. [CrossRef] [PubMed]
4. Hu, Y.; Guo, Y.; Sun, J.; Li, H.; Liu, W. Progress in MgO sorbents for cyclic CO_2 capture: A comprehensive review. *J. Mater. Chem. A* **2019**, *7*, 20103–20120. [CrossRef]
5. Bavarella, S.; Hermassi, M.; Brookes, A.; Moore, A.; Vale, P.; Moon, I.S.; Pidou, M.; McAdam, E.J. Recovery and concentration of ammonia from return liquor to promote enhanced CO_2 absorption and simultaneous ammonium bicarbonate crystallisation during biogas upgrading in a hollow fibre membrane contactor. *Sep. Purif. Technol.* **2020**, *241*, 116631. [CrossRef]
6. Al-Hamed, K.H.M.; Dincer, I. A novel multigeneration ammonia-based carbon capturing system powered by a geothermal power plant for cleaner applications. *J. Clean. Prod.* **2021**, *321*, 129017. [CrossRef]
7. Al-Hamed, K.H.M.; Dincer, I. A comparative review of potential ammonia-based carbon capture systems. *J. Environ. Manag.* **2021**, *287*, 112357. [CrossRef]
8. Kim, Y.; Lim, S.-R.; Jung, K.A.; Park, J.M. Process-based life cycle CO_2 assessment of an ammonia-based carbon capture and storage system. *J. Ind. Eng. Chem.* **2019**, *76*, 223–232. [CrossRef]
9. Ishaq, H.; Ali, U.; Sher, F.; Anus, M.; Imran, M. Process analysis of improved process modifications for ammonia-based post-combustion CO_2 capture. *J. Environ. Chem. Eng.* **2021**, *9*, 104928. [CrossRef]
10. Leung, D.Y.; Caramanna, G.; Maroto-Valer, M.M. An overview of current status of carbon dioxide capture and storage technologies. *Renew. Sustain. Energy Rev.* **2014**, *39*, 426–443. [CrossRef]
11. Rochelle, G.T. Amine scrubbing for CO_2 capture. *Science* **2009**, *325*, 1652–1654. [CrossRef] [PubMed]
12. Bai, H.; Yeh, A.C. Removal of CO_2 greenhouse gas by ammonia scrubbing. *Ind. Eng. Chem. Res.* **1997**, *36*, 2490–2493. [CrossRef]
13. Zhao, B.; Su, Y.; Tao, W.; Li, L.; Peng, Y. Post-combustion CO_2 capture by aqueous ammonia: A state-of-the-art review. *Int. J. Greenh. Gas Control* **2012**, *9*, 355–371. [CrossRef]
14. Mclarnon, C.R.; Duncan, J.L. Testing of Ammonia Based CO_2 Capture with Multi-Pollutant Control Technology. *Energy Procedia* **2009**, *1*, 1027–1034. [CrossRef]
15. Yu, H.; Qi, G.; Wang, S.; Morgan, S.; Allport, A.; Cottrell, A.; Do, T.; McGregor, J.; Wardhaugh, L.; Feron, P. Results from trialling aqueous ammonia-based post-combustion capture in a pilot plant at Munmorah Power Station: Gas purity and solid precipitation in the stripper. *Int. J. Greenh. Gas Control* **2012**, *10*, 15–25. [CrossRef]
16. Di Profio, G.; Tucci, S.; Curcio, E.; Drioli, E. Selective Glycine Polymorph Crystallization by Using Microporous Membranes. *Cryst. Growth Des.* **2007**, *7*, 526–530. [CrossRef]
17. Li, L.; Zhao, S.; Xin, Z.; Zhou, S. Nucleation kinetics of clopidogrel hydrogen sulfate polymorphs in reactive crystallization: Induction period and interfacial tension measurements. *J. Cryst. Growth* **2020**, *538*, 125610. [CrossRef]
18. Feng, D.; Guo, D.; Zhang, Y.; Sun, S.; Zhao, Y.; Chang, G.; Guo, Q.; Qin, Y. Adsorption-enrichment characterization of CO_2 and dynamic retention of free NH_3 in functionalized biochar with $H_2O/NH_3{\cdot}H_2O$ activation for promotion of new ammonia-based carbon capture. *Chem. Eng. J.* **2021**, *409*, 128193. [CrossRef]
19. Lenka, M.; Sarkar, D. Determination of metastable zone width, induction period and primary nucleation kinetics for cooling crystallization of l-asparaginenohydrate. *J. Cryst. Growth* **2014**, *408*, 85–90. [CrossRef]
20. Prasad, R.; Dalvi, S.V. Sonocrystallization: Monitoring and controlling crystallization using ultrasound. *Chem. Eng. Sci.* **2020**, *226*, 115911. [CrossRef]
21. Wang, M.; Zhu, J.; Zhang, S.; You, G.; Wu, S. Influencing factors for vulcanization induction period of accelerator / natural rubber composites: Molecular simulation and experimental study. *Polym. Test.* **2019**, *80*, 106145. [CrossRef]
22. Zhou, L.; Wang, Z.; Zhang, M.; Guo, M.; Xu, S.; Yin, Q. Determination of metastable zone and induction time of analgin for cooling crystallization. *Chin. J. Chem. Eng.* **2017**, *25*, 313–318. [CrossRef]
23. Šimon, P.; Nemčeková, K.; Jóna, E.; Plško, A.; Ondrušová, D. Thermal stability of glass evaluated by the induction period of crystallization. *Thermochim. Acta* **2005**, *428*, 11–14. [CrossRef]
24. He, D.; Ou, Z.; Qin, C.; Deng, T.; Yin, J.; Pu, G. Understanding the catalytic acceleration effect of steam on $CaCO_3$ decomposition by density function theory. *Chem. Eng. J.* **2020**, *379*, 122348. [CrossRef]
25. Söhnel, O.; Mullin, J.W. A method for the determination of precipitation induction periods. *J. Cryst. Growth* **1978**, *44*, 377–382. [CrossRef]
26. Yadav, A.; Labhasetwar, P.K.; Shahi, V.K. Membrane distillation crystallization technology for zero liquid discharge and resource recovery: Opportunities, challenges and futuristic perspectives. *Sci. Total Environ.* **2022**, *806*, 150692. [CrossRef] [PubMed]
27. Chabanon, E.; Mangin, D.; Charcosset, C. Membranes and crystallization processes: State of the art and prospects. *J. Membr. Sci.* **2016**, *509*, 57–67. [CrossRef]
28. Di Profio, G.; Curcio, E.; Drioli, E. Supersaturation Control and Heterogeneous Nucleation in Membrane Crystallizers: Facts and Perspectives. *Ind. Eng. Chem. Res.* **2010**, *49*, 11878–11889. [CrossRef]

29. You, S.; Zhang, Y.; Zhang, Y. Nucleation of ammonium aluminum sulfate dodecahydrate from unseeded aqueous solution. *J. Cryst. Growth* **2015**, *411*, 24–29. [CrossRef]
30. Wang, S.; Feng, M.; Du, H.; Weigand, J.J.; Zhang, Y.; Wang, X. Determination of metastable zone width, induction time and primary nucleation kinetics for cooling crystallization of sodium orthovanadate from NaOH solution. *J. Cryst. Growth* **2020**, *545*, 125721. [CrossRef]
31. Neumann, A.W.; Good, R.J.; Hope, C.J.; Sejpal, M. An equation-of-state approach to determine surface tensions of low-energy solids from contact angles. *J. Colloid Interface Sci.* **1974**, *49*, 291–304. [CrossRef]
32. Neumann, A.W. Haftfestigkeit an Grenzflächen. *Mater. Corros.* **1969**, *20*, 19–22. [CrossRef]
33. Yang, Y.; Liu, W.; Hu, Y.; Sun, J.; Tong, X.; Chen, Q.; Li, Q. One-step synthesis of porous Li_4SiO_4-based adsorbent pellets via graphite moulding method for cyclic CO_2 capture. *Chem. Eng. J.* **2018**, *353*, 92–99. [CrossRef]
34. Zhou, K.; Wang, H.P.; Chang, J.; Wei, B. Experimental study of surface tension, specific heat and thermal diffusivity of liquid and solid titanium. *Chem. Phys. Lett.* **2015**, *639*, 105–108. [CrossRef]
35. Correia, N.T.; Ramos, J.J.M.; Saramago, B.J.V.; Calado, J.C.G. Estimation of the Surface Tension of a Solid: Application to a Liquid Crystalline Polymer. *J. Colloid Interface Sci.* **1997**, *189*, 361–369. [CrossRef]
36. Sangwal, K. Novel Approach to Analyze Metastable Zone Width Determined by the Polythermal Method: Physical Interpretation of Various Parameters. *Cryst. Growth Des.* **2009**, *9*, 942–950. [CrossRef]
37. Gu, C.-H.; Young, V.; Grant, D.J.W. Polymorph screening: Influence of solvents on the rate of solvent-mediated polymorphic transformation. *J. Pharm. Sci.* **2001**, *90*, 1878–1890. [CrossRef]
38. Torkian, M.; Manteghian, M.; Safari, M. Caffeine metastable zone width and induction time in anti-solvent crystallization. *J. Cryst. Growth* **2022**, *594*, 126790. [CrossRef]
39. Ma, L.; Qin, C.; Pi, S.; Cui, H. Fabrication of efficient and stable Li_4SiO_4-based sorbent pellets via extrusion-spheronization for cyclic CO_2 capture. *Chem. Eng. J.* **2020**, *379*, 122385. [CrossRef]
40. Sangwal, K. Some features of metastable zone width of various systems determined by polythermal method. *CrystEngComm* **2011**, *13*, 489–501. [CrossRef]
41. Kashchiev, D.; Verdoes, D.; van Rosmalen, G.M. Induction time and metastability limit in new phase formation. *J. Cryst. Growth* **1991**, *110*, 373–380. [CrossRef]
42. Su, W.; Hao, H.; Glennon, B.; Barrett, M. Spontaneous Polymorphic Nucleation of d-Mannitol in Aqueous Solution Monitored with Raman Spectroscopy and FBRM. *Cryst. Growth Des.* **2013**, *13*, 5179–5187. [CrossRef]
43. Marichev, V.A. General thermodynamic equations for the surface tension of liquids and solids. *Surf. Sci.* **2010**, *604*, 458–463. [CrossRef]
44. Wang, Y.; Xiao, X.; Feng, X. An accurate and parallel method with post-processing boundedness control for solving the anisotropic phase-field dendritic crystal growth model. *Commun. Nonlinear Sci. Numer. Simul.* **2022**, *115*, 106717. [CrossRef]
45. Zhao, H.; Li, T.; Li, J.; Li, Q.; Wang, S.; Zheng, C.; Li, J.; Li, M.; Zhang, Y.; Yao, J. Excess polymer-assisted crystal growth method for high-performance perovskite photodetectors. *J. Alloys Compd.* **2022**, *908*, 164482. [CrossRef]
46. Wang, P.F.; Li, M.W.; Zhou, C.; Hu, Z.T.; Yin, H.W. Numerical simulations of flow and mass transfer during large-scale potassium dihydrogen phosphate crystal growth via three-dimensional motion growth method. *Int. J. Heat Mass Transf.* **2018**, *127*, 901–907. [CrossRef]
47. Sangwal, K. On the estimation of surface entropy factor, interfacial tension, dissolution enthalpy and metastable zone-width for substances crystallizing from solution. *J. Cryst. Growth* **1989**, *97*, 393–405. [CrossRef]
48. Jáger, G.; Tomán, J.J.; Juhász, L.; Vecsei, G.; Erdélyi, Z.; Cserháti, C. Nucleation and growth kinetics of $ZnAl_2O_4$ spinel in crystalline ZnO–amorphous Al_2O_3 bilayers prepared by atomic layer deposition. *Scr. Mater.* **2022**, *219*, 114857. [CrossRef]
49. Barata, P.A.; Serrano, M.L. Salting-out precipitation of potassium dihydrogen phosphate (KDP). I. Precipitation mechanism. *J. Cryst. Growth* **1996**, *160*, 361–369. [CrossRef]
50. Guo, K.; Lv, Y.; He, L.; Luo, X.; Zhao, J. Experimental study on the effect of spatial distribution and action order of electric field and magnetic field on oil-water separation. *Chem. Eng. Process.* **2019**, *145*, 107658. [CrossRef]
51. Ren, J.; Zhu, Z.; Qiu, Y.; Yu, F.; Ma, J.; Zhao, J. Magnetic field assisted adsorption of pollutants from an aqueous solution: A review. *J. Hazard. Mater.* **2021**, *408*, 124846. [CrossRef] [PubMed]
52. Cheng, N.; Guo, R.; Shuai, S.; Wang, J.; Xia, M.; Li, J.; Ren, Z.; Li, J.; Wang, Q. Influence of static magnetic field on the heterogeneous nucleation behavior of Al on single crystal Al2O3 substrate. *Materialia* **2020**, *13*, 100847. [CrossRef]
53. Taheri, M.H.; Mohammadpourfard, M.; Sadaghiani, A.K.; Kosar, A. Wettability alterations and magnetic field effects on the nucleation of magnetic nanofluids: A molecular dynamics simulation. *J. Mol. Liq.* **2018**, *260*, 209–220. [CrossRef]
54. Koza, J.A.; Uhlemann, M.; Gebert, A.; Schultz, L. Nucleation and growth of the electrodeposited iron layers in the presence of an external magnetic field. *Electrochim. Acta* **2008**, *53*, 7972–7980. [CrossRef]
55. Kadota, K.; Shirakawa, Y.; Wada, M.; Shimosaka, A.; Hidaka, J. Influence of clusters on the crystal surface of NaCl at initial growth stage investigated by molecular dynamics simulations. *J. Mol. Liq.* **2012**, *166*, 31–39. [CrossRef]
56. Galmarini, S.; Bowen, P. Atomistic simulation of the adsorption of calcium and hydroxyl ions onto portlandite surfaces—towards crystal growth mechanisms. *Cem. Concr. Res.* **2016**, *81*, 16–23. [CrossRef]
57. Separdar, L.; Rino, J.P.; Zanotto, E.D. Decoding crystal growth kinetics and structural evolution in supercooled ZnSe by molecular dynamics simulation. *Comput. Mater. Sci.* **2022**, *212*, 111598. [CrossRef]

Article

Quantitative Evaluation of CO_2 Storage Potential in the Offshore Atlantic Lower Cretaceous Strata, Southeastern United States

Dawod S. Almayahi [1,2,*], James H. Knapp [1] and Camelia Knapp [1]

1 Boone Pickens School of Geology, Oklahoma State University, Stillwater, OK 74075, USA; james.knapp@okstate.edu (J.H.K.); camelia.knapp@okstate.edu (C.K.)
2 Marine Geology Department, Marine Science Centre, University of Basrah, Basrah 61004, Iraq
* Correspondence: dalmaya@okstate.edu

Abstract: The geological storage of CO_2 in the Earth's subsurface has the potential to significantly offset greenhouse gas emissions for safe, economical, and acceptable public use. Due to legal advantages and vast resource capacity, offshore CO_2 storage provides an attractive alternative to onshore options. Although offshore Lower Cretaceous reservoirs have a vast expected storage capacity, there is a limited quantitative assessment of the offshore storage resource in the southeastern United States. This work is part of the Southeast Offshore Storage Resource Assessment (SOSRA) project, which presents a high-quality potential geological repository for CO_2 in the Mid- and South Atlantic Planning Areas. This is the first comprehensive investigation and quantitative assessment of CO_2 storage potential for the Lower Cretaceous section of the outer continental shelf that includes the Southeast Georgia Embayment and most of the Blake Plateau. An interpretation of 200,000 km of legacy industrial 2D seismic reflection profiles and geophysical well logs (i.e., TRANSCO 1005-1-1, COST GE-1, and EXXON 564-1) were utilized to create structure and thickness maps for the potential reservoirs and seals. We identified and assessed three target reservoirs isolated by seals based on their effective porosity values. The CO_2 storage capacity of these reservoirs was theoretically calculated using the DOE-NETL equation for saline formations. The prospective storage resources are estimated between 450 and 4700 Mt of CO_2, with an offshore geological efficiency factor of dolomite between 2% and 3.6% at the formation scale.

Keywords: carbon dioxide (CO_2); carbon capture and storage (CCS); offshore Atlantic; efficiency factors; southeastern United States

Citation: Almayahi, D.S.; Knapp, J.H.; Knapp, C. Quantitative Evaluation of CO_2 Storage Potential in the Offshore Atlantic Lower Cretaceous Strata, Southeastern United States. *Energies* **2022**, *15*, 4890. https://doi.org/10.3390/en15134890

Academic Editors: Dongdong Feng, Jian Sun and Zijian Zhou

Received: 2 June 2022
Accepted: 30 June 2022
Published: 4 July 2022

1. Introduction

Offshore geologic storage of carbon dioxide (CO_2) as a form of carbon capture and storage (CCS) technology has recently attracted considerable scientific attention. CCS technology is a potentially vital tool to reduce the levels of CO_2 emissions in the atmosphere and to prevent the most dangerous consequences of climate change [1–8]. The term offshore CO_2 storage refers to the injection of CO_2 into the geological strata beneath the seabed for safe and permanent storage [9–12]. Due to legal considerations and the vast resource storage capacity, offshore storage offers an attractive alternative to the onshore options. After many failed attempts, the Sleipner project in the North Sea was an early successful opportunity for commercial deployment of CO_2 storage. Although offshore Lower Cretaceous reservoirs have an extensive expected storage capacity, there has been no comprehensive assessment of the offshore storage resource estimate in the southeastern United States. An analysis of a 25,900 km^2 area of offshore Alabama and the western Florida Panhandle suggested that approximately 170 Gt of CO_2 could be stored in Miocene sandstone, whereas at least 30 Gt could be stored in the deeper Cretaceous formations [13]. Around 32 Gt of CO_2 could be stored within 190,000 km^2 of the Upper Cretaceous strata in the offshore southeastern United States [14]. A study offshore of the northeastern United States has recently concluded that the Cretaceous and Jurassic sandstone is able to store approximately 37–403 Mt of CO_2, with geological storage efficiency of 1–13% [15]. Realizing that the US Environmental Protection Agency estimates that about 40% of anthropogenic CO_2 emissions in the US are produced in the southeast, the lack of an offshore CO_2 assessment constitutes a significant gap in understanding of this prospective regional storage resource [16–19].

The research area is located offshore of the southeastern United States, covering the southern part of the Mid-Atlantic Planning Area (including the Carolina Trough) and the South Atlantic Planning Area (including the Southeast Georgia Embayment and Blake Plateau), as defined by the Bureau of Ocean Energy Management (BOEM) (Figure 1). Within the Mid-Atlantic Planning Area and South Atlantic Planning Area, there is a thick sequence of post-rift stratigraphy, which is considered to be a semi-closed saline aquifer, with sediments ranging in age from Jurassic to Pleistocene [20,21]. In these areas, the significant sedimentary deposits from north to south include the Carolina Trough, the Southeast Georgia Embayment, and the Blake Plateau Basin, with a range in sediment column thicknesses from 3048 to 7010 m [22–25]. A regional unconformity under the post-rift sediments known as the post-rift unconformity (PRU) cuts the entire region after rifting between Africa and North America ceases. This unconformity marks the transition to a widespread sediment deposition zone during the drifting stage.

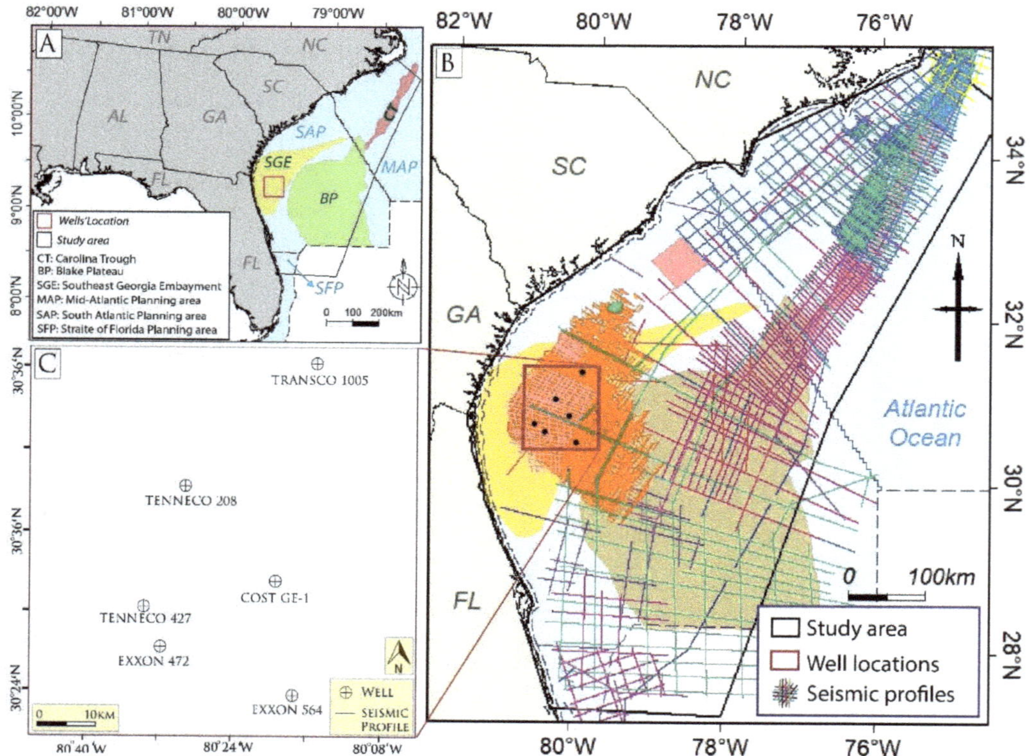

Figure 1. Location map of the study area. Panel (**A**) is the location of the study area in the regional map, panel (**B**) is the study area map, and panel (**C**) is the well locations.

This research evaluates the potential offshore CO_2 storage capacity within the Lower Cretaceous strata in the Mid- and South Atlantic Planning areas.

2. Geological Framework

The evolution of the Atlantic continental margin, including the study area, is broadly characterized by the terminal collision of the Laurentian and Gondwanan continents in the Late Paleozoic Era, followed by continental rifting beginning in the Early Triassic [9,20,23,25,26]. Mesozoic rifting involved local tectonic subsidence in early restricted extensional basins, followed by regional thermal subsidence along the eastern North American margin [21,24,26–28]. Stratigraphic sequences on this passive margin show extensive lateral continuity and relatively minor structural disturbance [25–27]. There is a thick sequence of post-rift stratigraphy ranging from Jurassic- to Pleistocene-aged sediments in the Mid-Atlantic and South Atlantic Planning Areas. The significant depositional centers in these areas from north to south include the Baltimore Canyon Trough, the Carolina Trough, the Southeast Georgia Embayment, and the Blake Plateau Basin. The oldest post-rift sediments are of Jurassic age, and are the product of rapid clastic sedimentation from erosion, followed by a period of evaporate deposition and then initiation of broad, shallow-water, carbonate deposition with some terrigenous intrusions [28,29]. The Jurassic section thickens seaward, and estimates from geophysical and stratigraphic studies suggest thicknesses of at least 7–8 km in the basins [20,21]. Typically, the Cretaceous section is characterized by more clastic sedimentation in the north and more carbonate deposition in the south, forming an extensive carbonate platform over the Blake Plateau and offshore Florida. From the Late Cretaceous to the Cenozoic, strong paleocurrents controlled the deposition of the clastic offshore sediments. The Late Cretaceous in the Blake Plateau exhibited a distinct facies change to the neighboring offshore Florida and Bahamas carbonate platforms [23–25]. The Suwannee Strait eventually evolved into the current Gulf Stream, providing strong erosive power that eroded most of the Paleogene sediments on the Blake Plateau and prevented deposition off the Florida–Hatteras slope, where it continues to the north along the shelf edge [25,28].

3. Materials and Methods

This study involves the integration of geological and geophysical data, combining regional 2D seismic reflection surveys, published sidewall core analyses, and well logs from commercial exploration wells. Seismic reflection data provide fundamental structural control over the subsurface geology, confirmed by accessible exploration wells. A total of 36 separate 2D seismic surveys was integrated and analyzed (Figure 1B), and a seismic mistie analysis was conducted to merge the individual surveys for this study. The well logs were then calibrated with the seismic reflection profiles. The seismic surveys were interpreted regionally for key stratigraphic horizons, and then porosities and permeabilities were derived from the log data and core reports. Subsequently, the porosity and permeability estimates were compared between the published sidewall core data and the derived values.

3.1. Well Sections

Seven commercial exploratory offshore wells (GETTY 913, TRANSCO 1005-1-1, TENNECO 208, COST GE-1, TENNECO 427, EXXON 472, and EXXON 564-1) (Figure 1C) were drilled in the Southeast Georgia Embayment from 1979 to 1980. These wells were stratigraphically correlated by Poppe et al. [25], and have also been seismic-stratigraphically interpreted and correlated with similar Mesozoic sedimentary sequences [30]. TRANSCO 1005-1, COST GE-1, and EXXON 564-1—the deepest three wells in the Southeast Georgia Embayment—were used in the present study. TRANSCO 1005-1 and COST GE-1 are the only two wells that penetrate the pre-rift sedimentary sequences from the Paleozoic Era. The EXXON 564-1 well penetrates only the post-rift sedimentary sequence from the Mesozoic Era.

The TRANSCO 1005-1-1 well encounters the pre-rift unconformity at a depth of ~2743 m, and bottoms in Paleozoic sedimentary rocks at a total depth (TD) of 3546 m [25]. The Paleozoic section in the TRANSCO 1005-1 well is a weakly metamorphosed shale and sandstone with meta-igneous intrusions (Figure 2). The log suite for this well includes mud logs, electric logs, drill cuttings, and biostratigraphic data. The Paleozoic rocks in the TRANSCO 1005-1 well have been correlated with Devonian strata in the COST GE-1 well [25].

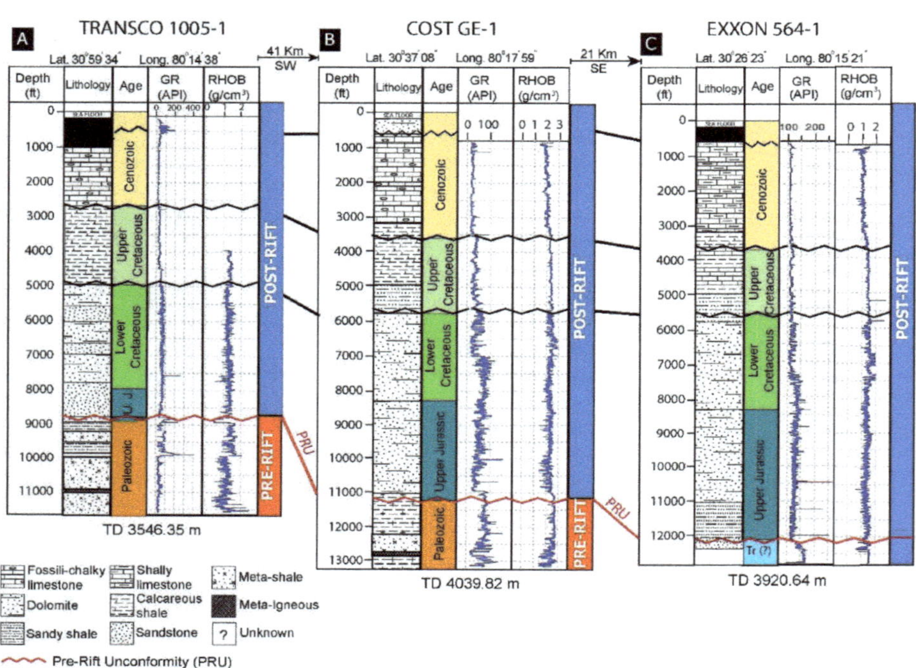

Figure 2. TRANSCO 1005-1-1, COST GE-1, and EXXON 564-1, stratigraphically and geophysically correlated by logs and lithology, modified and adapted with permission from Ref. [25]. Copyright 1995, copyright Poppe et al., and Ref. [31]. Copyright 2016, copyright Boote and Knapp.

The COST GE-1 well penetrates the pre-rift unconformity at 3200 m, drilling approximately 686 m of the Paleozoic sedimentary sequences and reaching 4040 m [9]. The COST GE-1 well showed a thick sequence from Paleozoic to Cenozoic (Figure 2). The Paleozoic section generally consists of non-fossiliferous quartzite, shale, and salt, underlain by metamorphic and meta-volcanic rocks [9].

Scholle [9] provided an analysis of the COST GE-1 well data, and described the stratigraphic units, porosity, and permeability measurements by both a conventional core and a sidewall core with respect to depth. The thickness of fossiliferous chalky limestone below the drill platform reaches 1006 m, corresponding to strata of Tertiary age. The Upper Cretaceous section is marked at a depth ranging from 1006 m to 1798 m, and is composed of calcareous shale, dolomite, and limestone. The section from 1798 to 2195 m is the Lower Cretaceous. Rocks encountered below 3353 m depth consist of highly indurated to weakly metamorphosed Paleozoic strata [9].

The EXXON 564-1 well encounters the pre-rift unconformity at a depth of 3737 m [30]. The last 183 m, under the post-rift unconformity in the EXXON 564-1 well, is a weakly metamorphosed shale and sandstone with meta-igneous intrusion of Triassic rocks. The EXXON 564-1 well has been correlated with the Devonian rocks in the COST GE-1 well (Figure 2).

3.2. Seismic Interpretation

The primary dataset consists of legacy 2D seismic reflection profiles offshore of the southeastern United States in the Atlantic Ocean. The dataset has been released by the Bureau of Ocean and Energy Management (Figure 1B). We interpreted and correlated a continuous surface stratigraphy of the storage elements (sinks and reservoir seals) along 200,000 km of the seismic profiles, covering approximately 200,000 km^2. Seismic interpretation started with picking high-frequency stratigraphy sequences targeted at creating three-dimensional maps of the reservoirs and seals. The TRANSCO 1005-1 well, COST GE-1 well, and EXXON 564-1 well were tied with the seismic profiles (Figure 3). The Schlumberger Petrel software can generate advanced velocity models using check-shot data. The velocity model uses input parameters such as tops and surfaces and the time–depth link [32]. The velocity model is created based on the time–depth relationship from the well data. The time–depth conversion uses linear velocity related to the depth functions ($V = V_0 + K \times Z$) and ($V = V_0 + K \times (Z - Z_0)$) for evaluating a velocity model [33]. Both K and the linear velocity slope indicate that the velocity increases with depth and reflects layer compaction. The compaction factor K is estimated with the fewest mistakes feasible and used to generate a V_0 surface; any modifications incorporated into the velocity model are reflected on the V_0 surface. The check-shot data of the three wells (COST GE-1, Exxon 564-1, and Transco 1005-1) were used to identify the depth of the upper and bottom surfaces of the Lower Cretaceous section.

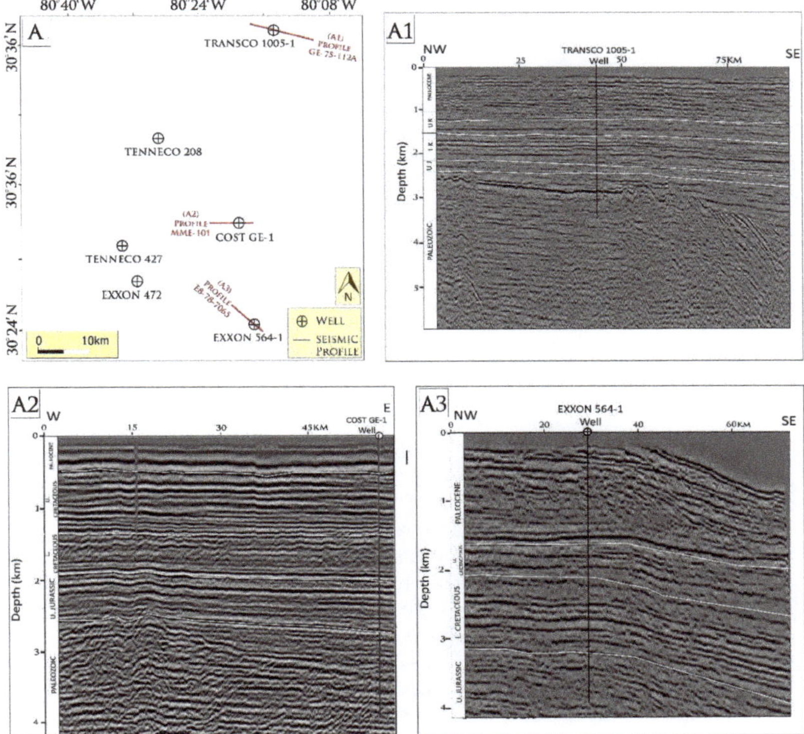

Figure 3. Seismic-well tie analysis for three wells: (**A**) The location map of the commercial wells in the Southeast Georgia Embayment. (**A1**) The seismic profile Number GE-75-112A intersecting the TRANSCO 1005-1-1 well. (**A2**) The seismic profile Number MME-101 intersecting the COST GE-1 well. (**A3**) The seismic profile Number E8-78-7065 intersecting the EXXON 564-1 well.

3.3. Calculation of CO_2 Storage Capacity

We developed an estimate of the regional CO_2 storage capacity offshore of the Lower Cretaceous section in the Mid–South Atlantic Ocean using the US Department of Energy (DOE) National Energy Technology Laboratory (NETL) method [16,34,35].

The US-DOE approach estimates CO_2 storage volume based on geological parameters such as formation area, thickness, and porosity [16,17,36]. Some articles such as the work of Teletzke et al. [37] criticize the DOE method or the Goodman method. However, the DOE method is the most comprehensive and well-documented storage method available at this time. The DOE method estimates carbon storage resources at the prospective scale in subsurface saline formations. This information plays an important role in establishing the scale of carbon capture and storage activities for governmental policy and commercial project decision making. When calculating the storage efficiency terms, the DOE method accounts for several parameters, as presented in Gorecki et al. [35]: reservoir width, reservoir length, thickness, domain discretization, rock properties, porosity, permeability (lateral), permeability anisotropy, relative permeability, capillary pressure, reservoir properties, initial pressure, pressure gradient, initial temperature, temperature gradient, brine concentration, pore compressibility, operation properties, injection rate, injection period, and perforation. Goodman et al. [16] used the static volumetric methodology and the CO_2 storage prospective resource estimation Excel analysis (CO_2-SCREEN) tool developed by the US Department of Energy National Energy Technology Laboratory (DOE-NETL) (Equation (1)). Equation (1) is mathematically expressed as follows:

$$GCO_2 = A \times \hbar g \times \varphi \times \rho \times E \tag{1}$$

where GCO_2 is the total mass of CO_2 in gigatons (Gt), A is the target area (in square meters), $\hbar g$ is the gross strata thickness (in meters), φ is the effective porosity, ρ is the CO_2 density in kilograms per cubic meter (kg/m^3), and ρCO_2 is the average CO_2 density evaluated at pressure and temperature, representing storage conditions anticipated for a specific deep saline aquifer. Ennis-King and Paterson [38] pointed out that the CO_2 density rises with a decrease in volume in the reservoir at depths ranging from 600 to 1000 m, depending on the specific geothermal conditions and pressure. Due to heat transfer, the average temperature in several geological formations increases by approximately 25–30.8 °C/km below the sea bed [39]. However, geothermal gradients vary significantly between local and global basins [40]. Nevertheless, the subsurface units suitable for geological carbon sequestration are 800 m or deeper below the seafloor, and seem to have higher pressure and temperature at depths greater than the CO_2 critical point [12]. The critical point indicates that CO_2 is injected at supercritical temperatures and pressures. CO_2 and certain other supercritical gases possess gas viscosity, which reduces resistance to flow compared to liquid and semi-liquid density, significantly reducing the volume required to store a given mass of fluid. Carbon dioxide behaves as a supercritical fluid at temperatures and pressures above the critical points of 30.85 °C and 7.38 MPa, respectively [39]. The 800 m depth requirement is a reasonable guess that varies based on the geothermal gradient and formation pressure at a given location [41]. The pressure in the pore spaces of sedimentary rocks is identical to hydrostatic pressure. This pressure is generated by a water column at a corresponding elevation to the depth of the pore space, since the pore space is frequently filled with water and, although in a convoluted manner, is connected to the ground surface. When the pore space is not connected with the surface at equilibrium, the hydrostatic pressure can be exceeded, and overpressure occurs [38].

Scholle [9] pointed out that pressure and temperature data for the COST GE-1 well were identified based on three temperature logs. The Lower Cretaceous temperature was estimated as 72.3 °C at the top and 81.4 °C at the bottom, with a geothermal gradient of 16 °C/km, based on a geothermal gradient that was only available at the COST GE-1 well. The parameters (A, $\hbar$, and φ) are the yield of the total pore volume of the studied section. The ρ parameter is the volume conversion to the mass of CO_2, and the efficiency factor

(*E*) reduces the total CO_2 mass for storage to an accurate, realistic value [41,42]. The CO_2 storage efficiency factor is the portion of rock suitable for CO_2 storage, and is defined as the fraction of pore space where injected CO_2 can permanently displace formation fluids [43]. Table 1 shows efficiency factors for different lithologies and estimated with different methods, including numerical and Monte Carlo simulations [11,16,35,44]. According to the US DOE approach, storage efficiency is a function of aquifer characteristics such as area, thickness, and porosity—the product of which represents the aquifer pore volume—as well as displacement efficiency components such as areal, vertical, and microscopic components, and is expressed as the product of these individual efficiencies [16,35,45].

Table 1. Numerical and Monte Carlo estimates for saline formation efficiency factors at the formation scale. Adapted with permission from Ref. [16]. Copyright 2011, Copyright Goodman et al., Ref. [35]. Copyright 2009, copyright Goreckie et al., and Ref. [46]. Copyright 2009, copyright Preston et al.

Lithology	Monte Carlo Method (E%)			Numerical Method (E%)		
	P_{10}	P_{50}	P_{90}	P_{10}	P_{50}	P_{90}
Clastic	1.86	2.7	6	1.2	2.4	4.1
Dolomite	2.58	3.26	5.54	2	2.7	3.6
Limestone	1.41	2.04	3.27	1.3	2	2.8

Goodman et al. [16] used Monte Carlo sampling to calculate local- and regional-scale storage efficiency values based on statistical properties (i.e., mean values, standard deviation, ranges, and distributions) that describe geological and displacement parameters for three lithologies: clastics, dolomite, and limestone. They obtained slightly lower values for storage efficiency (E) than Gorecki et al. [35]. Efficiency in saline formations can be calculated using Equation (2):

$$E = E_{An/At} \times E_{Hn/Hg} \times E_{Øe/Øt} \times E_A \times E_v \times E_d \quad (2)$$

where $E_{An/At}$ is the percentage-to-total-area ratio ideal for CO_2 storage; $E_{Hn/Hg}$ is the fraction-to-gross-thickness ratio that meets the porosity and permeability standards for CO_2 storage; $E_{Øe/Øt}$ represents the ratio of linked porosity to total porosity; E_A is the effective aquifer area; E_v is the volumetric displacement; and E_d is the microscopic displacement. The net-to-total-area ratio $E_{An/At}$ is the proportion of the aquifer area suitable for CO_2 storage, expressed as a net-to-gross-thickness ratio. $E_{Hn/Hg}$ is the fraction of the geological formation in the vertical dimension that meets the porosity and permeability requirements for CO_2 injection and storage, and $E_{Øe/Øt}$ is the effective (interconnected)-porosity-to-total-porosity ratio. The storage efficiency factor reflects the total pore volume filled with CO_2. There is no comparison established between the CO_2 stored by different processes. For a 15–85% certainty value, Monte Carlo simulations generate an E range between 1 and 4% of the bulk volume of a deep saline aquifer, with an average of 2.4% for 50% confidence. The Monte Carlo simulated by USDOE-NETL [47] that established the proposed range for E varied several calculation factors, i.e., from 0.20 to 0.80 of the saline aquifer appropriate for CO_2 storage; 0.25 to 0.75 of the geological unit has the porosity and permeability required for CO_2 injection; the interconnected porosity fraction ranges from 0.6 to 0.95; the areal displacement efficiency ranges between 0.5 and 0.8, while the vertical displacement efficiency ranges between 0.6 and 0.9. Due to CO_2 buoyancy, CO_2 occupies between 0.2 and 0.6 percent of the net aquifer thickness. The effectiveness of pore-scale displacement ranges from 0.5 to 0.8. The maximum and minimum values for each parameter were calculated to reflect varied lithologies and geological depositional systems in North America, with the maximum and minimum values representing reasonably high and low values, respectively.

Several analogies are found in the US DOE method [48]; the effect of total water saturation is included in the efficiency factor E through the pore-scale displacement efficiency. The salty aquifers with TDS more than 10,000 ppm and deeper than 800 m should be considered, as this is the minimum depth required to assure that CO_2 is in a dense liquid or supercritical phase confined by aquitards or aquicludes (caprock), which include shale, anhydrite, and evaporite. The considerations of the US DOE method introduce storage efficiency coefficient calculations through Monte Carlo simulations of CO_2 storage within deep saline aquifers in North America. The US DOE obtained a range of values for these storage efficiency coefficients for the 0.15 and 0.85 confidence intervals, ranging between 0.1 and 0.4 for deep saline aquifers. Based on the IPCC [1] report, the US DOE screening requirements were assumed, and these coefficients have a value of unity in local-scale assessments or, more broadly, when the effective aquifer area, thickness, and porosity are known. In this study, we used the value of the efficiency factor suggested by the US DOE [35] (Table 1).

Goodman et al. [16] demonstrated that geological uncertainty has a greater impact on storage estimation than the approach used. Thus, it is critical to determine the geological estimates and ranges of storage efficiency factors for certain geological parameters to improve or refine storage estimates. In addition, due to the legacy of seismic data and the relatively limited well data available over the 200,000 km^2 study area, uncertainty associated with the subsurface data gap must be assumed in the storage resource assessment. The potential capacity of the several reservoirs of the Lower Cretaceous section was calculated using all parameters in Equation (1).

4. Results and Discussions

4.1. Well Data Analysis

The well log interpretation is the most fundamental geophysical approach for geological and geophysical reservoir characterizations. The density log (RHOB) provides lithology interpretation, porosity calculation, and petrophysical properties. The gamma ray (GR) log is used to interpret lithology, porosity, and permeability. The spontaneous potential (SP) log is useful for lithology identification and permeability calculation. Both GR and SP have a similar response to porous layers, and can determine lithology and correlate stratigraphy. Density logs (RHOB) provide a continuous record of the bulk density, determined by the porosity of the formation and the fluid content of the pore spaces. GR and ROHB logs for the TRANSCO 1005-1 well, COST GE-1 well, and EXXON 564-1 well were stratigraphically interpreted and correlated in this study to obtain similar equivalents in the sedimentary sequences (Figure 2). Related to the sidewall core analysis on the COST GE-1 well, the porosity was valued between 0.12 and 0.35, and the permeability was estimated between 9.87×10^{-18} and 5.4×10^{-13} m^2, within the Lower Cretaceous strata [9].

The Lower Cretaceous strata, between depths of 1798 m and 2195 m, have 14 lithological intervals, which are mainly composed of varying proportions of calcite, clay, shale, sandstone, limestone, and dolomite with carbonite materials [9] (Table 2).

Table 2. The 14 lithological intervals of the Lower Cretaceous strata, between depths of 1798 and 2195 m in the COST GE-1 well, based on [9]. Adapted with permission from Ref. [9]. Copyright 1979, copyright Scholle.

Unit	Depth		Lithology	Porosity
	ft	m		
1	5900	1798	Shale, gray, silty, calcareous, micaceous, and sandstone.	Low
2	5990	1826	Shale, silty, calcareous, micaceous, non-calcareous sandstone.	Very low
3	6080	1853	More shale, slightly calcareous, carbonaceous, fossiliferous.	Low to moderate
4	6320	1926	Coarse-to-medium crystals, dense, and fossil fragments.	Low to high
5	6500	1981	Partly sandy, dense silty, hard, calcareous to non-calcareous.	Low
6	6800	2073	Sandstone, shell, sandstone, anhydrite, and gypsum.	Low to high
7	6890	2100	Limestone, shale, very fine-grained calcareously cemented sandstone, and anhydrite with dense dolomite.	Moderate
8	7020	2140	Dolomite, finely crystalline to dolomite, limestone increasing with depth, shale, and sandstone.	Low to high
9	7070	2155	Limestone, fossiliferous, dolomite, and non-calcareous.	Low to high
10	7160	2182	Shale and sandstone, much calcareous cement.	Moderate to low

Table 2. *Cont.*

Unit	Depth		Lithology	Porosity
	ft	m		
11	7200	2195	Shale, sandstone, and silty shale with calcareous cement. Limestone, some dolomite, and fossiliferous to non-fossiliferous.	High
12	7400	2256	Shale, some gravel trace, dolomite, and fossiliferous to non-fossiliferous.	High
13	7490	2283	Lithology like unit 12 with decreasing shale, increasing dolomite.	High
14	7910	2411	Shale to fine sandstone, gravel, faintly calcareous and non-calcareous shale, dolomite with some clayey coatings, non-fossiliferous, much coal, anhydrite, and sandy dolomite.	Moderate

For the COST GE-1 well, the net porosity was geophysically derived by the ratio of the density log (RHOB) and the gamma ray log (GR). The values calculated from the log data were then compared with the measured values from the conventional and sidewall cores for the Lower Cretaceous lithological units to identify potential reservoirs and seals (Figure 4). Significant CO_2 storage was predicted where high primary and secondary porosity values ranged from 0.20 to 0.33, accounting for the greatest permeability range of 1.97×10^{-13}–5.43×10^{-13} m^2 recorded in the Lower Cretaceous section.

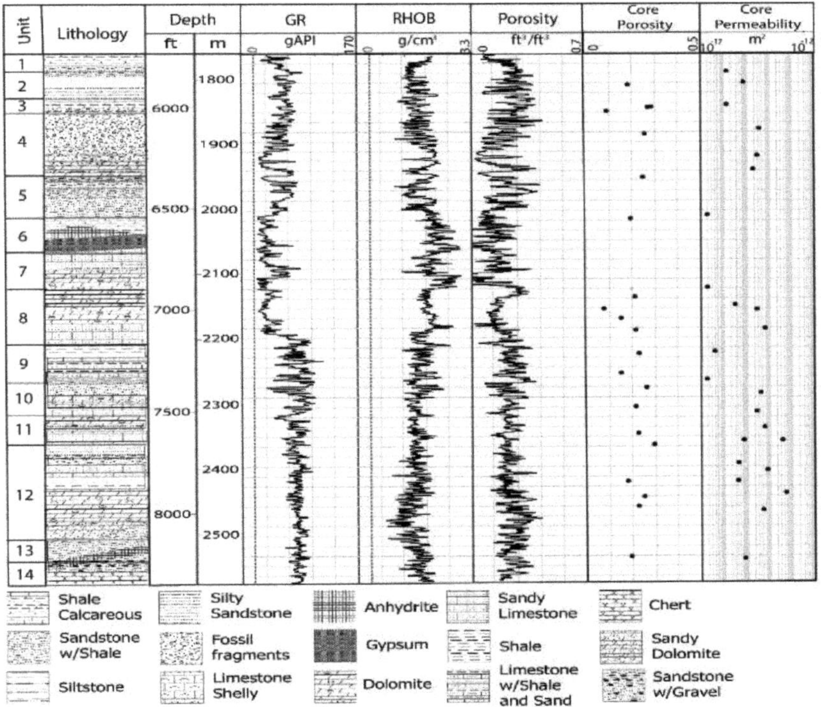

Figure 4. Analysis of density logs and gamma ray logs to generate porosity logs, compared with estimated porosity from analysis of core samples from the COST GE-1 well.

4.2. Geological CO_2 Storage

In a practical CO_2 storage evaluation, Chadwick et al. [42] concluded that a generic case may be associated with injecting large quantities—reaching 400 Mt—of CO_2 into a single anticlinal aquifer structure. The effective size of the reservoir is a critical parameter to determine water displacement, with high pressure leading to increased porosity of 20% or higher, which is a requirement for safe injection into porous strata. While the CO_2 injection pressure significantly increases from 5 to 10 MPa in the injection zone, the effective storage capacity in the caprocks increases when the pressure of the injected CO_2 exceeds the initial formation pressure [49]. Table 3 demonstrates atypical geological criteria of the static and dynamic storage capacities for the reservoir depth of 1000–2500 m with a thickness of 21–50 m, where the porosity is greater than 0.20 and the permeability is greater than 1.978×10^{-13} m^2. The ideal seal thickness is 100 m with lateral continuity, and non-faulted strata or capillary entry pressure is present.

Table 3. Ideal geological CO_2 storage criteria for reservoir properties and caprocks. Reprinted with permission from Ref. [42]. Copyright 2008, copyright Chadwick et al., and Ref. [50]. Copyright 2017, copyright Chadwick et al.

Media	Properties	Positive Indicators	Cautionary Indicators
Reservoir	Static storage capacity	Evaluated effective CO_2 storage capacity greater than total injected CO_2	Evaluated effective CO_2 storage capacity equal to total injected CO_2
	Dynamic storage capacity	Predicted injection-induced pressures below the rate of inducing geomechanical damage to the reservoir or caprock.	Geomechanical instability limits reaching the predicted injection-induced pressures.
	Depth (m)	Greater than 800	Less than 800
	Thickness (m)	Greater than 50	Less than 20
	Porosity	Greater than 0.20	Less than 0.10
	Permeability (m^2)	Greater than 4.93×10^{-12}	Less than 1.97×10^{-13}
	Stratigraphy	Capacity much larger than total injected CO_2	Capacity $\leq$ total injected CO_2
Caprocks	Lateral stratigraphy	Uniform and small or no fault	Lateral variations and medium-to-large fault
	Thickness (m)	Greater than 20	Less than 20
	Capillary entry pressure	Greater than the maximum predicted injection-induced pressure increase	Equal to the maximum predicted injection-induced pressure increase

The Lower Cretaceous section from the COST GE-1 well was described in terms of lithology and rock properties through 14 core samples [9]. Dolomite rocks are the most dominant rocks in this section. The porosity and permeability of the different stratigraphic intervals were the primary basis for identifying the main storage units. The reservoirs and seals were classified and evaluated with the observance of the positive indicators of the CO_2 storage criteria described by Chadwick et al. [42] (Figure 5). Three reservoirs separated by three seals were identified within the Lower Cretaceous section. Figure 3 illustrates the 14 intervals that appear most suitable for permanent offshore CO_2 storage. Figure 6 reveals structural maps for the top and bottom topographic surfaces and the thickness of the Lower Cretaceous strata. The Lower Cretaceous section ranges in depth between 1798 m and 2539 m, and consists of dolomite interbedded with sandstones and calcareous silty shales. Based on the rock composition and physical rock properties, this section (Table 2) records the lithological description and porosity values with depth for the COST GE-1 well, based on core analyses and geophysical logs. Figure 5 shows the potential CO_2 storage reservoirs and seals based on the rock properties compared with the favorable conditions for CO_2 storage [42]. Scholle [9] noted impermeable shale with calcareous shale layers interbedded with highly permeable dolomite in the COST GE-1 well. However, a few samples of sandstone were marked between 1768 m and 2530 m, with high primary and secondary porosity and high permeability, suitable as reservoir rock for CO_2 storage. This section is dominated by dolomite with porosities that vary widely and unsystematically, with depth from 0.17 to 0.32 and permeability between 2.096×10^{-13} and 5.43×10^{-13} m^2. The porosity log was derived and calculated from well logs to fill in the gaps between the core intervals (Figure 4).

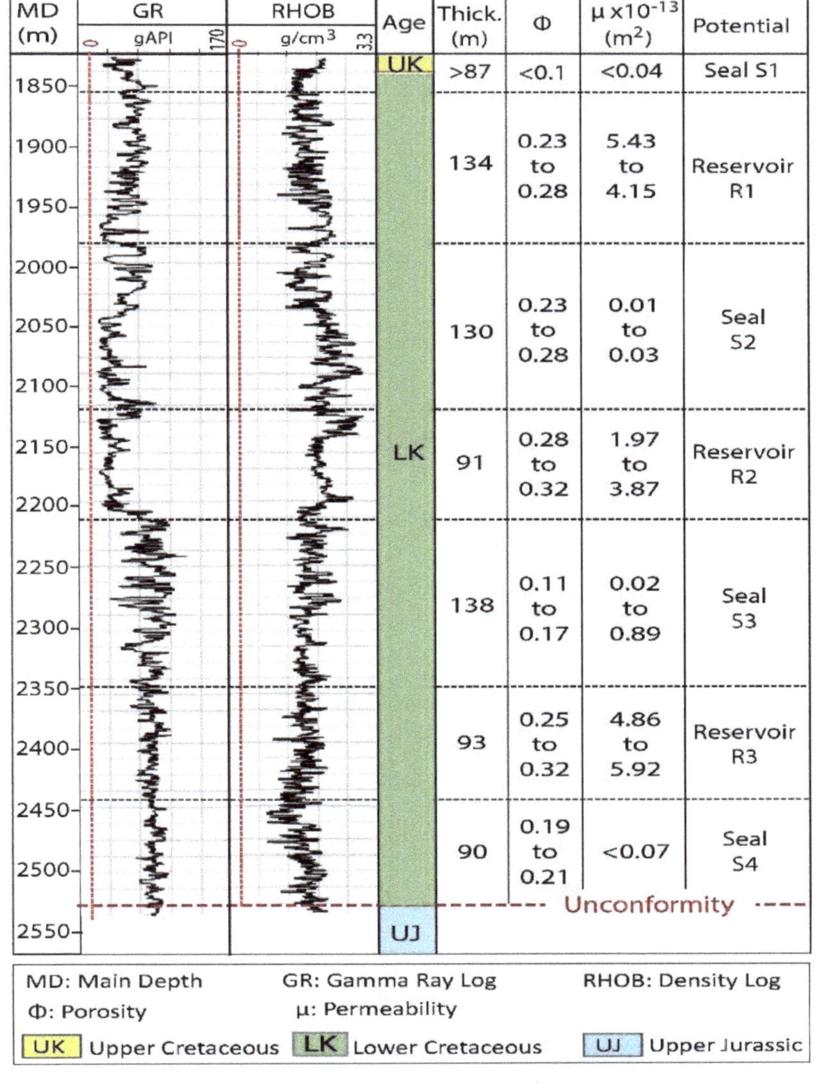

Figure 5. Characterizations of the storage elements, seals, and reservoirs identified based on the geological and geophysical data at the COST GE-1 well.

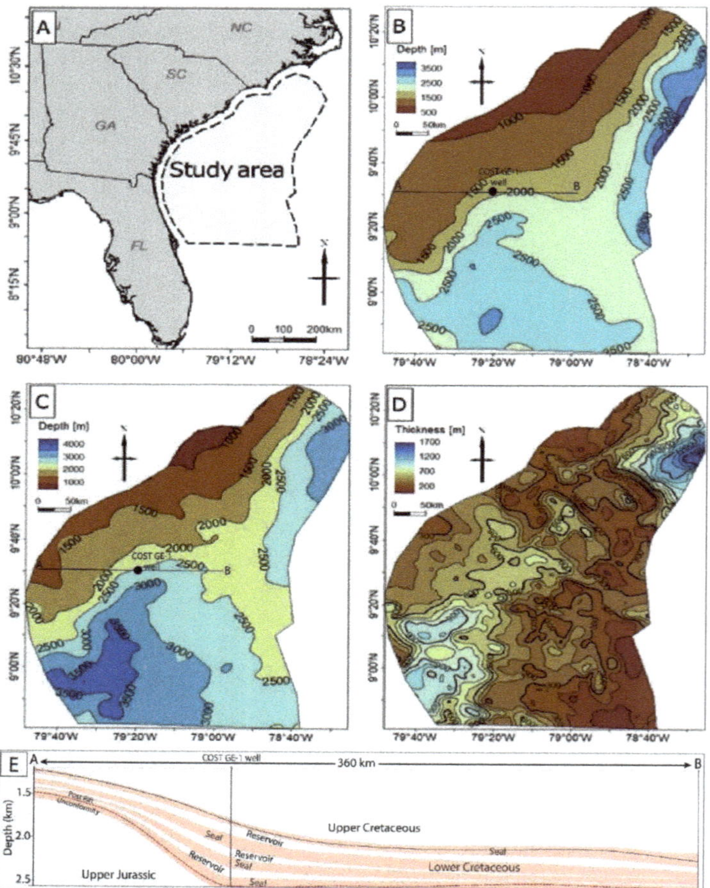

Figure 6. Structural maps of the study area: (**A**) the location map of the study area, (**B**) the depth map for the top topographic surface of the Lower Cretaceous section, (**C**) the depth map for the bottom topographic surface of the Lower Cretaceous section, (**D**) the thickness map of the Lower Cretaceous section, and (**E**) is the cross-section AB shown in panels (**B**,**C**) in this figure. The cross-section AB demonstrates the lateral extension and thickness of the reservoirs and seals across the Lower Cretaceous strata in the study area.

4.3. CO_2 Storage Capacity Calculations

The capacity for CO_2 storage potential of the Lower Cretaceous section was calculated based on the rock compositions and petrophysical properties at the COST GE-1 well. Three potential reservoirs were associated with four potential seals characterized and assessed in the Lower Cretaceous section. The three reservoirs are sealed by thick caprocks mainly composed of shale, siltstone, anhydrite, and limestone. These reservoirs are marked as R1, R2, and R3, and their seals are marked as S1, S2, S3, and S4 (Figure 5). According to Scholle [9], the trapping mechanism, characterized as an overlying seal, involves stratigraphic trapping through lateral facies variations. Figure 5 shows that reservoir R1 ranges in depth from 1855 to 1989 m, reservoir R2 ranges in depth from 2119 to 2210 m, and reservoir R3 ranges in depth from 2349 to 2442 m. The three reservoirs are composed as follows: (1) R1 is composed of calcareous shale, anhydrite, and gypsum; (2) R2 is composed of limestone and shale; and (3) R3 is composed of calcareous shale, anhydrite, fossil fragments, and gypsum. The average porosities of the reservoirs (R1, R2, and R3) are 0.23–0.28, 0.28–0.32, and 0.25–0.32, respectively. We used Equation (1), which was developed earlier by Goodman et al. [16], for applying the dolomite efficiency factor (Equation (2)) at the formation scale to give 2.58%, 3.26%, and 5.54% for probabilities of 0.10, 0.50, and 0.90, respectively. This work considers the probabilities for the area (A) parameter with a wide range of thickness values over the area (Table 4) to apply Equation (1). The uncertainty of CO_2 density with depth and thickness in the Lower Cretaceous section can be reduced by calculating the density of CO_2 based on the depth of each reservoir (Figure 6). For accuracy of CO_2 density values, the Lower Cretaceous section is divided into three depth zones: (1) the shallow depth (SLK) is in the 300–1000 m range, (2) the depth of the COST GE-1 well (GLK) is in the 1600–2450 m range, and (3) the deep reservoir (DLK) ranges from 3000 to 3500 m. To identify the temperature of the reservoirs in the three depth zones, we assumed that the geothermal gradient at COST GE-1 well (16 °C) is constant across the study area (Figure 7A; Table 5). Based on the temperature–pressure–density graph [41] (Figure 7B), the density values of supercritical CO_2 were estimated based on the depth and temperature considered in the NETL CO_2 calculation method [47].

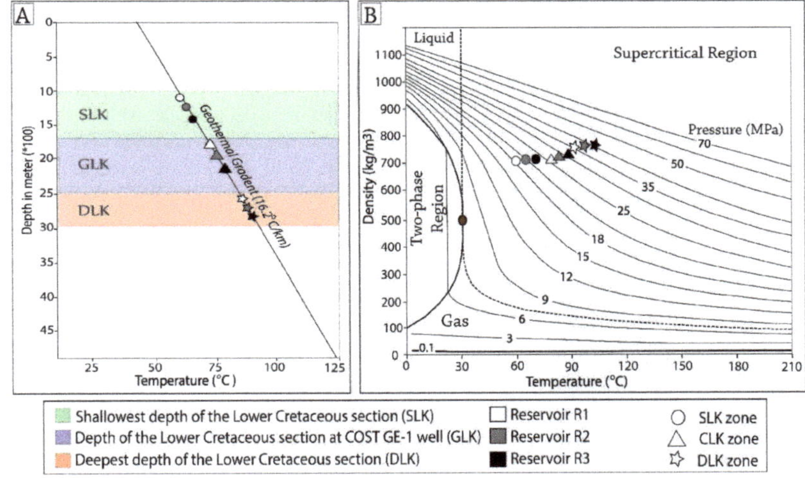

Figure 7. (**A**) The three reservoirs in three depth zones plotted in the geothermal gradient at the COST GE-1 well. (**B**) The time–depth–pressure graph to identify the density of supercritical CO_2 for three reservoirs in the three depth zones of the Lower Cretaceous section.

Table 4. The physical parameters for the three reservoirs applied in the NETL method (Equation (1)) in the local and regional zones.

Zone	Reservoir	Area (km^2)	Gross Thickness (m)		Total Porosity (%)		Pressure (MPa)		Temperature (°C)	
			Mean	Std Dev	Mean	Std Dev	Mean	Std Dev	Mean	Std Dev
Local	1	10,000	134	0.0093	0.28	0.0012	26	0.0004	72.8	0.06
	2	10,000	91	0	0.32	3.2×10^{-18}	29	0.0193	75.4	0.006
	3	10,000	93	0	0.32	0	32	0.0385	76.2	0.06
Regional	1	200,000	83	0.193	0.245	0.012	26.3	0.1925	70	1.6
	2	200,000	60	0.0001	0.3	3.2×10^{-17}	29.3	0.1925	73.4	1.6
	3	200,000	63	0.0001	0.285	0	32.6	0.3849	75.5	0.06

Table 5. CO_2 density values estimated based on depth, temperature, and overburden pressure for the reservoirs in the three depth zones.

Zone	Reservoir	Depth (m)	Temp. (°C)	Pressure (MPa)	Density (kg/m^3)
SLK	R1	1100	56.2	18	700
	R2	1300	58.6	20	708
	R3	1550	60.2	22	712
GLK	R1	1855	72.8	26	722
	R2	2120	76.2	29	732
	R3	2350	75.4	32	740
DLK	R1	2550	81.1	35	760
	R2	2680	85.4	39	768
	R3	2860	90.9	44	778

We estimated the potential storage resources of the three reservoirs for local and regional areas. The local area was detected where seismic profiles and well data were densely concentrated in the Southeast Georgia Embayment, which covers approximately 10,000 km^2. The regional storage resource covers approximately 200,000 km^2, which we detected based on the abundance and density of the data. We considered three probabilities (P10, P50, and P90) for each reservoir to determine the geological storage efficiency factor in both areas. For the integrity and safety of CO_2 storage, we interpreted and evaluated impermeable rock units that were denoted as seals. Although the seismic interpretation indicated no significant faults in the Lower Cretaceous section, the uniform lateral stratigraphy was a considerable concern due to the lack of wells in the study area. Table 4 demonstrates the required parameters that were applied in Equation (1) to identify the probabilistic estimate of the efficiency factor from P10 to P90 in Equation (2), as shown in Table 6.

Table 6. The probabilities from P10 to P90 of the ratio parameters that were considered in Equation (2) to identify the efficiency factor (E) at the formation scale for dolomite reservoirs of the Lower Cretaceous section.

Lithology	Net-to-Total Area ($E_{An/At}$)		Net-to-Gross Thickness ($E_{Hn/Hg}$)		Effective-to-Total Porosity ($E_{Øe/Øt}$)		Volumetric Displacement (E_v)		Microscopic Displacement (E_d)	
	P_{10}	P_{90}	P_{10}	P_{90}	P_{10}	P_{90}	P_{10}	P_{90}	P_{10}	P_{90}
Dolomite	0.2	0.8	0.17	0.68	0.53	0.71	0.26	0.43	0.57	0.64

Table 7 shows the results of the quantitative estimates of CO_2 storage capacity in the local and regional potential storage resources for the Lower Cretaceous potential reservoirs. The total capacity of the three storage resources with a geological storage efficiency (E) of dolomite between 0.65 and 5.40% ranges between 48.98 and 376.70 Mt of CO_2 at P10 to P90 for the local area (Figure 8), and the E ranges between 450.85 and 4705.46 Mt of CO_2 for the regional area (Figure 9).

Table 7. Volumetric CO_2 storage capacity (GCO_2) in Mt with the storage efficiency factor (E%) at P10, P50, and P90 for the three Lower Cretaceous reservoirs within local and regional zones in the Mid–South Atlantic Ocean.

Zone	Reservoir	Storage Resource (Mt)			Storage Efficiency (%)		
		P_{10}	P_{50}	P_{90}	P_{10}	P_{50}	P_{90}
Local	1	19.12	60.45	146.57	0.69	2.19	5.31
	2	14.85	47.77	117.57	0.67	2.17	5.34
	3	15.01	51.08	122.56	0.65	2.2	5.28
Regional	1	182.63	635.83	1628.72	0.65	2.18	5.25
	2	88.54	414.75	1574.37	0.67	2.22	5.4
	3	179.67	597.97	1502.37	0.67	2.17	5.3

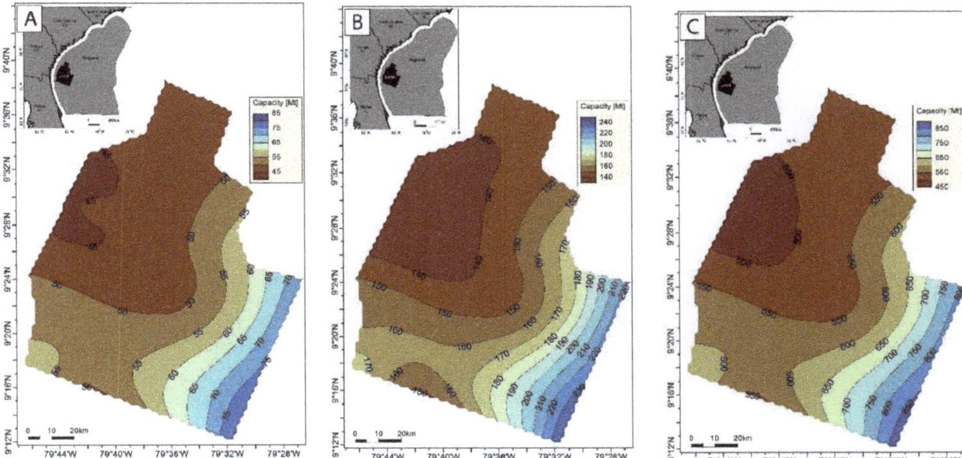

Figure 8. The total CO_2 storage capacity (in gigatons) for the Lower Cretaceous section: (A) the total capacity at P10, (B) P50, and (C) P90 in the local area.

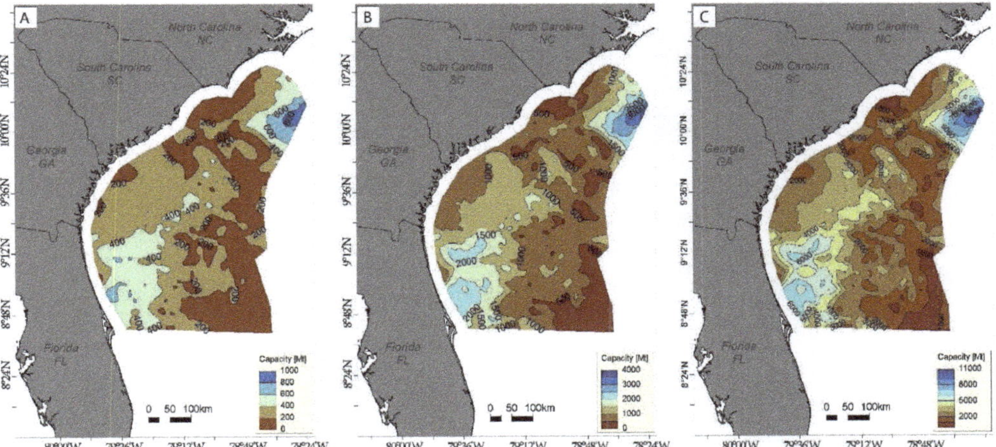

Figure 9. The total CO_2 storage capacity (in gigatons) for the Lower Cretaceous section: (A) the total capacity at P10, (B) P50, and (C) P90 in the regional area.

At P50, the average CO_2 storage in the Lower Cretaceous section is approximately 1.65 Gt. Reservoir R1 has a maximum storage capacity value of 0.63 Gt of CO_2. R2 is greater than 0.42 Gt of CO_2, and R3 is greater than 0.59 Gt of CO_2 at P50.

5. Summary and Conclusions

This work presents the first comprehensive study to identify and evaluate the CO_2 storage potential of the Lower Cretaceous section of the Mid–South Atlantic offshore southeastern United States. Based on the analysis of three wells in the Southeast Georgia Embayment, the CO_2 geological storage capacity estimate provides evidence of three significant permeable storage strata that are isolated by impermeable seals in the depth interval of 1767.84–2529.84 m. Based on an analysis of the COST GE-1 well, we identified wide and unsystematic porosity ranges from 0.17 to 0.32, and permeabilities between 2.1×10^{-13} and 5.43×10^{-13} m^2 (low percentage of sandstone and high percentage of dolomite). These layers are suitable reservoir rocks qualified for permanent CO_2 storage.

The US DOE methodology was used for calculating pore volume spaces to estimate the geological CO_2 storage potential capacity in megatons (Mt). The CO_2 storage potential of the Lower Cretaceous section was calculated based on the rock compositions and petrophysical properties at the COST GE-1 well. Three potential reservoirs were associated with four potential seals that were characterized and assessed. According to Scholle [9], the trapping mechanism indicated by an overlying seal involves stratigraphic variations. The prospective storage resources of the three reservoirs were calculated locally, where seismic profiles and well data were densely concentrated in the Southeast Georgia Embayment (10,000 km^2), and regionally, where we suggested a regional storage resource of 200,000 km^2. We considered three probability values (P10, P50, and P90) of each reservoir for determining the geological storage efficiency factor in both areas. This study suggests that the CO_2 storage capacity ranges approximately from 48.98 to 376.70 Mt locally, and from 450.85 to 4705.46 Mt regionally, in the three Lower Cretaceous reservoirs, with geological storage efficiencies from 0.65% to 5.4%.

The average storage potential is approximately 82 tons of CO_2, which could be safely stored per 1 km^2 offshore of the Lower Cretaceous section, at a probability of 0.5. The largest CO_2 storage capacity value for reservoir R1 was >3.2 tons/km^2. The intermediate and lowest values at P50 in reservoirs R3 and R2 were less than or equal to 2.7 tons/km^2. Since the reservoir heterogeneity impacts the pressure distribution, and the CO_2 plume migration is significantly affected by the permeability, we suggest that reservoir R2 and seals S2 and S3 can provide additional protection for safe injection and storage in case unpredictable leakage occurs due to unexpected natural hazards.

The uncertainty associated with subsurface data gaps was incorporated into the storage resource evaluation due to the legacy of seismic data and the relatively limited well data available over the study area.

Author Contributions: Formal analysis, D.S.A.; Investigation, D.S.A.; Supervision, J.H.K.; Writing—original draft, D.S.A.; Writing—review & editing, J.H.K. and C.K. All authors have read and agreed to the published version of the manuscript.

Funding: This research received no external funding.

Institutional Review Board Statement: This study did not require ethical.

Informed Consent Statement: Not applicable.

Data Availability Statement: https://walrus.wr.usgs.gov/namss/search/ (accessed on 1 June 2022).

Acknowledgments: We thank the Iraqi Ministry of Higher Education and Scientific Research (MOHESR), the Marine Science Center (MSC)–Basrah University, and the Iraqi Cultural office in Washington DC for funding and supporting the first author. This material is based upon work supported by the US Department of Energy National Energy Technology Laboratory, through a Cooperative Agreement DE-FE0026086 with the Southern States Energy Board. Cost sharing and research support were provided by the Project Partners and an Advisory Committee. The SOSRA team also includes the Virginia Polytechnic Institute and State University/Virginia Center for Coal and Energy Research; the Virginia Department of Mines, Minerals, and Energy; the South Carolina Geological Survey; Oklahoma State University; the Geological Survey of Alabama; the State Oil and Gas Board of Alabama; and Advanced Resources International, Inc. In addition, we would like to thank Schlumberger for providing access to Petrel software, and we are grateful to the Schlumberger technical team for their support. We would like to thank the US Department of Energy National Energy Technology Laboratory and Angela Goodman for providing further information about the DOE approach and CO_2-SCREEN techniques. We also thank the University of South Carolina and Susannah Boote for assistance in downloading and organizing the database at the University of South Carolina.

Conflicts of Interest: The authors declare no conflict of interest.

References

1. Metz, B.; Davidson, O.; De Coninck, H.; Loos, M.; Meyer, L. *IPCC Special Report on Carbon Dioxide Capture and Storage*; Intergovernmental Panel on Climate Change: Geneva, Switzerland, 2005.
2. Thomas, D.C.; Benson, S.M. *Carbon Dioxide Capture for Storage in Deep Geologic Formations-Results from the CO_2 Capture Project: Vol 1-Capture and Separation of Carbon Dioxide from Combustion, Vol 2-Geologic Storage of Carbon Dioxide with Monitoring and Verification*; Elsevier: Amsterdam, The Netherlands, 2005.
3. Hart, P.E. Namss—A National Archive of Marine Seismic Surveys. In Proceedings of the 2007 GSA Denver Annual Meeting, Denver, CO, USA, 28–31 October 2007.
4. Solomon, S.; Carpenter, M.; Flach, T.A. Intermediate storage of carbon dioxide in geological formations: A technical perspective. *Int. J. Greenh. Gas Control.* **2008**, *2*, 502–510. [CrossRef]
5. Hertel, T.W.; Golub, A.A.; Jones, A.D.; O'Hare, M.; Plevin, R.J.; Kammen, D.M. Effects of US maize ethanol on global land use and greenhouse gas emissions: Estimating market-mediated responses. *BioScience* **2010**, *60*, 223–231. [CrossRef]
6. Hortle, A.; Michael, K.; Azizi, E. Assessment of CO_2 storage capacity and injectivity in saline aquifers–comparison of results from numerical flow simulations, analytical and generic models. *Energy Procedia* **2014**, *63*, 3553–3562. [CrossRef]
7. Cumming, L.; Gupta, N.; Miller, K.; Lombardi, C.; Goldberg, D.; ten Brink, U.; Schrag, D.; Andreasen, D.; Carter, K. Mid-Atlantic US Offshore Carbon Storage Resource Assessment. *Energy Procedia* **2017**, *114*, 4629–4636. [CrossRef]
8. Okwen, R.; Yang, F.; Frailey, S. Effect of geologic depositional environment on CO_2 storage efficiency. *Energy Procedia* **2014**, *63*, 5247–5257. [CrossRef]
9. Scholle, P.A. *Geological Studies of the COST GE-1 Well, United States South Atlantic outer Continental Shelf Area*; USGS: Denver, CO, USA, 1979.
10. Smyth, R.C.a.H.; Susan, D.; Meckel, T.; Breton, C.; Paine, J.G.; Hill, G.R.; Andrews, J.R.; Lakshminarasimhan, S.; Herzog, H.; Zhang, H.H.; et al. Potential Sinks for Geologic Storage of Carbon Dioxide Generated in the Carolinas. US Bureau of Econ Geol, South Carolina, United States Summary Report 2007. Available online: http://citeseerx.ist.psu.edu/viewdoc/download?doi=10.1.1.470.1152&rep=rep1&type=pdf (accessed on 15 May 2016).
11. Zhou, Q.; Birkholzer, J.T.; Tsang, C.-F.; Rutqvist, J. A method for quick assessment of CO_2 storage capacity in closed and semi-closed saline formations. *Int. J. Greenh. Gas Control* **2008**, *2*, 626–639. [CrossRef]
12. Schrag, D.P. Storage of carbon dioxide in offshore sediments. *Science* **2009**, *325*, 1658–1659. [CrossRef]
13. Esposito, R.A.; Pashin, J.C.; Hills, D.J.; Walsh, P.M. Geologic assessment and injection design for a pilot CO_2-enhanced oil recovery and sequestration demonstration in a heterogeneous oil reservoir: Citronelle Field, Alabama, USA. *Environ. Earth Sci.* **2010**, *60*, 431–444. [CrossRef]
14. Almutairi, K.F. Assessment of upper cretaceous strata for Offshore CO_2 storage: Southeastern United States. Ph.D. Thesis, University of South Carolina, Columbia, SC, USA, 2018. Available from ProQuest Dissertations and Theses Full Text. Available online: https://scholarcommons.sc.edu/etd/4811 (accessed on 14 January 2019).
15. Fukai, I.; Keister, L.; Ganesh, P.R.; Cumming, L.; Fortin, W.; Gupta, N. Carbon dioxide storage resource assessment of Cretaceous- and Jurassic-age sandstones in the Atlantic offshore region of the northeastern United States. *Environ. Geosci.* **2020**, *27*, 25–47. [CrossRef]
16. Goodman, A.; Hakala, A.; Bromhal, G.; Deel, D.; Rodosta, T.; Frailey, S.; Small, M.; Allen, D.; Romanov, V.; Fazio, J. US DOE methodology for the development of geologic storage potential for carbon dioxide at the national and regional scale. *Int. J. Greenh. Gas Control* **2011**, *5*, 952–965. [CrossRef]
17. Gray, K. *Carbon Utilization and Storage Atlas*; Southern States Energy Board: Peachtree Corners, GA, USA, 2012.
18. Warwick, P.D.; Blondes, M.S.; Brennan, S.T.; Corum, M.D.; Merrill, M.D. US Geological survey geologic carbon dioxide storage resource assessment of the United States. *Energy Procedia* **2013**, *37*, 5275–5279. [CrossRef]
19. Levine, J.S.; Fukai, I.; Soeder, D.J.; Bromhal, G.; Dilmore, R.M.; Guthrie, G.D.; Rodosta, T.; Sanguinito, S.; Frailey, S.; Gorecki, C. US DOE NETL methodology for estimating the prospective CO_2 storage resource of shales at the national and regional scale. *Int. J. Greenh. Gas Control* **2016**, *51*, 81–94. [CrossRef]
20. Dillon, W.P.; Paull, C.K.; Buffler, R.T.; Fail, J.-P. Rifted Margins. In *Structure and Development of the Southeast Georgia Embayment and Northern Blake Plateau: Preliminary Analysis*; USGS: Denver, CO, USA, 1979.
21. Dillon, W.P.; Klitgord, K.D.; Paull, C.K. *Mesozoic Development and Structure of the Continental Margin off South Carolina*; USGS: Denver, CO, USA, 1983; Volume 1313.
22. Maher, J.C.; Applin, E.R. *Geologic Framework and Petroleum Potential of the Atlantic Coastal Plain and Continental Shelf*; 2330-7102; United States Government Publishing Office: Washington, DC, USA, 1971; p. 98.
23. Poag, C.W. Stratigraphy of the Atlantic continental shelf and slope of the United States. *Annu. Rev. Earth Planet. Sci.* **1978**, *6*, 251–280. [CrossRef]
24. Pinet, P.R.; Popenoe, P. A scenario of Mesozoic-Cenozoic ocean circulation over the Blake Plateau and its environs. *Geol. Soc. Am. Bull.* **1985**, *96*, 618–626. [CrossRef]
25. Poppe, L.J.; Popenoe, P.; Poag, C.W.; Swift, B.A. Stratigraphic and palaeoenvironmental summary of the south-east Georgia Embayment: A correlation of exploratory wells. *Mar. Pet. Geol.* **1995**, *12*, 677–690. [CrossRef]
26. Dalziel, I.W.; Dalla Salda, L.H.; Gahagan, L.M. Paleozoic Laurentia-Gondwana interaction and the origin of the Appalachian-Andean mountain system. *Geol. Soc. Am. Bull.* **1994**, *106*, 243–252. [CrossRef]

27. Badley, M.; Price, J.; Dahl, C.R.; Agdestein, T. The structural evolution of the northern Viking Graben and its bearing upon extensional modes of basin formation. *J. Geol. Soc.* **1988**, *145*, 455–472. [CrossRef]
28. Dillon, W.P.; Popenoe, P. The Blake Plateau Basin and Carolina Trough. *Geol. N. Am.* **1988**, *2*, 291–328.
29. Dillon, W.P.; Popenoe, P.; Grow, J.A.; Klitgord, K.D.; Swift, B.A.; Paull, C.K.; Cashman, K.V. Rifted Margins: Field Investigations of Margin Structure and Stratigraphy. In *Growth Faulting and Salt Diapirism: Their Relationship and Control in the Carolina Trough, Eastern North America*; American Association of Petroleum Geologists: Tulsa, OK, USA, 1982.
30. Lizarralde, D.; Holbrook, W.S.; Oh, J. Crustal structure across the Brunswick magnetic anomaly, offshore Georgia, from coincident ocean bottom and multi-channel seismic data. *J. Geophys. Res. B* **1994**, *99*, 21741–21757. [CrossRef]
31. Boote, S.K.; Knapp, J.H. Offshore extent of Gondwanan Paleozoic strata in the southeastern United States: The Suwannee suture zone revisited. *Gondwana Res.* **2016**, *40*, 199–210. [CrossRef]
32. Roth, W.M. Problem-centered learning for the integration of mathematics and science in a constructivist laboratory: A case study. *Sch. Sci. Math.* **1993**, *93*, 113–122. [CrossRef]
33. Schlumberger. Schlumberger Petrel Manual 2014. 2016. Available online: http://isegunque.ddns.net/236.html2016 (accessed on 11 April 2016).
34. Gorecki, C.D.; Holubnyak, Y.; Ayash, S.; Bremer, J.M.; Sorensen, J.A.; Steadman, E.N.; Harju, J.A. A new classification system for evaluating CO_2 storage resource/capacity estimates. In Proceedings of the SPE International Conference on CO_2 Capture, Storage, and Utilization, San Diego, CA, USA, 2–4 November 2009.
35. Gorecki, C.D.; Sorensen, J.A.; Bremer, J.M.; Knudsen, D.; Smith, S.A.; Steadman, E.N.; Harju, J.A. Development of storage coefficients for determining the effective CO_2 storage resource in deep saline formations. In Proceedings of the SPE International Conference on CO_2 Capture, Storage, and Utilization, San Diego, CA, USA, 2–4 November 2009.
36. Brennan, S.T.; Burruss, R.C.; Merrill, M.D.; Freeman, P.A.; Ruppert, L.F. A probabilistic assessment methodology for the evaluation of geologic carbon dioxide storage. *US Geol. Surv. Open-File Rep.* **2010**, *1127*, 31. [CrossRef]
37. Teletzke, G.; Palmer, J.; Drueppel, E.; Sullivan, M.B.; Hood, K.; Dasari, G.; Shipman, G. Evaluation of practicable subsurface CO_2 storage capacity and potential CO_2 transportation networks, onshore North America. In Proceedings of the 14th Greenhouse Gas Control Technologies Conference Melbourne, Melbourne, Australia, 21–26 October 2018; pp. 21–26.
38. Ennis-King, J.; Paterson, L. Reservoir engineering issues in the geological disposal of carbon dioxide. In Proceedings of the Fifth International Conference on Greenhouse Gas Control Technologies, Cairns, Australia, 13–16 August 2000; pp. 290–295.
39. Holloway, S. Carbon dioxide capture and geological storage. *Philos. Trans. R. Soc. A* **2007**, *365*, 1095–1107. [CrossRef] [PubMed]
40. Tissot, B.P.; Welte, D.H. An Introduction to Migration and Accumulation of Oil and Gas. In *Petroleum Formation and Occurrence*; Springer: Berlin/Heidelberg, Germany, 1978; pp. 257–259.
41. Bachu, S. Screening and ranking of sedimentary basins for sequestration of CO_2 in geological media in response to climate change. *Environ. Geol.* **2003**, *44*, 277–289. [CrossRef]
42. Chadwick, A.; Arts, R.; Bernstone, C.; May, F.; Thibeau, S.; Zweigel, P. *Best Practice for the Storage of CO_2 in Saline Aquifers-Observations and Guidelines from the SACS and CO_2 STORE Projects*; British Geological Survey: Nottingham, UK, 2008; Volume 14.
43. DOE-NETL. Best Practices for Risk Analysis and Simulation for Geologic Storage of CO_2; DOE/NETL-2011/1459; March 2011. National Energy Technology Laboratory. Available online: https://www.academia.edu/13753393/Best_Practices_for_Risk_Analysis_and_Simulation_for_Geologic_Storage_of_CO2 (accessed on 14 January 2019).
44. van der Meer, L.B.; Yavuz, F. CO_2 storage capacity calculations for the Dutch subsurface. *Energy Procedia* **2009**, *1*, 2615–2622. [CrossRef]
45. Gray, K. *Carbon Sequestration Atlas of the United States and Canada*; Southern States Energy Board: Peachtree Corners, GA, USA, 2010.
46. Preston, C.; Whittaker, S.; Rostron, B.; Chalaturnyk, R.; White, D.; Hawkes, C.; Johnson, J.; Wilkinson, A.; Sacuta, N. IEA GHG Weyburn-Midale CO_2 monitoring and storage project–moving forward with the Final Phase. *Energy Procedia* **2009**, *1*, 1743–1750. [CrossRef]
47. Sanguinito, S.M.; Goodman, A.; Levine, J. *NETL CO_2 Storage ProspeCtive Resource Estimation Excel aNalysis (CO_2-SCREEN) User's Manual*; National Energy Technology Lab.: Pittsburgh, PA, USA, 2017.
48. Sanguinito, S.; Goodman, A.L.; Sams, J.I., III. CO_2-SCREEN tool: Application to the oriskany sandstone to estimate prospective CO_2 storage resource. *Int. J. Greenh. Gas Control.* **2018**, *75*, 180–188. [CrossRef]
49. Liu, E.; Li, X.Y.; Chadwick, A. Multicomponent Seismic Monitoring of CO_2 Gas Cloud in the Utsira Sand: A feasibility Study: Saline Aquifer CO_2 Storage Phase 2 (SACS2): Work Area 5 (Geophysics): Feasibility of Multicomponent Seismic Acquisition. 2001; Unpublished.
50. Chadwick, R.; Williams, G.; Noy, D. CO_2 storage: Setting a simple bound on potential leakage through the overburden in the North Sea Basin. *Energy Procedia* **2017**, *114*, 4411–4423. [CrossRef]

 MDPI

Article

System Performance Analyses of Supercritical CO_2 Brayton Cycle for Sodium-Cooled Fast Reactor

Min Xie [1], Jian Cheng [2], Xiaohan Ren [3,*], Shuo Wang [4], Pengcheng Che [1] and Chunwei Zhang [1]

1 HE National Engineering Research Center of Power Generation Equipment, Harbin 150028, China; xiemin@harbin-electric.com (M.X.); chepc@harbin-electric.com (P.C.); zhangcw@harbin-electric.com (C.Z.)
2 Harbin Electric International Company Limited, Harbin 150028, China; chengjian@china-hei.com
3 Institute of Thermal Science and Technology, Shandong University, Jinan 250061, China
4 State Key Laboratory of Efficient and Clean Coal-Fired Utility Boilers, Harbin Boiler Company Limited, Harbin 150046, China; 57626@163.com
* Correspondence: renxh@sdu.edu.cn

Abstract: The system performance of the supercritical CO_2 Brayton cycle for the Sodium Fast Reactor with a partial-cooling layout was studied, and an economic analysis was carried out. The energetic, exergetic, and exergoeconomic analyses are presented, and the optimized results were compared with the recompression cycle. The sensitivity analyses were conducted by considering the variations in the pressure ratios and inlet temperatures of the main compressor and the turbine. The exergy efficiency of the partial-cooling cycle reached 63.65% with a net power output of 34.39 MW via optimization. The partial-cooling cycle obtained a minimum total cost rate of 2230.36 USD/h and exergy efficiency of 63.65% when the pressure ratio was equal to 3.50. The inlet temperature of the main compressor was equal to 35 °C, and the inlet temperature of the turbine was equal to 480 °C. The total cost of recuperators decreased with the increase in the pressure ratio and the inlet temperatures of the main compressor. In addition, the total cost of recuperator could be reduced by increasing the outlet temperature of the turbine. The change in cost from exergy loss and destruction with the pressure ratio was substantially larger than with the inlet temperature of the turbine or the main compressor. Manipulating the pressure ratio is an essential method to guarantee good economy of the system. Moreover, capital investment, operation, and maintenance costs normally accounted for large proportions of the total cost rate, being almost double the cost from the exergy loss and destruction occurring in each condition.

Keywords: partial-cooling cycle; supercritical CO_2 Brayton cycle; advanced fast reactors; exergoeconomic analyses; CO_2 utilization

Citation: Xie, M.; Cheng, J.; Ren, X.; Wang, S.; Che, P.; Zhang, C. System Performance Analyses of Supercritical CO_2 Brayton Cycle for Sodium-Cooled Fast Reactor. *Energies* **2022**, *15*, 3555. https://doi.org/10.3390/en15103555

Academic Editor: Francesco Frusteri

Received: 17 March 2022
Accepted: 28 April 2022
Published: 12 May 2022

1. Introduction

The supercritical CO_2 (sCO_2) cycle was proposed in a patent that used CO_2 as the working fluid [1] in 1968, but this cycle slowly developed due to the technology limitation at that time. The sCO_2 cycle is currently regarded as having the potential to work effectively with multiple heat sources. The first known business unit of the CO_2 cycle is EPS-100, and many studies on the waste heat recovery have also been conducted [2,3]. Numerous countries have also launched different scales of experimental projects. The R&D of sCO_2 Brayton technology is generally divided into several phases from the system design to the pilot station operation. In addition to EPS-100, the SunShot Program in the USA, the Hero-sCO_2 Program and the flex-sCO_2 Project in the European Union, and the 5-MWe sCO_2 pilot station in China are available. The application of the sCO_2 Brayton cycle has advantages of a high energy density, compactness, and high efficiency due to the good thermophysical properties of sCO_2 [4]. In addition to pure CO_2, using a CO_2-based binary mixture may be used for raising the critical temperature [5,6].

The sCO_2 cycle is believed to be applied for fossil combustion power [7,8], heat recovery [9,10], solar thermal power [11,12], and generation IV nuclear reactors [13,14], and can even be combined with other cycles [15]. Among these heat sources, the new-generation nuclear reactor is regarded as a major source of sustainable energy, with a tremendous capacity, near-zero emissions, and multipurpose products [16]. Previous studies [17] revealed that the sCO_2 Brayton cycle can be combined with the helium-cooled fast reactor [18,19], the sodium-cooled fast reactor (SFR) [20], the lead–bismuth fast reactor, and the high-temperature gas reactor. The SFR, which is the fastest-growing reactor type, has several demonstration reactors in different countries. Therefore, it is regarded as the quickest route to applying the sCO_2. As a heat source, the SFR has a limited range at moderate temperatures. The sCO_2 Brayton cycle has been demonstrated to be well-suited to a combination with the SFR in previous studies [21]. The specifications considering the temperature of the secondary sodium loop are as follows: 550 °C/395 °C with the Advanced Fast Reactor (AFR) [22], 520 °C/335 °C with the Japan Atomic Energy Agency Sodium Fast Reactor (JSFR) [23], and 520 °C/350 °C with the China Experimental Fast Reactor (CEFR). Considering power class, CEFR-1200 has four cycles with 300 MW per cycle, AFR-100 can output 100 MWe, and the electricity output of JSFR is 1500 MWe with two loops. A certain number of kWe/Mwe-class space reactors are also currently being developed. The sCO_2 Brayton cycle fits the temperature gap of the secondary sodium inlet and outlet at 155–185 °C. Moreover, sCO_2 could avoid sodium–water interactions considering the safety aspect and operational advantages. Thus, replacing the steam Rankine cycle with the sCO_2 Brayton cycle with high conventional loop pressure could increase the safety of the system. In general, combined with the SFR, the sCO_2 Brayton cycle is used as a conventional loop, which is well-adapted to the reactor [21] regarding the temperature gaps and ranges of heat exchange processes [17]. The AFR–100 can reach a higher efficiency of 42.3% and 104.9 MW net power with the sCO_2 Brayton cycle [21].

The recuperation cycle is the preliminary layout of the sCO_2 Brayton cycle, as shown in Figure 1. The conventional loop includes the Na-sCO_2 heat exchanger as the heat source, turbine, recuperator, cooler, and compressor. Unlike the steam Rankine cycle, the sCO_2 Brayton cycle must have a recuperation process to avoid heat waste of the high-temperature working fluid from the turbine outlet. The system performance in terms of the cycle efficiencies of the recuperation cycle remains unsatisfactory. A recuperation cycle with a concise structure was adopted as a reference for the system design and performance analysis. The recompression cycle, the partial-cooling cycle, and even substantially complex structures were considered to realize an improved system performance [24–26]. Aside from layouts, operation parameters, such as working fluid pressures, temperatures, and recuperated points, affect the system efficiency remarkably [27].

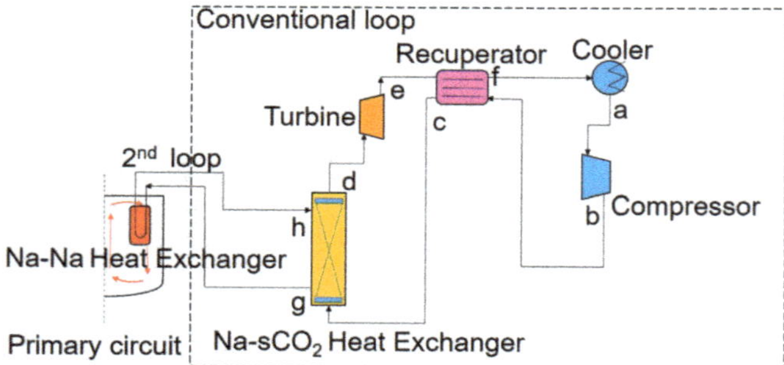

Figure 1. Layouts of the SFR-sCO_2 recuperation cycle. a–h represent state points of working fluid in the cycle.

Moreover, applying the sCO_2 Brayton cycle is expected to reduce the capital cost per unit output electrical power of the nuclear power plant or the levelized cost of electricity [17]. Exergy analysis shows that the highest exergy loss takes place in the heliostat field, at nearly 42.5% of incident solar exergy, with a recompression Brayton cycle with partial-cooling and an improved heat recovery (RBC-PC-IHR) configuration [28]. Exergoeconomic analysis was applied to sCO_2 power systems with several configurations of the turbine inlet temperature and cooling system. The recompression concept and the novel cycle showed the best performance [29]. With the recompression cycle, the maximum exergy efficiency reached about 55% in previous reports, and that using the CO_2-C_7H_8 binary mixture was nearly 60% [5]. Previous research has already performed exergoeconomic investigation of the sCO_2-SFR with the recompression cycle [30]. However, the exergoeconomics of the partial-cooling cycle of the SFR has not been presented.

The system performance and economic analyses of the sCO_2 Brayton cycle, particularly for the SFR with the partial-cooling layout, are discussed in the current research. The energetic, exergetic, and exergoeconomic analyses are presented, and the optimized results are compared with the recompression cycle. The sensitivity analyses mainly focus on the effect of the pressure and temperature ranges.

2. Analysis Methods

2.1. Energetic Analysis

The sCO_2 power cycle includes heat exchangers, turbomachinery, pipes, and valves. The balance equation of heat exchangers is as follows:

$$\dot{Q}_{HX} = \dot{m}_{hot}\Delta h_{hot} = \dot{m}_{cold}\Delta h_{cold} \tag{1}$$

The balance equation of turbomachinery is as follows:

$$\dot{W} = \dot{m}_{CO_2}\Delta h \tag{2}$$

The thermal efficiency of the cycle is as follows:

$$\eta_{cyc} = \frac{\dot{W}_{net}}{\dot{Q}_{InCycle}} \tag{3}$$

where $\dot{Q}_{InCycle}$ represents the heat transferred to the sCO_2 via the Na-sCO_2 heat exchanger, and $\dot{W}_{net}$ is the net power outputted from the cycle. The pressure loss of pipes and valves is not considered.

2.2. Exergetic Analysis

The exergetic analysis elucidates the route and the amount of the exergy lost with the same simulation results [4]. The exergy of the working fluid stream can be expressed as follows:

$$e_{sCO_2} = h - h_a - T_a(s - s_a) \tag{4}$$

The exergy rate of the working fluid stream can be expressed as follows:

$$\dot{E}_{sCO_2} = \dot{m}_{sCO_2} e_{sCO_2} \tag{5}$$

The exergy loss for heat exchangers can be calculated by the following:

$$\dot{E}_{loss} = \Delta\dot{E}_{hot} - \Delta\dot{E}_{cold} = \left(\dot{E}_{hot,in} - \dot{E}_{hot,out}\right) - \left(\dot{E}_{cold,out} - \dot{E}_{cold,in}\right) \quad (6)$$

For the turbine, the exergy loss can be calculated as follows:

$$\dot{E}_{loss,turbine} = \Delta\dot{E}_{turbine} - \dot{W}_{turbine} = \dot{E}_{turbine,in} - \dot{E}_{turbine,out} - \dot{W}_{turbine} \quad (7)$$

The exergy loss for compressors can be calculated as follows:

$$\dot{E}_{loss,comp} = \dot{W}_{comp} - \Delta\dot{E}_{comp} = \dot{W}_{comp} - \left(\dot{E}_{comp,out} - \dot{E}_{comp,in}\right) \quad (8)$$

The exergy efficiency of the cycle is computed as follows:

$$\eta_{exe,cyc} = \frac{\dot{W}_{net}}{\dot{E}_{InCycle}} \quad (9)$$

where $\dot{E}_{InCycle}$ represents the exergy transferred to the sCO_2 via the Na-sCO_2 heat exchanger.

2.3. Exergoeconomic Analysis

Economic analysis is an essential aspect of evaluating technical applications. Exergoeconomic analysis is used for the establishment of a guideline of the cost per exergy unit. Specifically, the exergy costing method, which is simple and direct, is widely used in the field of exergoeconomic analysis [31–33]. The exergy costing method was selected for this study, and the results were analyzed in combination with energetic and exergetic calculations.

The cost balance equation for the *k*th component in a power conversion system is as follows [30]:

$$\sum\left(c_{out}\dot{E}_{out}\right)_k + c_{W,k}\dot{W}_k = \sum\left(c_{in}\dot{E}_{in}\right)_k + c_{q,k}\dot{E}_k + \dot{Z}_k \quad (10)$$

The cost rate of the overall system ($\dot{C}_{total}$) is the sum of capital investment, operation and maintenance costs ($\dot{Z}$, in Table 1), and the cost derived from exergy loss and destruction ($\dot{C}_{L+D}$), as follows:

$$\dot{C}_{total} = \dot{Z} + \dot{C}_{L+D} \quad (11)$$

Table 1. Cited data for cost of components.

Component	Cited Data for Cost of Components
Reactor core [34]	$\dot{Z}_k = c_{core} \times \dot{Q}_r$
Turbine [35]	$\dot{Z}_k = 479.34\left(\frac{\dot{m}_{in}}{0.93-\eta_{Turb}}\right)\cdot ln(\beta)\cdot(1+e^{0.036T_{in}-54.4})$
Compressors [35]	$\dot{Z}_k = 71.1\left(\frac{\dot{m}_{in}}{0.93-\eta_{Comp}}\right)\cdot\beta\cdot ln(\beta)$
Recuperators [36]	$\dot{Z}_k = 2681A^{0.59}$
Coolers [36]	$\dot{Z}_k = 2143A^{0.514}$
Reactor core [33,37–39]	c_{fuel} = 7.4 USD/(MW·h)

The definitions of "fuel" and "product" considering the partial-cooling cycle are summarized in Table 2. In addition, the solution for a set of equations is provided in Table 3.

Table 2. Fuel–product definition of the partial-cooling cycle.

Component	Fuel	Product
Reactor core (including Na-sCO_2 Heat Exchanger)	$\dot{E}_c + \dot{E}_{fuel}$	$\dot{E}_d$
Turbine	$\dot{E}_d - \dot{E}_e$	$\dot{W}_T$
Recuperator HT	$\dot{E}_e - \dot{E}_j$	$\dot{E}_c - \dot{E}_i$
Recuperator LT	$\dot{E}_j - \dot{E}_f$	$\dot{E}_n - \dot{E}_m$
Main Cooler	$\dot{E}_f - \dot{E}_a$	$\Delta\dot{E}_{Mcooler,water}$
Subcooler	$\dot{E}_k - \dot{E}_l$	$\Delta\dot{E}_{Scooler,water}$
Main Compressor	$\dot{W}_{MC}$	$\dot{E}_b - \dot{E}_a$
Recompressor LT	$\dot{W}_{RCLT}$	$\dot{E}_m - \dot{E}_l$
Recompressor HT	$\dot{W}_{RCHT}$	$\dot{E}_h - \dot{E}_g$

Table 3. Exergetic cost rate balance and auxiliary equations for the partial-cooling cycle.

Component	Balance Equation	Auxiliary Equation(s)
Reactor core (including Na-sCO_2 Heat Exchanger)	$\dot{C}_d = \dot{C}_{fuel} + \dot{C}_c + \dot{Z}_{core}$	-
Turbine	$\dot{C}_e + \dot{C}_T = \dot{C}_d + \dot{Z}_T$	$\dot{C}_e/\dot{E}_e = \dot{C}_d/\dot{E}_d$
Recuperator HT	$\dot{C}_j + \dot{C}_c = \dot{C}_e + \dot{C}_i + \dot{Z}_{RHT}$	$\dot{C}_j/\dot{E}_j = \dot{C}_e/\dot{E}_e$ $\dot{C}_i = \dot{C}_n + \dot{C}_h$
Recuperator LT	$\dot{C}_f + \dot{C}_n = \dot{C}_j + \dot{C}_m + \dot{Z}_{RLT}$	$\dot{C}_f/\dot{E}_f = \dot{C}_j/\dot{E}_j$
Main Cooler	$\dot{C}_a + \dot{C}_{Mcooler} = \dot{C}_f + \dot{Z}_{Mcooler}$	$\dot{C}_a/\dot{E}_a = \dot{C}_f/\dot{E}_f$
Subcooler	$\dot{C}_l + \dot{C}_{Scooler} = \dot{C}_k + \dot{Z}_{Scooler}$	$\dot{C}_l = \dot{C}_k$ $\dot{C}_k = x\dot{C}_b$ $\dot{C}_g = (1-x)\dot{C}_b$
Main Compressor	$\dot{C}_b = \dot{C}_{MC} + \dot{C}_a + \dot{Z}_{MC}$	$\dot{C}_{MC}/\dot{W}_{MC} = \dot{C}_T/\dot{W}_T$
Recompressor LT	$\dot{C}_m = \dot{C}_{RCLT} + \dot{C}_l + \dot{Z}_{RCLT}$	$\dot{C}_{RCLT}/\dot{W}_{RCLT} = \dot{C}_T/\dot{W}_T$
Recompressor HT	$\dot{C}_h = \dot{C}_{RCHT} + \dot{C}_g + \dot{Z}_{RCHT}$	$\dot{C}_{RCHT}/\dot{W}_{RCHT} = \dot{C}_T/\dot{W}_T$

Following the previous research [30], the primary circuit, the second loop, and the Na-sCO_2 heat exchanger were categorized as the "Reactor core". $\dot{C}_{fuel}$ represents the fuel cost rate of the entire part and can be calculated as follows:

$$\dot{C}_{fuel} = c_{fuel}\dot{Q}_{core}, \tag{12}$$

where c_{fuel} is the fuel cost (USD/(MW·h)). c_{fuel} was set as 7.4 USD/(MW·h) based on the mean value comparison of the constant dollar unit costs between the fast reactors and the gas-cooled reactor [33,37–39].

2.4. Solution Procedures

A system simulation program (The original program was developed by Harbin Electric Company Limited, and the Software Copyright Registration Number is 2018SR349409) was developed with the functions energy and exergy balance calculations, parameter sensitivity analyses, and system performance optimizations. Multiobjective optimization was performed using a genetic algorithm developed in Python 3 (Figure 2), which was crosschecked with an enumeration algorithm. A genetic algorithm begins with randomly selected values of parameters (as the initial population). Then, each parameter is evaluated via fitness calculation to test its ability to solve a problem. The selection operation selects some of the parameters for reproduction. The biological crossing over and recombination of parameters are conducted in coupling and mutation operation. Then, a new population is used in a new calculation until the operation objectives are attained. Table 4 shows the manipulated parameters in the optimization. The optimization aims to obtain the maximum value of $\eta_{exe,cyc}$ or the minimum value of C_{total}. The split ratios in the cycle are adjusted based on the optimization process in the range of 0.1–0.9 to obtain the best result. It is one of the optimization parameters, and it is manipulated as other parameters in the genetic algorithm and the enumeration algorithm.

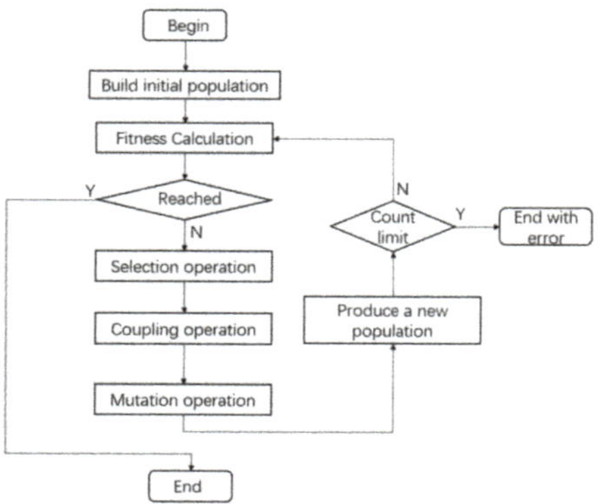

Figure 2. Flow diagram of genetic algorithm method.

Table 4. Manipulated parameters in the optimization.

Parameter	Symbol	Range	Step Size
Maximum pressure of the cycle	p_H	[15.00, 28.00] MPa	0.50 MPa
Minimum pressure of the cycle	p_L	[8.00, 12.00] MPa	0.50 MPa
Mass flow rate of the sCO_2	$\dot{m}_{sCO2}$	[300.00, 450.00] kg/s	1.00 kg/s
Maximum temperature of the cycle	T_{max}	[430.00, 480.00] °C	1.00 °C
Minimum temperature of the cycle	T_{min}	[32.00, 38.00] °C	1.00 °C
Split ratios	r	[0.1, 0.9]	0.05

3. Results and Discussions

3.1. Energy and Exergy Analyses with Different Layouts

The heat and work distribution of the recuperation cycle based on the energy and exergy analyses in previous studies is shown in Figure 3a. The heat flow map indicates that the heat flow rate into the cycle ($\dot{Q}_{in}$) was set as the reference quantity of energy. The power input was $0.137\dot{Q}_{in}$ when $1\dot{Q}_{in}$ energy came into the cycle to support the sCO_2 compression. As the output energy, the turbine output $0.473\dot{Q}_{in}$ as power, while the energy released $0.664\dot{Q}_{in}$ from the cooler. The heat flow rate of the recuperator in the recuperation process, which is an enormous heat exchange process, was $1.497\dot{Q}_{in}$. The cooler wasted a considerable amount of energy according to the heat flow map. However, the discharging energy from the cooler could not be reused due to the low pressure and temperature.

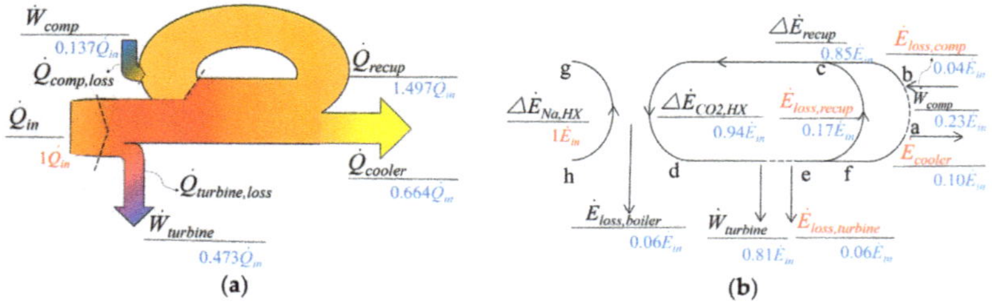

Figure 3. Energy and exergy distributions of the recuperation cycle [25]. (**a**) Heat and power, (**b**) Exergy and exergy loss. A–h represent state points of working fluid in the cycle.

The exergy flow map and the exergy losses of the process to improve the cycle structure are illustrated and shown in Figure 3b. The exergy flow rate into the cycle ($\dot{E}_{in}$) was set as the reference quantity of exergy in the exergy flow map. The input exergies were $1\dot{E}_{in}$ and $0.23\dot{E}_{in}$ from the heat source and the compression process, respectively. The main output exergy was $0.81\dot{E}_{in}$ via the turbine. Exergy losses mainly occurred in the recuperation, cooling, and heat exchange in the heat source processes as $0.17\dot{E}_{in}$, $0.10\dot{E}_{in}$, and $0.06\dot{E}_{in}$, respectively. Most exergy losses occurred in the heat exchange processes; thus, many types of cascade cycles could be adopted to reduce the exergy losses. Among these cycles, the recompression cycle, which is known as the Feher cycle, had good performance application in the sCO_2 Brayton cycle with the SFR, as shown in Figure 4a. Moreover, the partial cooling in Figure 4b had a relatively simple structure and good performance compared to the compression cycle. The energy and exergy performance of the compression cycle were analyzed in the previous research. The current research focuses on the study of the partial-cooling cycle and completes its comparison with the compression cycle.

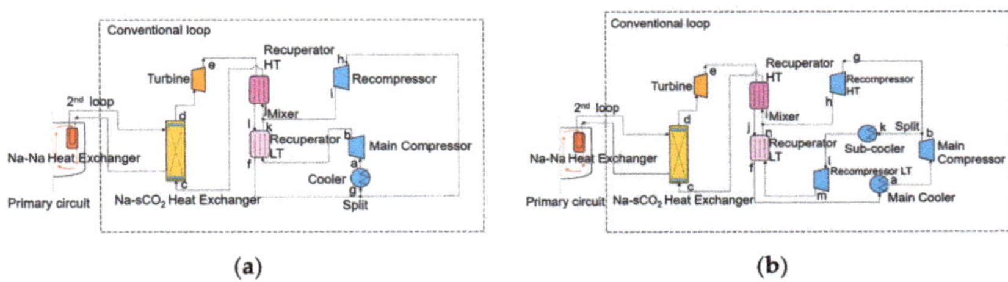

Figure 4. Advanced cycle layouts of sCO_2-SFR. (**a**) Recompression cycle, (**b**) partial cooling cycle. a–n represent state points of working fluid in the cycle.

3.2. System Performance of the Partial-Cooling Cycle

The optimization of the sCO_2-SFR with the partial-cooling layout at the highest exergy efficiency reaching 63.65% with a net power output of 34.39 MW is shown in Figure 5. A series of parameter limitations was set during the optimization based on the engineering practice, as listed in Table 4.

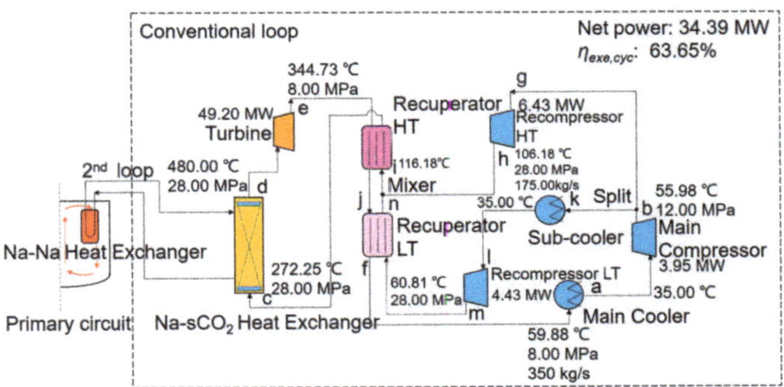

Figure 5. Optimal simulation results of the partial-cooling cycle with sCO_2-SFR. a–n represent state points of working fluid in the cycle.

Figure 6 shows the *T-s* diagram of the recompression [30] and partial-cooling cycle with optimization results. Compared with the recompression cycle, the compression processes are more complex, including three parts. In the partial-cooling cycle, the pressure of the working fluid was raised in the main compressor (a–b). A part of sCO_2 was cooled in the subcooler (k–l), and the recompressor LT further raised the working fluid pressure (l–m), and then the temperature of sCO_2 was increased in the recuperator LT (m–n). The other part of sCO_2 from the main compressor went into the recompressor HT (g–h) without the cooling process. Both sCO_2 streams entered the mixer with roughly comparable temperatures. sCO_2 went into the recuperator HT to recycle the heat (i–c) after the mixing, and then the working fluid absorbed heat from the heat source of the conventional loop (c–d). Afterward, the working fluid stream with a high temperature and pressure expanded in the turbine (d–e) outputting power. Processes e–j, j–f, and f–a are the heat discharges in the recuperator HT, the recuperator LT, and the main cooler, respectively.

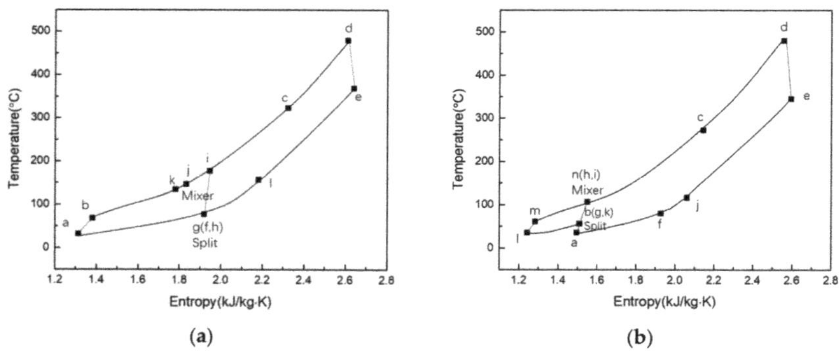

Figure 6. *T-s* diagram at the optimum point with the advanced cycles. (**a**) Recompression cycle, (**b**) partial cooling cycle. a–n represent state points of working fluid in the cycle.

The following observations are presented in Table 5: a 3.57 MW exergy loss from the heat source of the conventional loop (6.61%$\Delta\dot{E}_{in}$), a 5.82 MW exergy loss from the recuperators (10.77%$\Delta\dot{E}_{in}$), a 4.38 MW exergy loss from the coolers (8.10%$\Delta\dot{E}_{in}$), a 3.60 MW exergy loss from the turbine (6.67%$\Delta\dot{E}_{in}$), and a 2.27 MW exergy loss from the compressors (4.21%$\Delta\dot{E}_{in}$). Table 5 also reveals that the exergy losses from the Na-sCO_2 heat exchanger were greater in the partial-cooling cycle than in the recompression cycle due to the wide temperature gap between the two sides of the exchanger. Moreover, additional exergy loss was observed from these coolers due to the subcooler application in the partial-cooling cycle.

Table 5. Comparisons of exergy losses between the partial-cooling and recompression cycles.

Items	Partial Cooling		Recompression	
	Power (MW)	Ratio (%)	Power (MW)	Ratio (%)
$\dot{E}_{loss,Na-sCO2}$	3.57	6.61	2.06	4.23
$\dot{E}_{loss,recup}$	5.82	10.77	5.02	10.28
$\dot{E}_{cooler}$	4.38	8.10	3.17	6.49
$\dot{E}_{loss,turbine}$	3.60	6.67	3.48	7.13
$\dot{E}_{loss,comp}$	2.27	4.21	2.37	4.85

3.3. *Effects of Operation Parameters on Cycle Efficiencies*

The cycle efficiencies are discussed on the basis of sensitivity analysis results. The sensitivity of efficiencies to the main system parameters, including the pressure ratio ($\beta = p_H/p_L$), the inlet temperature of the main compressor (T_{min}), and the inlet temperature of the turbine (T_{max}), was analyzed.

3.3.1. Pressure Ratio

β was set in the range of [2.00, 3.50] with the $p_{comp,in}$ = 8.00 MPa, T_{min} = 35.00 °C, T_{max} = 480.00 °C, and $\dot{m}_{CO_2}$ = 350.00 kg/s. Theoretically, the maximum β could be reached at a high p_H. However, realizing a high p_H is difficult under the current situation of compressor technology. Therefore, only the calculation with the β range in [2.00, 3.50] is shown in this section.

The cycle efficiencies (η_{cyc} and $\eta_{exe,cyc}$) of the partial-cooling cycle were generally lower than those of the recompression cycle, as shown in Figure 7, but the gap was narrowed with the increase in β. The effect tendencies of β on η_{cyc} and $\eta_{exe,cyc}$ were based on the partial-cooling cycle rise with the increase in β from 32.51% to 36.53% and from 54.90% to 63.65%, respectively. If the value of β can be raised, then superior system efficiencies could be reached. Unlike the partial-cooling cycle, the maximum value of η_{cyc} at β = 2.75 was 37.96%, and that of $\eta_{exe,cyc}$ at β = 3.00 was 64.84% as observed in the specified β range in the recompression cycle. Therefore, a high β would not elicit a superior system performance in the recompression cycle case.

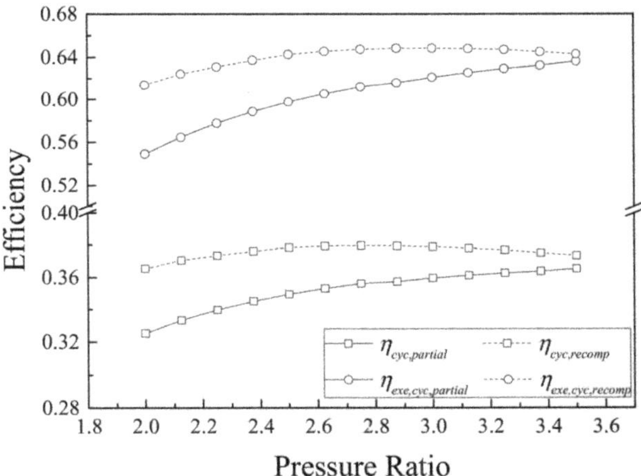

Figure 7. Effects of β on η_{cyc} and $\eta_{exe,cyc}$ with the partial-cooling cycle compared with the recompression cycle.

3.3.2. Inlet Temperature of the Main Compressor

T_{min} cannot be less than 32.00 °C (close to the critical point) because only the supercritical Brayton cycle was considered in this study. The range of T_{min} was decided to fall within [32.00, 36.00] °C with β = 2.75, $p_{comp,in}$ = 8.00 MPa, T_{max} = 480.00 °C, and $\dot{m}_{CO_2}$ = 428.57 kg/s. Figure 8 shows that the cycle efficiencies of the partial-cooling cycle were better stabilized under T_{min} than the recompression cycle. η_{cyc} increased initially and decreased from 35.94% to 36.06% (maximum value, at T_{min} = 35.00 °C), then to 35.37% under the partial-cooling cycle. Similarly, $\eta_{exe,cyc}$ initially rose from 61.68% to 62.31% (maximum value, at T_{min} = 35.00 °C) and then dropped to 61.70%. As discussed in Section 3.3.1, the decreasing T_{min} not only reduced the outlet temperature of the compressor, leading to high efficiencies, but also enhanced the heat loss from the subcooler, resulting in energy and exergy losses of steam due to the application of the subcooler. Optimized T_{min}, which could maximize the efficiencies, was elicited by the aforementioned influences.

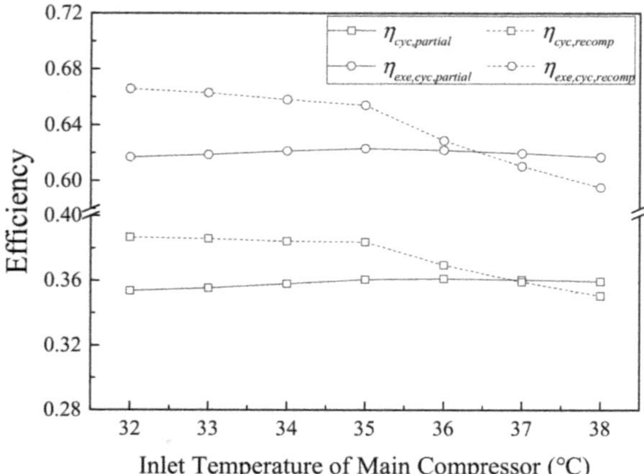

Figure 8. Effects of T_{min} on η_{cyc} and $\eta_{exe,cyc}$ with the partial-cooling cycle compared with the recompression cycle.

3.3.3. Inlet Temperature of Turbine

Figure 9 shows that the range of T_{max} was decided to fall within [430.00, 480.00] °C with β = 3.5, $p_{comp,in}$ = 8.00 MPa, T_{min} = 35.00 °C, and $\dot{m}_{CO_2}$ = 350.00 kg/s. η_{cyc} and $\eta_{exe,cyc}$ linearly increased with T_{max} regardless of the partial-cooling or the recompression cycle. η_{cyc} increased by 2.79% from 33.74% to 36.53% under the partial-cooling cycle, whereas $\eta_{exe,cyc}$ dramatically rose by 5.19% from 58.46% to 63.65%. These variations indicate that T_{max} is essential to obtain high η_{cyc} and $\eta_{exe,cyc}$. The highest T_{max} reached in the current research was 480 °C due to the temperature limit of the SFR core (the outlet temperature of the core is close to 550 °C).

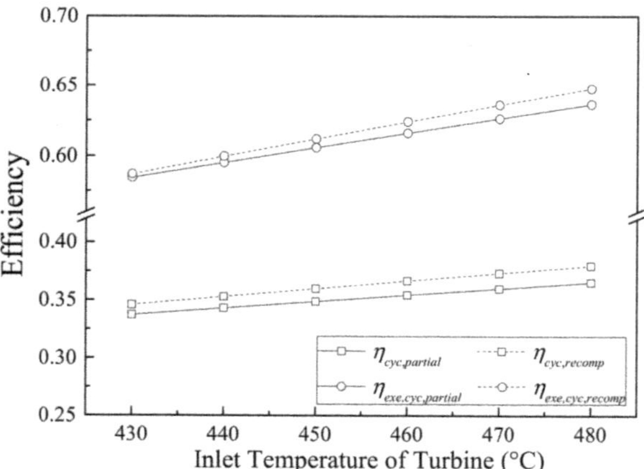

Figure 9. Effects of T_{max} on η_{cyc} and $\eta_{exe,cyc}$ with the partial-cooling cycle compared with the recompression cycle.

The partial-cooling cycle showed a good system performance with the optimization results and the sensitivity analysis. The exergoeconomic of the two layouts will be analyzed in the next section for further comparison.

3.4. Exergoeconomic Discussion

The effects of system parameters on exergoeconomic are initially analyzed in this section. The exergoeconomic optimized results of the partial-cooling cycle are then summarized and compared with the recompression cycle.

3.4.1. Pressure Ratio

Pressure ratio (β) is a major factor causing cost variation. A large β results in a high outlet pressure of the compressor with the same inlet pressure. Thus, a high outlet pressure would provide additional power support to the compressor. Moreover, the cost of the compressor and the turbine increased in accordance with the equations in Table 1. Figure 10 shows that β was within the range [2.00, 3.50] with $p_{comp,in}$ = 8.00 MPa, T_{min} = 35.00 °C, T_{max} = 480.00 °C, and $\dot{m}_{CO_2}$ = 428.57 kg/s. $\dot{C}_{total}$ dropped 261.19 USD/h from 2491.55 USD/h to 2230.36 USD/h with the partial-cooling cycle. Meanwhile, both efficiencies continuously increased.

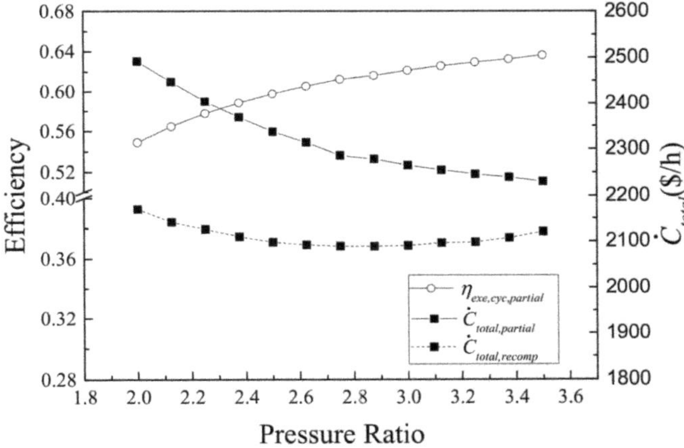

Figure 10. Effects of β on $\dot{C}_{total}$ with the partial-cooling cycle compared with the recompression cycle.

$\dot{C}_{total}$ comprises capital investment, operation, and maintenance costs ($\dot{Z}$), and the cost from exergy loss and destruction ($\dot{C}_{L+D}$). The effect of β on $\dot{C}_{L+D}$ is generally larger than that on $\dot{Z}$, and the variation in $\dot{C}_{L+D}$ with β introduces changes in $\dot{C}_{total}$. Considering $\dot{Z}$, the increase in β indicates that the maximum cycle pressure would be high, further causing additional $\dot{Z}$ of compressors and turbines. Therefore, $\dot{Z}$ slightly increased with the rise in β, as shown in Figure 11. Unlike $\dot{Z}$, the variation in $\dot{C}_{L+D}$ is complicated due to exergy loss and destruction in the partial-cooling cycle, and $\dot{C}_{L+D}$ significantly dropped initially and then declined from 984.35 USD/h to 660.20 USD/h.

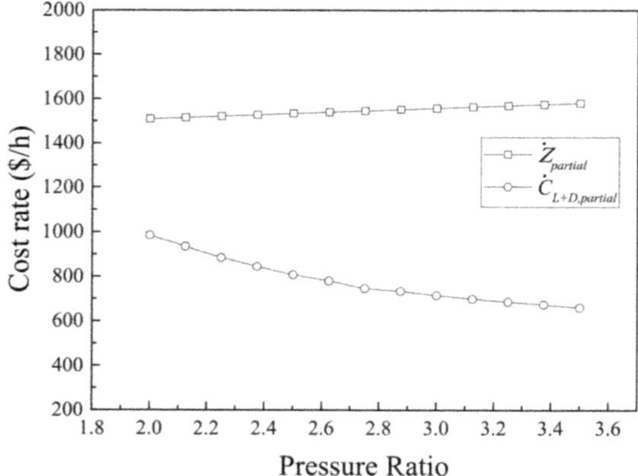

Figure 11. Effects of β on $\dot{Z}$ and $\dot{C}_{L+D}$ with the partial-cooling cycle.

$\dot{C}_{component}$ is the total cost of each component, such as the compressor, the cooler, the turbine, the recuperator, and the Na-sCO_2 heat exchanger. As seen in Figure 12, the total cost of the Na-sCO_2 heat exchanger barely changed with β maintaining a value of around 1570–1580 USD/h. The total costs of other components are shown in Figure 12. Turbines and compressors had a lower cost at lower β. By contrast, the total cost of recuperators decreased with increasing β. With higher β, the exergy of the working fluid at the outlet of turbine decreased, as did the energy needed for recuperation. Therefore, the total cost of recuperator was reduced.

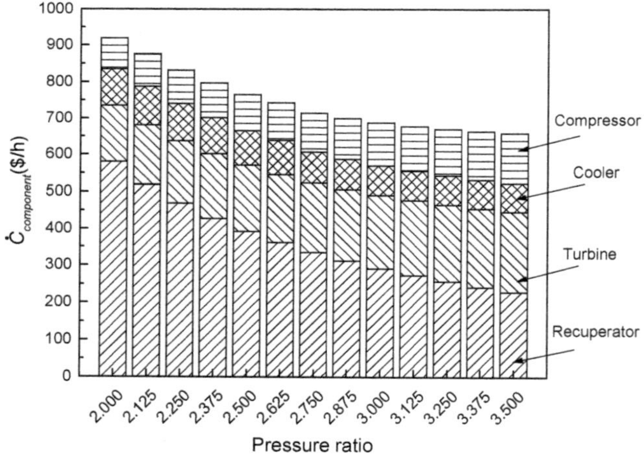

Figure 12. Effects of β on $\dot{C}_{component}$ with the partial-cooling cycle.

3.4.2. Inlet Temperature of the Main Compressor

Considering the effect of T_{min}, the range of T_{min} was decided to fall within [32.00, 38.00] °C when β = 3.00, $p_{comp,in}$ = 8.00 MPa, T_{max} = 480.00 °C, and $\dot{m}_{CO_2}$ = 350.00 kg/s. The $\dot{C}_{total}$ of the partial-cooling cycle was significantly higher than that of the recompression cooling cycle. This value decreased from 2308.44 USD/h to 2249.50 USD/h when T_{min} increased from 32.00 °C to 36.00 °C, and then remained around 2250 USD/h with the increase in T_{min}, as shown in Figure 13.

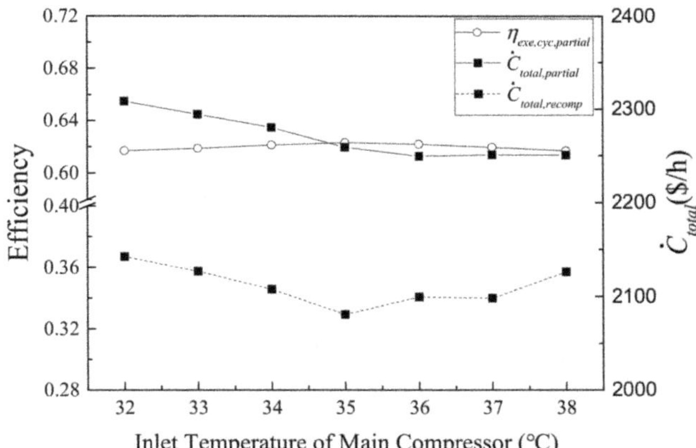

Figure 13. Effects of T_{min} on $\dot{C}_{total}$ with the partial-cooling cycle compared with the recompression cycle.

Figure 14 shows that the variation in $\dot{C}_{total}$ was caused by the change in $\dot{C}_{L+D}$, and $\dot{Z}$ barely changed with T_{min}. As the inlet of the compressor, T_{min} increased the outlet temperature of the compressor with the same pressure ratio in the range of [32.00, 38.00] °C in the partial-cooling cycle. However, temperatures were not the major influencing factors on $\dot{Z}$ according to the equations in Table 1. By contrast, T_{min} can affect the variations in exergy loss and destruction in the entire system. $\dot{C}_{L+D}$ decreased by 60.55 USD/h from 759.31 USD/h to 698.76 USD/h when T_{min} varied from 32.00 °C to 36.00 °C, and then remained at approximately 700 USD/h. $\dot{C}_{L+D}$ continued to decrease with the increase in T_{min} when T_{min} was less than 35.00 °C.

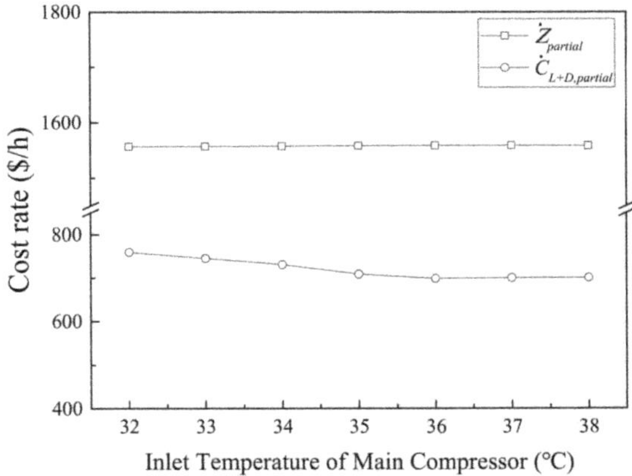

Figure 14. Effects of T_{min} on $\dot{Z}$ and $\dot{C}_{L+D}$ with the partial-cooling cycle.

Figure 15 shows the variation in component total cost with T_{min}. T_{min} hardly affected the total cost of turbines and compressors. The total cost of the Na-sCO_2 heat exchanger barely changed with β maintained around 1564–1584 USD/h. The total cost of the recuperator continued to decrease with the increase in T_{min}. During the compression process, a higher outlet temperature of the compressor could be obtained with higher T_{min}. The former narrowed the temperature gap of the recuperators. Therefore, the heat exchange capacity and the cost of recuperators were reduced.

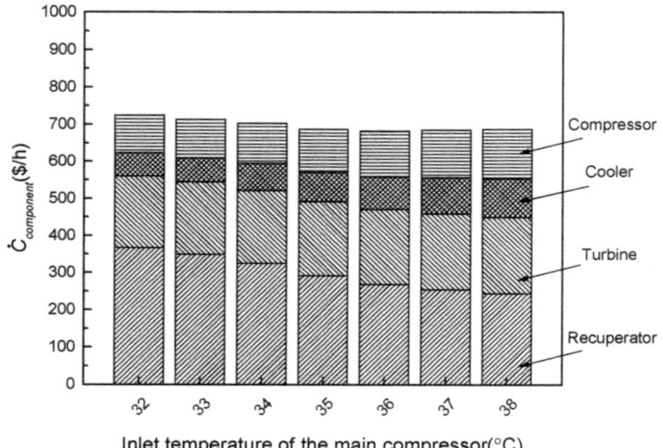

Figure 15. Effects of T_{min} on $\dot{C}_{component}$ with the partial-cooling cycle.

3.4.3. Inlet Temperature of Turbine

Figure 16 shows that the range of T_{max} was decided to fall within [430.00, 480.00] °C when $\beta = 3.50$, $p_{comp,in} = 8.00$ MPa, $T_{min} = 35.00$ °C, and $\dot{m}_{CO_2} = 350.00$ kg/s. The $\dot{C}_{total}$ variation lines linearly declined with the increase in T_{max}, from 2251.67 USD/h to 2088.88 USD/h (for the recompression cycle) [30] and from 2392.85 USD/h to 2230.36 USD/h (for the partial-cooling cycle). Increasing T_{max} was one of the most effective approaches to reducing the total cost considering system efficiencies. However, the heat source temperature for SFR limited the T_{max} of the sCO_2 cycle. High T_{max} could be reached in the future with the optimization of the SFR core design and the performance enhancements of the Na-sCO_2 heat exchanger.

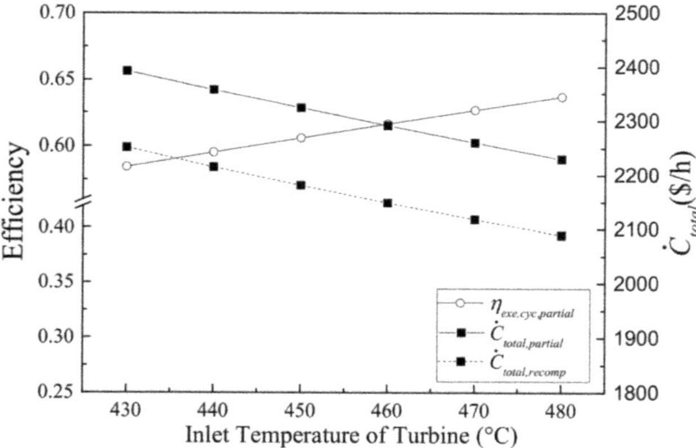

Figure 16. Effects of T_{max} on η_{cyc}, $\eta_{exe,cyc}$, and $\dot{C}_{total}$ with the partial-cooling cycle compared with the recompression cycle.

Similar to the effect of β and T_{min}, Figure 17 shows that $\dot{Z}$ barely changed with T_{max}. $\dot{C}_{L+D}$ dropped because the increase in T_{max} reduced the exergy loss and the destruction of the cycle from 823.42 USD/h to 660.20 USD/h for the partial-cooling cycle.

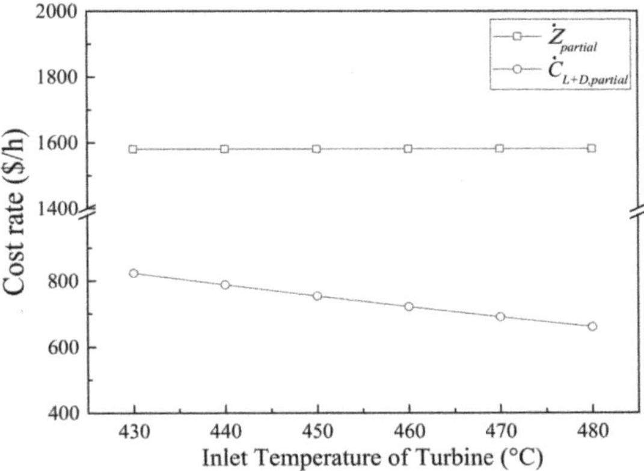

Figure 17. Effects of T_{max} on $\dot{Z}$ and $\dot{C}_{L+D}$ with the partial-cooling cycle.

Figure 18 shows the variation in component total costs with T_{max}. T_{max} barely affected the total cost of turbines and compressors. The total cost of the compressor, cooler, turbine, and recuperator barely changed with T_{max}. However, the total cost of the Na-sCO_2 heat exchanger was significantly affected by T_{max}. The total cost of the recuperator decreased by increasing T_{max}; the former dropped about 155.36 USD/h when T_{max} changed from 430 °C to 480 °C.

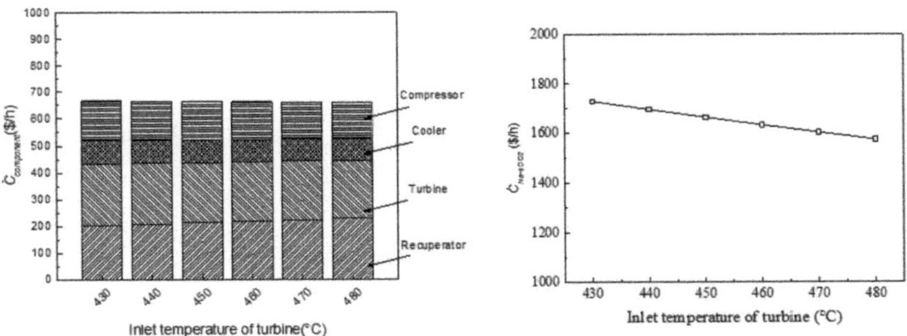

Figure 18. Effects of T_{max} on $\dot{C}_{component}$ with the partial-cooling cycle. (**a**) Multiple components, (**b**) Na-sCO_2 heat exchanger.

The comparison among β, T_{min}, and T_{max} revealed that the change in $\dot{C}_{L+D}$ with β was substantially larger than that with T_{max} or T_{min}. Thus, manipulating β is important to guarantee the good economy of the system. Moreover, the sum of capital investment, operation, and maintenance costs ($\dot{Z}$) accounted for large proportions of the total cost rate, which was almost double that of $\dot{C}_{L+D}$ in each condition.

3.5. Comparison Summary of the Two Layouts

The partial-cooling cycle showed good system performance with high efficiency and net power via the optimization process based on the operation conditions of the SFR in this study.

Compared with the recompression cycle, the exergy losses from the Na-sCO_2 heat exchanger were greater in the partial-cooling cycle due to the wide temperature gap between the two sides of the exchanger. Moreover, the coolers demonstrated additional exergy losses due to the subcooler application in the partial-cooling cycle.

Furthermore, the recompression cycle could slightly improve the exergoeconomic performance compared to the partial-cooling cycle. The minimum $\dot{C}_{total}$ of 2230.36 USD/h with $\eta_{exe,cyc}$ = 63.65% could be obtained for the partial-cooling cycle, where β = 3.50, T_{min} = 35.00 °C, and T_{max} = 480.00 °C. Similarly, the minimum $\dot{C}_{total}$ reached 2088.88 USD/h with $\eta_{exe,cyc}$ = 64.72% at β = 2.75 for the recompression cycle [30]. Meanwhile, the partial-cooling cycle showed good potential for reducing the total cost by increasing the pressure ratio with the development of sCO_2 technology.

4. Conclusions

The current research investigated the system performance and carried out economic analysis of the sCO_2 Brayton cycle for the SFR with a partial-cooling layout. The exergy efficiency of the partial-cooling cycle reached 63.65% with a net power output of 34.39 MW after optimization, and the minimum $\dot{C}_{total}$ was 2230.36 USD/h. Similarly, the minimum $\dot{C}_{total}$ reached 2088.88 USD/h with $\eta_{exe,cyc}$ = 64.72% at β = 2.75 for the recompression cycle. Compared with the recompression cycle, the exergy losses from the Na-sCO_2 heat exchanger were greater in the partial-cooling cycle due to the wide temperature gap between its two sides. Moreover, the coolers had additional exergy losses due to the subcooler application in the partial-cooling cycle.

The cost derived from exergy loss and destruction in the selected ranges of parameters were significantly affected by the pressure ratio and the inlet temperatures of the main compressor and the turbine. The change in cost from exergy loss and destruction with the pressure ratio was substantially greater than that from the inlet temperature of the turbine or the inlet temperature of the main compressor. The total cost of recuperators decreased with the increase in β, T_{min}. In addition, the total cost of the recuperator could be reduced by increasing T_{max}. Therefore, manipulating β is important to guarantee the good economy of the system. Moreover, the capital investment and operation and maintenance costs normally accounted for large proportions of the total cost rate, which was almost double the cost from exergy loss and destruction in each condition. The partial-cooling cycle obtained the minimum total cost rate when the pressure ratio was equal to 3.50. The inlet temperatures of the main compressor and the turbine were 35 °C and 480 °C, respectively.

Author Contributions: Writing—original draft preparation and investigation, M.X.; investigation and resource, J.C.; supervision, X.R.; writing—review and editing, S.W.; Visualization, P.C.; project administration, C.Z. All authors have read and agreed to the published version of the manuscript.

Funding: This research was funded by the National Key Research and Development Program of China grant number 2020YFF0218100 and Natural Science Foundation of Heilongjiang Province of China grant number YQ2020E032. And the APC was funded by the National Key Research and Development Program of China grant number 2020YFF0218100.

Institutional Review Board Statement: Not applicable.

Informed Consent Statement: Not applicable.

Data Availability Statement: Not applicable.

Acknowledgments: We thank Zhongyuan Ma of the Nuclear Power Division, Harbin Electric Company Limited, for his assistance with the Sodium Fast Reactor System.

Conflicts of Interest: The authors declare no conflict of interest.

Nomenclature

Symbols

A	heat transfer area (m^2)
β	compressor pressure ratio
c	cost rate per exergy unit (USD/GJ)
$\dot{C}$	cost rate (USD/h)
e	specific exergy (kJ/kg)
$\dot{E}$	rate of exergy (MW)
h	specific enthalpy (kJ/kg)
$\dot{m}$	mass flow rate (kg/s)
p	pressure (MPa)
$\dot{Q}$	rate of heat (MW)
$\dot{W}$	rate of work (MW)
s	specific entropy (kJ/(kg·K))
T	temperature (°C)
η	efficiency
$\dot{Z}$	capital cost rate (USD/h)

Subscript

a	setting of environment for analysis
cyc	cycle
comp	compressor
exe	exergy
MC	main compressor
Na-sCO2	Na-sCO_2 heat exchanger
partial	partial-cooling cycle
RC	recompressor
recomp	recompression cycle
recup	recuperators
RHT	recuperator HT
RLT	recuperator LT
Turb	turbine

References

1. Gianfranco, A. Method for Obtaining Mechanical Energy from a Thermal Gas Cycle with Liquid Phase Compression. U.S. Patent No. 3376706, 9 April 1968.
2. Nami, H.; Mahmoudi, S.M.S.; Nemati, A. Exergy, ecomomic and environmental impact assessment and optimization of a novel cogeneration system including a gas turbine, a supercritical CO_2 and an organic Rankine cycle (GT-HRSG/SCO_2). *Appl. Therm. Eng.* **2017**, *110*, 1315–1330. [CrossRef]
3. Chaudhary, A.; Mulchand, A.; Dave, P. Feasibility Study of Supercritical CO2 Rankine Cycle for Waste Heat Recovery, Proceedings. In Proceedings of the 6th International Symposium-Supercritical CO2 Power Cycles, Pittsburgh, PA, USA, 27–29 March 2018.
4. Xie, M.; Xie, Y.H.; He, Y.C.; Dong, A.H.; Zhang, C.W.; Shi, Y.W.; Zhang, Q.H.; Yang, Q.G. Thermodynamic analysis and system design of the supercritical CO_2 Brayton cycle for waste heat recovery of gas turbine. *Heat Transf. Res.* **2020**, *51*, 129–146. [CrossRef]
5. Siddiqui, M.E. Thermodynamic Performance Improvement of Recopression Brayton Cycle Utilizing CO2-C7H8 Binary Mixture. *Mechanika* **2021**, *27*, 259–264. [CrossRef]
6. Manzolini, G.; Binotti, M.; Bonalumi, D.; Invernizzi, C.; Iora, P. CO_2 mixtures as innovative working fluid in power cycles applied to solar plants. Techno-economic assessment. *Sol. Energy* **2019**, *181*, 530–544. [CrossRef]
7. Li, H.; Zhang, Y.; Yang, Y.; Han, W.; Yao, M.; Bai, W.; Zhang, L. Preliminary design assessment of supercritical CO2 cycle for commercial scale coal-fired power plants. *Appl. Therm. Eng.* **2019**, *158*, 113785. [CrossRef]
8. Shi, Y.; Zhong, W.; Shao, Y.; Xiang, J. System Analysis on Supercritical CO2 Power Cycle with Circulating Fluidized Bed Oxy-Coal Combustion. *J. Therm. Sci.* **2019**, *28*, 505–518. [CrossRef]
9. Siddiqui, M.E.; Almitani, K.H. Proposal and Thermodynamic Assessment of S-CO2 Brayton Cycle Layout for Improved Heat Recovery. *Entropy* **2020**, *22*, 305. [CrossRef]
10. Haroon, M.; Sheikh, N.A.; Ayub, A.; Tariq, R.; Sher, F.; Beheta, A.T.; Imran, M. Exergetic, Economic and Exergo-Environmental Analysis of Bottoming Power Cycles Operating with CO2-Based Binary Mixture. *Energies* **2020**, *13*, 5080. [CrossRef]
11. Garg, P.; Srinivasan, K.; Dutta, P.; Kumar, P. Comparison of CO_2, and Steam in Transcritical Rankine Cycles for Concentrated Solar Power. *Energy Procedia* **2014**, *49*, 1138–1146. [CrossRef]

12. Dunima, S.; Jahn, I.; Hooman, K.; Lu, Y.; Veeraragavan, A. Comparison of direct and indirect natural draft dry cooling tower cooling of the sCO_2 Brayton cycle for concentrated solar power plants. *Appl. Therm. Eng.* **2018**, *130*, 1070–1080. [CrossRef]
13. Handwerk, C.S.; Driscoll, M.J.; Hejzlar, P. Optimized Core Design of a Supercritical Carbon Dioxide-Cooled Fast Reactor. *Nucl. Technol.* **2007**, *164*, 3320–3336. [CrossRef]
14. Pope, M.A. Thermal hydraulic design of a 2400 MWth Direct Supercritical Co-Cooled Fast Reactor. Ph.D. Thesis, Institute of Technology, Cambridge, MA, USA, 2006.
15. Gao, W.; Yao, M.Y.; Chen, Y.; Li, H.; Zhang, Y.; Zhang, L. Performance of S-CO_2 Brayton Cycle and Organic Rankine Cycle (ORC) Combined System Considering the Diurnal Distribution of Solar Radiation. *J. Therm. Sci.* **2019**, *28*, 463–471. [CrossRef]
16. *Nuclear Development Report, Risks and Benefits of Nuclear Energy*; NEA No. 6242; OECD Nuclear Energy Agency: Paris, France, 2007.
17. Sienicki, J.J.; Moisseytsev, A. Nuclear Power. In *Fundamentals and Applications of Supercritical Carbon Dioxide (sCO_2) Based Power Cycles*; Brun, K., Friedman, P., Dennis, R., Eds.; Woodhead Publishing: Sawston, UK, 2017; pp. 339–391.
18. Hoffmann, J.R.; Feher, E.G. 150 kWe Supercritical Closed Cycle System. *ASME J. Eng. Power.* **1971**, *93*, 70–80. [CrossRef]
19. Feher, E.G.; Harvey, R.A.; Milligan, H.H. *150-kWe Supercritical Cycle Turbo-pump Design Layout and Performance Demonstration Final Technical Report*; McDonnell Douglas Astronautics Company-West: St. Louis, MO, USA, 1971.
20. Dievoet, V.J.P. The sodium-CO_2 fast breeder reactor concept. In Proceedings of the International Conference on Sodium Technology and Large Fast Reactor Design, Bologna, Italy, 7–9 November 1968.
21. Sienicki, J.J.; Moisseytsev, A.; Krajtl, L. A Supercritical CO_2 Brayton Cycle Power Converter for a Sodium-Cooled Fast Reactor Small Modular Reactor. In Proceedings of the ASME 2015 Nuclear Forum collocated with the ASME 2015 Power Conference, the ASME 2015 9th International Conference on Energy Sustainability, and the ASME 2015 13th International Conference on Fuel Cell Science, Engineering and Technology. American Society of Mechanical Engineers, San Diego, CA, USA, 28 June–2 July 2015. V001T01A004.
22. Grandy, C.; Sienicki, J.J.; Moisseytsev, A.; Krajtl, L.; Farmer, M.T.; Kim, T.K.; Middleton, B. *Advanced Fast Reactor—100 (AFR-100) Report for the Technical Review Panel*; U.S. Department of Commerce National Technical Information Service: Alexandra, VA, USA, 2014.
23. Masakazu, I.; Tomoyasu, M.; Shoji, K. A Next Generation Sodium-cooled Fast Reactor Concept and Its R&D Program. *Nucl. Eng. Technol.* **2007**, *3*, 171–186.
24. Moisseytsev, A.; Sienicki, J.J. Investigation of alternative layouts for the supercritical carbon dioxide Brayton cycle for a sodium-cooled fast reactor. *Nucl. Eng. Des.* **2009**, *239*, 1362–1371. [CrossRef]
25. Xie, M.; Xie, Y.H.; Zhang, Q.H.; He, Y.; Dong, A.; Zhang, C.; Shi, Y.; Yang, Q. Thermodynamic Analysis and Multiple System Layouts Comparison of the Supercritical CO2 Brayton Cycle for Sodium-cooled Fast Reactor with Offshore Scenes. In Proceedings of the International Conference on Energy, Ecology and Environment (ICEEE 2018) and International Conference on Electric and Intelligent Vehicles (ICEIV 2018), Melbourne, Australia, 21–25 November 2018.
26. Neises, T.; Turchi, C. Supercritical CO2 Power Cycles: Design Considerations for Concentrating Solar Power. In Proceedings of the The 4th International Symposium-Supercritical CO2 Power Cycles, Pittsburgh, PA, USA, 9–10 September 2014.
27. Lu, S.P.; Liu, Y.W.; Lin, Y.S. Parameters optimization of supercritical carbon dioxide Brayton cycle based on CEFR. *Energy Conserv.* **2018**, *7*, 21–26.
28. Siddiqui, M.E.; Almitani, K.H. Energy and Exergy Assessment of S-CO2 Brayton Cycle Coupled with a Solar Tower System. *Processes* **2020**, *8*, 1264. [CrossRef]
29. Noaman, M.; Saade, G.; Morosuk, T.; Tsataronis, G. Exergoeconomic Analysis Applied to Supercrtical CO_2 Power Systems. *Energy* **2019**, *183*, 756–765. [CrossRef]
30. Xie, M.; Xie, Y.; Zhang, Q.; Zhang, C.; Dong, A.; Shi, Y.; Zhang, Y. Exergy and Exergoeconomic Analyses Based on Recompression Cycle of the Supercritical CO_2 Brayton Cycle for Sodium-cooled Fast Reactor. In Proceedings of the International Conference on Energy, Ecology and Environment (ICEEE 2019) and International Conference on Electric and Intelligent Vehicles (ICEIV 2019), Stavanger, Norway, 23–27 July 2019.
31. Shokati, N.; Mohammadkhani, F.; Yari, M.; Mahmoudi, S.M.S.; Rosen, M.A. A comparative exergoeconomic analysis of waste heat recovery from a gas turbine-modular helium reactor via organic Rankine cycles. *Sustainability* **2013**, *6*, 2474–2489. [CrossRef]
32. Mohammadkhani, F.; Shokati, N.; Mahmoudi, S.M.S.; Yari, M.; Rosen, M.A. Exergoeconomic assessment and parametric study of a gas turbine-modular helium reactor combined with two organic rankine cycles. *Energy* **2014**, *65*, 533–543. [CrossRef]
33. Khaljani, M.; Saray, R.K.; Bahlouli, K. Comprehensive analysis of energy, exergy and exergo-economic of cogeneration of heat and power in a combined gas turbine and organic rankine cycle. *Energy Convers. Manag.* **2015**, *97*, 154–165. [CrossRef]
34. Schultz, K.R.; Brown, L.C.; Besenbruch, G.E.; Hamilton, C.J. *Large Scale Production of Hydrogen by Nuclear Energy for the Hydrogen Economy*; GA-A24265; General Atomics: San Diego, CA, USA, 2003.
35. Ghaebi, H.; Amidpour, M.; Karimkashi, S.; Rezayan, O. Energy, exergy and thermoeconomic analysis of a combined cooling, heating and power (CCHP) system with gas turbine prime mover. *Int. J. Energy Res.* **2011**, *35*, 697–709. [CrossRef]
36. Guo, Y.Z.; En, W.; Shan, T.T. Techno-economic study on compact heat exchangers. *Int. J. Energy Res.* **2008**, *32*, 1119–1127.
37. Dixon, B.; Ganda, F.; Hoffman, E.A.; Hansen, J. *Advanced Fuel Cycle Cost Basis—2017 Edition*; U.S. Department of Energy Fuel Cycle Options Campaign: Washington, DC, USA, 2017.

38. Wang, X.; Yang, Y.; Zheng, Y.; Dai, Y. Exergy and exergoeconomic analyses of a supercritical CO_2 cycle for a cogeneration application. *Energy* **2017**, *119*, 971–982. [CrossRef]
39. Zare, V.; Mahmoudi, S.M.S.; Yari, M. An exergoeconomic investigation of waste heat recovery from the Gas Turbine-Modular Helium Reactor (GT-MHR) employing an ammonia–water power/cooling cycle. *Energy* **2013**, *61*, 397–409. [CrossRef]

Article

Decoupling Investigation of Furnace Side and Evaporation System in a Pulverized-Coal Oxy-Fuel Combustion Boiler

Zixue Luo *, Zixuan Feng, Bo Wu and Qiang Cheng

State Key Laboratory of Coal Combustion, Huazhong University of Science and Technology, Wuhan 430074, China; m202171260@hust.edu.cn (Z.F.); m201470995@hust.edu.cn (B.W.); chengqiang@hust.edu.cn (Q.C.)
* Correspondence: luozixue@hust.edu.cn; Tel.: +86-27-8754-2417; Fax: +86-27-8754-5526

Abstract: A distributed parameter model was developed for an evaporation system in a 35 MW natural circulation pulverized-coal oxy-fuel combustion boiler, which was based on a computational fluid dynamic simulation and in situ operation monitoring. A mathematical model was used to consider the uneven distribution of working fluid properties and the heat load in a furnace to predict the heat flux of a water wall and the wall surface temperature corresponding to various working conditions. The results showed that the average heat flux near the burner area in the air-firing condition, the oxy-fuel combustion with dry flue gas recycling (FGR) condition, and the oxy-fuel combustion with wet flue-gas recycle condition were 168.18, 154.65, and 170.68 kW/m^2 at a load of 80%. The temperature and the heat flux distributions in the air-firing and the oxy-fuel combustion with wet FGR were similar, but both were higher than those in the oxygen-enriched combustion conditions with the dry FGR under the same load. This study demonstrated that the average metal surface temperature in the front wall during the oxy-fuel combustion condition was 3.23 °C lower than that under the air-firing condition. The heat release rate from the furnace and the vaporization system should be coordinated at a low and middle load level. The superheating surfaces should be adjusted to match the rising temperature of the flue gas while shifting the operation from air to oxy-fuel combustion, where the distributed parameter analytical approach could then be applied to reveal the tendencies for these various combustion conditions. The research provided a type of guidance for the design and operation of the oxy-fuel combustion boiler.

Keywords: oxy-fuel combustion; evaporation system; distributed parameter modeling; pulverized coal boiler

Citation: Luo, Z.; Feng, Z.; Wu, B.; Cheng, Q. Decoupling Investigation of Furnace Side and Evaporation System in a Pulverized-Coal Oxy-Fuel Combustion Boiler. *Energies* **2022**, *15*, 2354. https://doi.org/10.3390/en15072354

Academic Editors: Dongdong Feng, Andres Siirde and Antonio Crespo

Received: 19 February 2022
Accepted: 22 March 2022
Published: 23 March 2022

1. Introduction

Oxy-fuel combustion is a promising power production technology being actively investigated [1]. This operation method could lead to retrofitting of the existing pulverized-coal combustion power plant boiler, based on flue gas recirculation and pure oxygen injection rather than using air. The medium flow and heat transfer characteristics in the furnace should be regulated by flue gas recycling (FGR) in order to obtain a higher CO_2 concentration. This technique could be easily implemented for large-scale emission reductions of CO_2, due to the reduced flue gas exhaust and the significant heat loss reduction, especially in the context of carbon neutralization.

Pulverized coal combustion in ordinary air and CO_2 atmospheric conditions differ in their heat and mass transfer characteristics. The furnace heat exchange is dependent on the flame temperature, flue gas composition, particle composition, and radiation absorption rate between the water wall temperature and the interior flow field of the furnace. The radiation heat transfer in the furnace primarily occurs through CO_2, moisture, char, soot, and ash. The flue gas and the heat transfer law in the design and operation of the boiler focus on exploring the radiation characteristics of oxy-fuel combustion. Guo et al. [2]

constructed a 200 MW tangential pulverized-coal boiler numerical simulation with oxy-fuel and air combustion conditions to show the radiation characteristics of these fluid properties and compositions. The gray and non-gray radiation models over various performances have been adopted to obtain the characteristics of heat transfer, the temperature distribution in the furnace, and the composition of exhaust gas in air-firing and oxygen-enriched wet/dry FGR conditions [3]. Black et al. investigated the difference in heat transfer between the air and the oxy-fuel combustion environment within a 500 MW pulverized-coal power plant boiler [4], where the radiation heat transfer over various concentrations of the oxy-fuel combustion conditions were analyzed. Gani et al. [5] carried out a modeling study on coal combustion in air and oxy-fuel mode by using Fluent, a CFD tool, and investigated the effects of balance gas N_2 and CO_2 and the effects of oxygen concentration on coal burning. Ahn et al. [6] studied the heat transfer and gas radiation characteristics in a 0.5 MW class oxy-fuel boiler under different loads with or without FGR. The natural circulation boiler furnace water wall was the primary heat-absorbing surface, where the fire-facing side was the radiant endothermic process, and a specific velocity of hot water and steam flowed inside the water wall. The evaporator was the worst within the operating equipment. The working fluid status was fickle in the part. De Kerret et al. [7] proposed a new methodology based on the understanding of key contributions to vertical two-phase flow pattern maps in tube bundles, leading to a more complete flow pattern map. The different two-phase flow patterns can be bubbly, intermittent, or annular. The heat transfer mode of the working fluid varies along the tube length corresponding to the different flow patterns. Guzella et al. [8] put forward two-dimensional simulations of nucleation, growth and departure of bubbles and boiling heat transfer at different reduced temperatures. Weise et al. described the heat transfer coefficient during flow boiling in horizontal tubes with circumferentially varying heat flux [9], adapting flow-pattern-based heat transfer models using local heat flux. The model of a gas–liquid two-phase flow in a helically coiled tube was established by Wu et al. [10] based on the separated phase flow model. The model can be used to solve gas-film velocity, gas-film thickness and heat transfer coefficient, and predict gas–liquid two-phase boiling heat transfer in helically coiled tubes. Studying the working fluid and the heat transfer in the heat exchanger of oxy-fuel combustion boilers is important when reducing the flow resistance, preventing the deterioration of heat transfer, and avoiding the unstable working medium flow of the heat-absorbing surface.

Boiler evaporation system modeling is generally constructed by the following three methods: the lumped parameter model, the linear and nonlinear parameter model, and the distributed parameter model [11]. The lumped parameter model is a model established under a simplified assumption that the state parameters of the medium in the heating pipes are uniform. Then, a representative point is selected in space, and the medium parameters of the point are used as the lumped parameters of the link. The distributed parameter model can fully reflect the linear/nonlinear distributed parameter characteristics of thermal objects and can also accurately reflect the variation law of metal wall temperature. Li et al. [12] improved the traditional thermodynamic method, where the evaporator heat-absorbing surface was divided into the preheater, the evaporator, and the superheater. The established model could conduct static and dynamic simulation analyses. Yu et al. [13] established a lumped-parameter model for natural circulation drum-boilers to calculate the heat flux transferred into water wall tubes, which can be used to describe the complicated dynamics of the entire evaporation system. Laubscher et al. [14] proposed a computational fluid dynamics modeling methodology used to evaluate the thermal performance of the water wall evaporator, the platen, and the final stage superheaters of a subcritical pulverized coal-fired boiler at full and reduced boiler loads. Tang et al. [15] developed a thermal–hydraulic model for the evaporator system of a 660 MWe ultra-supercritical CFB boiler. Pressure drop, mass flux distribution and metal temperature in the evaporator system were estimated. Chu et al. [16] proposed a 2D accurate, distributed parameter model for the evaporation system of a controlled natural circulation boiler, based on the 3D temperature distribution and the emissivity of the particle phase. The evaporation system is divided into a subcooled region, a boiling region, and a super-heated region, according to the secondary water/steam status. The boundaries of these regions are movable. The steam–water separation takes place at the dryout location in the evaporating tubes. Zheng et al. [17] presented the distributed parameter model for the evaporation system based on 3D combustion monitoring in the furnace. The mathematical model was formulated to predict the transient distribution of the parameters, such as heat flux, the metal surface temperature, and steam quality. A distributed parameter model was built in a supercritical W-shaped once-through boiler, which accounted for the non-uniform distributions of the surface heat transfer and the frictional resistance coefficient [18]. The heat transfer characteristics of the W-shaped water wall at the near-critical and supercritical pressures were analyzed. Laubscher et al. [19,20] developed a computational fluid dynamics model and a steady-state 1D computer model, simulating the heat transfer in the evaporation system of grate-fired biomass boilers. Many scholars have carried out evaporation system modeling research studies on various types of boilers, but there are relatively few studies on the heating surfaces of oxy-fuel combustion boilers. At the same time, most of the studies above were based on boiler side mechanisms and lack of furnace side combustion information.

In this paper, a medium temperature and medium pressure natural circulation air/oxy-fuel combustion boiler was investigated. The conventional air-firing and oxy-fuel combustion boiler was simulated numerically under various load conditions by using Fluent to establish a distribution parameter model on the wall temperature and heat flux. This demonstration unit was proven to be an effective strategy for the design, operation, and combustion diagnosis of an oxy-fuel combustion boiler.

2. Materials and Methods

2.1. Numerical Simulation of Oxy-Fuel Combustion Boiler

The boiler was a pulverized-coal combustion unit single-furnace, with a slightly positive pressure, a π-type arrangement, and swirl burners mounted on the front wall. The steam mass flow rate was 38 t/h, the main steam pressure was 4.3 MPa, and the feed water temperature was 105 °C. The structure of the boiler and the thermodynamic system diagram are shown in Figure 1.

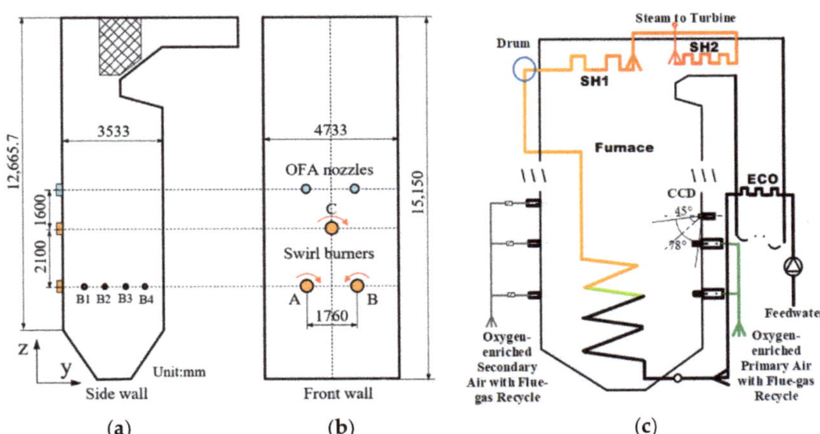

Figure 1. Structure of the boiler: (**a**) z–y plane of the furnace, (**b**) z–x plane of the furnace, and the thermodynamic system diagram (**c**). OFA nozzles: over-fire air nozzles; SH1, 2: 1-stage superheater, 2-stage superheater; CCD: charge-coupled device; ECO: economizer.

As a part of our work on oxy-fuel combustion [21], the cross-section of the boiler was 4733 mm wide, 3533 mm deep, and 15,150 mm tall. The designed chamber volume heating load was 163.3 kW/m^3 and the cross-section heating load was 1.993 MW/m^2. The front wall was arranged to have three swirl burners in a triangular shape. B1–B4 (Figure 1a) are four measurement openings arranged symmetrically for measuring the temperature and heat flux. The whole furnace was composed of the entire steel framework and the suspension structure.

The oxy-fuel combustion system included the swirl burner, the over-fire air, the oxygen injectors, and the auxiliary equipment. The staged air/flue gas was used for combustion-supporting organization of the flow field, which ensured the stability of the coal combustion. Pure oxygen should be mixed previously with flue gas as the primary air and then conveyed with pulverized-coal into the furnace via burners. Table 1 shows the proximate and the ultimate analysis of coal. The coal had low sulfur, high ash content and was easily ignited. The inlet flow rates of the air/flue gas at all stages under different working conditions are shown in Table 2.

Table 1. Proximate and ultimate analyses of the coal.

Proximate Analysis/%				Ultimate Analysis/%					$Q_{net,ar}$/(MJ/kg)
M_{ar}	A_{ar}	V_{ar}	FC_{ar}	S_{ar}	N_{ar}	C_{ar}	H_{ar}	O_{ar}	
9.22	22.77	16.93	51.08	0.32	0.50	54.14	3.65	9.40	21.53

Table 2. Velocity of the flow rates at the inlet of the boiler.

Items	Velocity, m/s								
	60% Load			80% Load			100% Load		
	Air/1	Dry FGR/2	Wet FGR/3	Air/4	Dry FGR/5	Wet FGR/6	Air/7	Dry FGR/8	Wet FGR/9
Prim. air	12.8	11.0	11.0	17.1	14.6	14.9	23.0	20.0	20.4
Sec. air	18.9	17.8	18.0	25.3	23.7	25.0	37.0	38.6	40.0
OFA	22.9	0	0	30.6	0	0	32.2	0	0
Cool. air	4.8	4.8	4.8	7.2	7.2	7.2	8.8	11.1	11.4

Air: conventional combustion situation; Dry FGR: oxygen-enriched combustion with dewatering flue gas recirculation; Wet FGR: oxygen-enriched combustion without dewatering flue gas recirculation. The concentration of the oxygen in the dry/wet FGR oxygen-enriched combustion was 28%. 1–9: case 1–case 9. Prim. air-Primary air; Sec. air-Secondary air; OFA-Over Fire Air; Cool. air-Cooling air.

2.2. Distributed Mathematical Model

The natural circulation oxy-fuel combustion boiler had 4 common down-comers, 28 connecting pipes, 202 rising water wall tubes, and 28 steam-water leading pipes. The boiler water wall was designed with a 12-water circulation circuit. The evaporation zone between the ash hopper (6.33 m) and the furnace arch (15.33 m) was divided into 18 layers. The horizontal cross-section was uniformly divided into 10 × 10 grids. There were 720 surface meshes in total.

The inner water wall heat flux adopted the average surface temperature of the inner tube wall, and the flue gas model adopted the temperature outside the fire-facing tube wall. The water/steam model considered the inner wall temperature as the input for calculating the working fluid enthalpy, whereas the external wall temperature was used to validate the model and evaluate the safe operation of the boiler. Neither the average temperature of the inner wall nor that of the outer wall can be used as the lumped parameter to meet the requirements at the same time. Therefore, a dynamic lumped model with the average temperature of the inner wall of the tube as the lumped parameter combined with a steady-state two-dimensional heat conduction model that can reflect the temperature of the outer wall of the tube was adopted. The lump-parameter model is:

$$MC_p\frac{dT_{in}}{d\tau} = Q_{out} - Q_{in} \quad (1)$$

The two-dimensional steady-state heat conduction model is:

$$\frac{\partial}{\partial r}\left(r\frac{\partial T}{\partial r}\right)+\frac{1}{r}\frac{\partial^2 T}{\partial \varphi^2}=0 \quad (2)$$

The dynamic model of the water/steam in the risers, coupled with mass, momentum, and energy conservation equations is shown as:

$$D_{i1}-D_{i2}=V_i\frac{d\rho_{i2}}{d\tau} \quad (3)$$

$$Q_{iin}+D_{i1}(h_{i1}-h_{i2})=V_i\rho_{i2}\frac{d\mu_{i2}}{d\tau} \quad (4)$$

$$p_{i1}-p_{i2}-\rho_{i1}gH=\xi\frac{1}{2A_i^2}\frac{D_{i1}^2}{\rho_{i1}} \quad (5)$$

The single-phase water pressure variations under subcritical conditions and the surface heat transfer coefficient α were calculated with the Dittuse–Boelter correlation [17]:

$$\alpha=0.023\frac{\lambda}{D}\left(\frac{\mu C_p}{\lambda}\right)^{0.4}\left(\frac{\rho w D}{\mu}\right)^{0.8} \quad (6)$$

The vapor–liquid two-phase flow expressing the correlation of heat transfer coefficient between the two-phase flow and liquid is [17]:

$$\alpha=0.00122\left[\frac{k_L^{0.79}C_{PL}^{0.45}\rho_L^{0.49}}{\sigma^{0.45}\mu_L^{0.29}h_{LG}^{0.24}\rho_G^{0.24}}\right]\Delta T_S^{0.24}\Delta P_S^{0.75}S+0.023\left[\frac{G(1-x)D}{\mu_L}\right]^{0.8}\left(\frac{\mu_L C_{PL}}{k_L}\right)\frac{k_L}{D}F \quad (7)$$

3. Results and Discussion

The in situ operating conditions were performed to verify the results from the air and oxy-fuel combustion boiler. Table 3 shows the typical cases during the mode transition of the study.

Table 3. Operation parameters during the mode transition.

Case	Load/MW	Feedwater Pres./MPa	Feedwater Temp/°C	Feedwater Flow/(t/h)
1/2/3	21	4.59	103	24.37
4/5/6	28	6.13	105	32.44
7/8/9	35	7.66	112	38.50

Figure 2 shows the online measurement of the external wall temperatures and the simulation results at 60%, 80%, and 100% load rates during the dry FGR oxy-fuel combustion at a height of 11.5 m. The maximum temperature relative error in case 5 was 2.14% at a width of 2.58 m. The wall temperature rose as the load increased, and the metal wall temperature in the middle of the front wall was higher than that on both sides, due to the arrangement of the swirl burners. The tube wall temperature predicted by the mathematical model was in agreement with that in the in situ operation. The highest temperature of the front wall was higher than that of the left, and the average temperature of the left wall close to the front was higher than that near the rear wall. The swirl burners were installed on the front wall in a triangular configuration; thus, the heat flux of the front wall was higher, corresponding to the higher wall metal surface temperature.

Figures 3 and 4 depict heat flux distributions on the front water wall and the metal wall temperature on the rear water wall under three different combustion conditions with a 100% load. The heat flux in the air-firing condition was more non-uniform than that in the oxy-fuel dry/wet FGR conditions at the 100% load. The front wall heat flux in the air-firing condition was more uneven than in the other two oxy-fuel combustions. The wall heat flux in the dry FGR condition was well distributed, due to the three atomic gases (CO_2 and H_2O) present in the specific heats, diffusion coefficients, and radiation characteristics. The wall heat flux in the wet FGR oxy-fuel condition was higher than that in the dry FGR oxy-fuel condition on account of the higher gas emission rate and temperature.

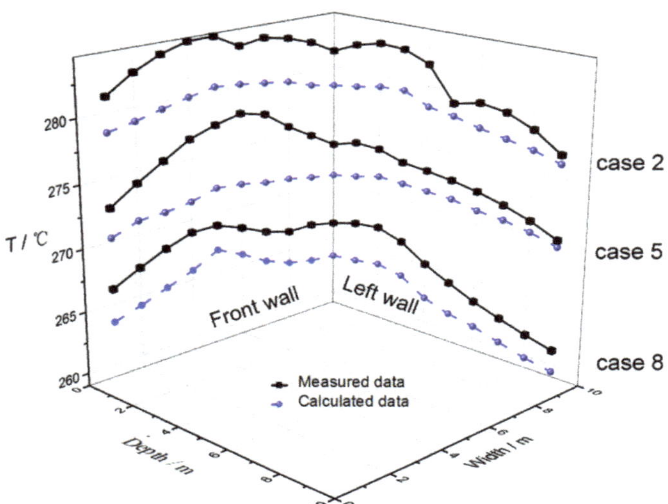

Figure 2. Comparison of temperatures on the front and left wall in case 2, case 5, and case 8.

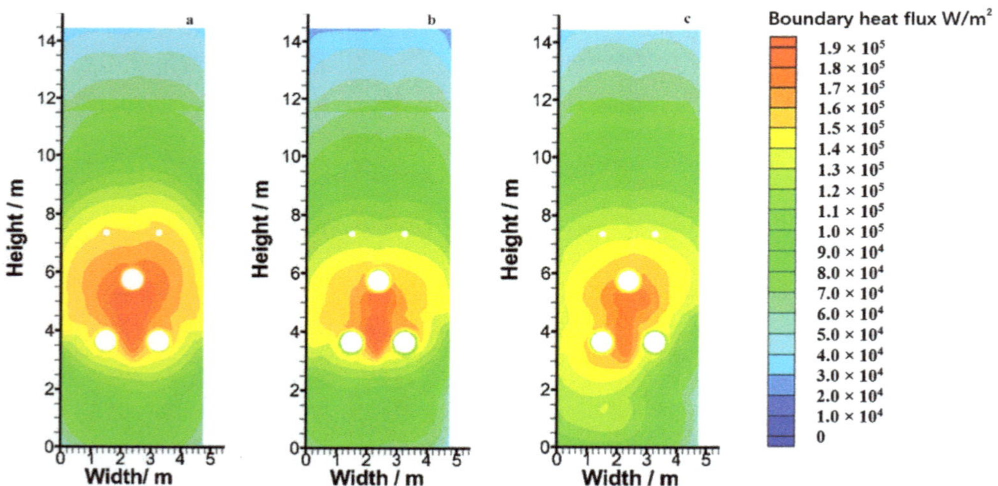

Figure 3. Distributions of heat flux in the front wall: (**a**) case 7, (**b**) case 8; and (**c**) case 9.

As shown in Figure 4, the distributions of the metal wall temperature on the rear water wall had similar tendencies with the heat flux distributions under the same combustion conditions. There was a high temperature region near the central position with symmetrical arrangement of the swirl burners, and the high temperature flue gas aggregated in the combustion area. The wall temperature in the air-firing, oxy-fuel dry/wet FGR conditions showed that the metal temperature was consistent with the uneven wall heat flux during the mode transition. The temperature difference of the front wall under the air-firing condition was 7.37 °C higher than that under the oxy-fuel wet FGR condition. The metal wall temperature distribution in the oxy-fuel dry/wet FGR condition was more uniform than that in the air-firing condition at the same load, which was consistent with the result that the wall heat flux distribution was more uniform under the oxy-fuel condition than under the air condition.

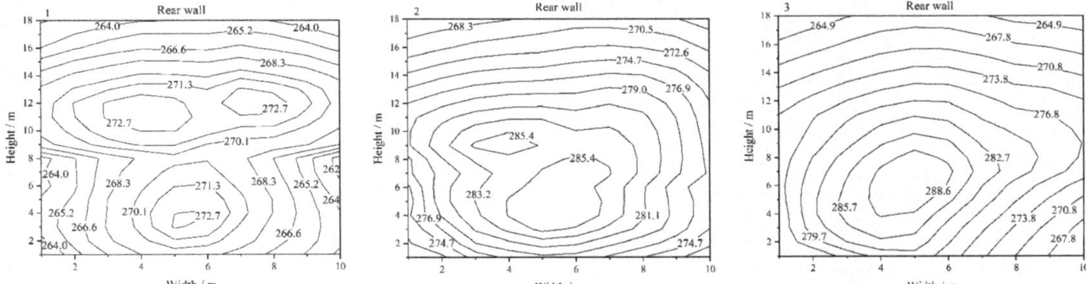

Figure 4. Distributions of the metal temperature (°C) on the rear wall: (**1**) case 7, (**2**) case 8, and (**3**) case 9.

Figure 5 shows the metal temperature distribution on the front water wall, which was similar to the distribution of heat flux in the corresponding cases. The highest water wall temperature in the oxy-fuel dry/wet FGR condition was lower than that in the air-firing condition. It can also be found that when the load rose from 60% to 80%, the wall temperature rose. This is because when the load increased, the corresponding wall heat flux grew, and the wall temperature grew. The oxy-fuel dry/wet FGR water wall temperature and heat flux varied with the in situ operation conditions, which means the model can produce correct predictions.

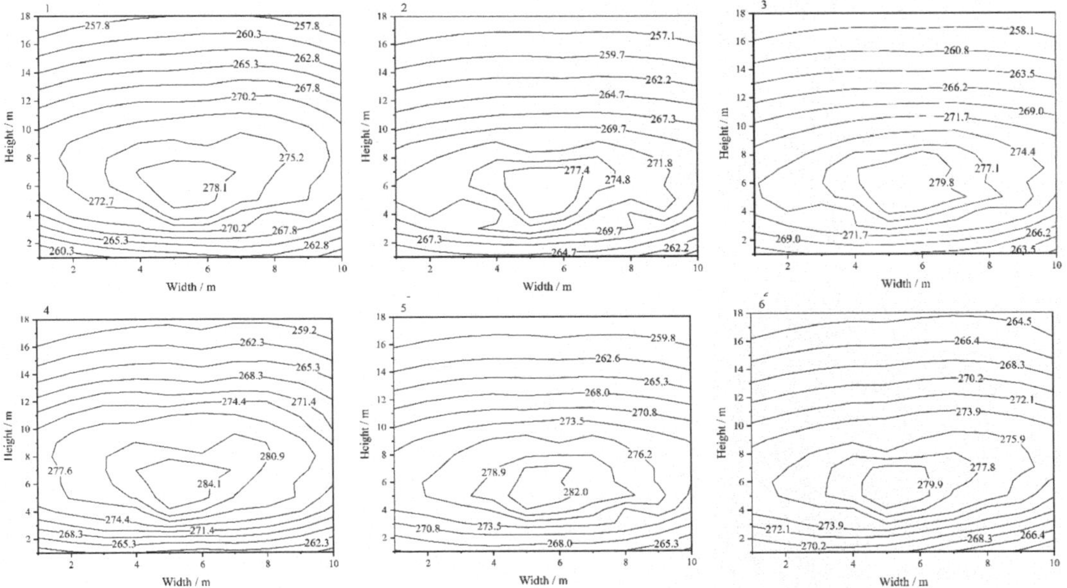

Figure 5. Distributions of metal temperature (°C) on the front wall. **1–3**: case **1**–case **3**, **4–6**: case **4**–case **6**.

Figure 6 illustrates the variation at a specific location of the boiler. The heat flux rose with the height increment, reached its maximum at 4 to 5 m, and then decreased with the increase in height. This maximum position was slightly over the burner combustion region. The pulverized-coal was mixed with the secondary air at the center of the three-swirl burner region; thus, the flue gas temperature was relatively higher, and the radiant heat flux was the highest. The flue gas temperature and the heat flux decreased with the rising height. The highest heat fluxes in the air-firing and the oxy-fuel wet FGR conditions were significantly higher than that in the oxy-fuel dry FGR condition at the same load. The maximum heat fluxes in the three cases were 192.6, 187.3, and 201.8 kW/m^2 at 100% load. The wall heat flux in the oxy-fuel dry FGR condition near the burner region increased more smoothly than that in the other two working conditions at the same load. The heat flux had a smooth change in the variable operation conditions.

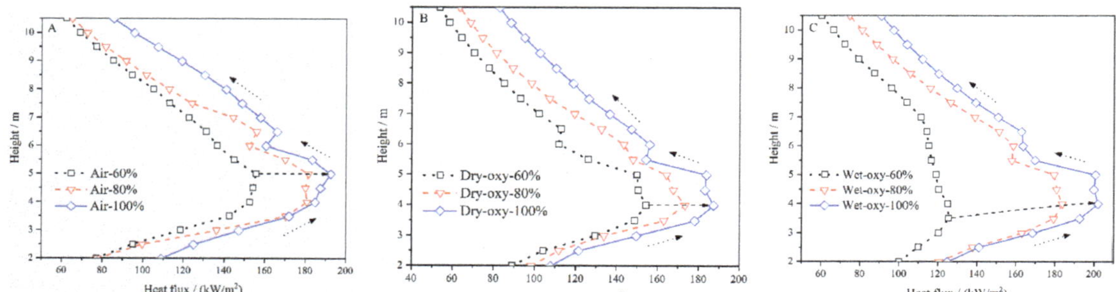

Figure 6. Heat flux distributions at 2.36 m wide at the three typical loads in air-firing (**A**), dry FGR (**B**), and wet FGR (**C**) conditions.

The oxy-fuel dry FGR heat flux in the burner region was significantly lower than the other two conditions at 80% load, as seen in Figure 7. The front wall average heat flux and metal temperature are illustrated in Table 4. The wall heat flux in the oxy-fuel dry FGR condition varied with the changing load. When the load rose from 60% to 80%, and then from 80% to 100%, the oxy-fuel dry FGR condition wall heat flux increased by 14.054 and 14.586 kW/m^2, the relative change rate of the average wall heat flux was 1.038, and the relative change rates of the average wall heat flux in air-firing and oxy-fuel wet FGR conditions were 1.455 and 1.477. These values were higher than that in the oxy-fuel dry FGR condition (1.038).

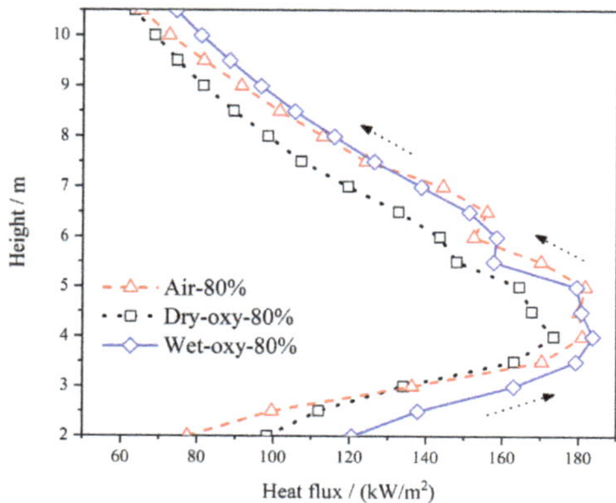

Figure 7. Heat flux distributions at the width of 2.36 m at 80% load in various combustion environments.

Table 4. Comparison of the average heat flux and temperature of the front wall in different cases.

	Air-Fired			**Dry FGR**			**Wet FGR**		
Case	1	4	7	2	5	8	3	6	9
Heat flux/(W/m^2)	104.78	117.52	136.03	96.94	110.99	125.52	106.63	119.58	138.71
Water wall temperature/°C	267.99	271.03	279.45	263.66	269.47	276.94	268.43	271.47	280.22

Figure 8 illustrates the steam quality distribution on the rear wall in the three different combustion environments at 100% load. The steam quality under the three different combustion modes showed that the middle value was higher than both sides at the same height, due to the high density of the heat flux in the middle area of the furnace. The water wall steam quality distribution in the dry FGR oxy-fuel condition was more uniform than that in the air-firing condition under the same load. The outlet steam quality of the rear wall in the air-firing condition was 0.063. The heating surface should be rearranged taking into account both the air-firing and the oxy-fuel combustion in the boiler.

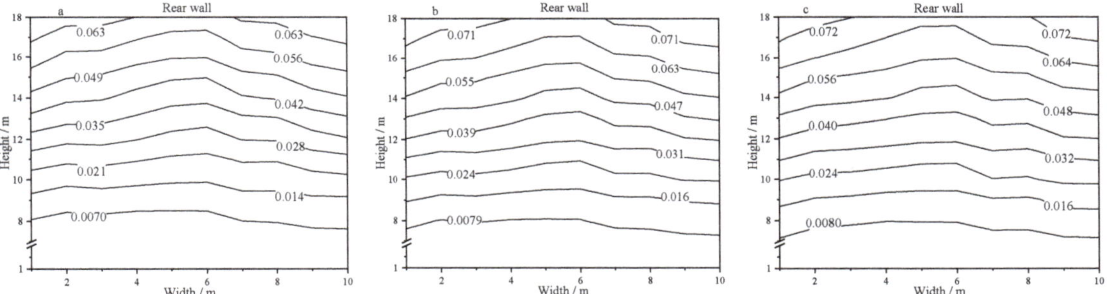

Figure 8. Distributions of the steam quality on the rear wall. **(a)**—case 7, **(b)**—case 8, and **(c)**—case 9.

As seen in Figure 9, the steam quality varied at the width of 2.36 m on the front water wall under the three typical loads in the oxy-fuel dry FGR combustion condition. The steam quality rose with the increment of height. As the load increased, the starting point of the gas vaporization advanced, and the steam quality increased. The present study demonstrates that the heat transfer to the membrane wall has a similar property under a 28% concentration of dry FGR oxy-fuel combustion.

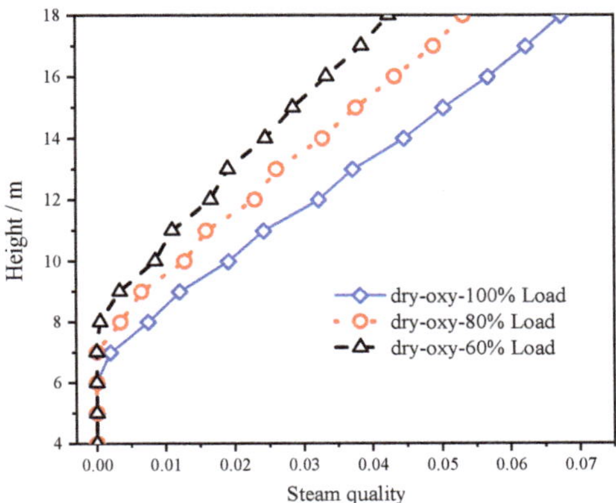

Figure 9. Steam quality on the front water wall under the dry FGR condition.

4. Conclusions

(1) A distribution parameter model for the evaporation system was developed in a 35 MW natural circulation air/oxy-fuel combustion boiler. Comparing the calculated value of the wall temperature with the measured value, it was found that the maximum relative error was 2.14%, which was located at the width of 2.58 m under the condition of oxy-fuel dry FGR combustion at 80% load, and the error was within a reasonable range. Therefore, the model proposed in this paper can produce correct predictions.

(2) When compared to the conventional air combustion, the water wall heat flux in the oxy-fuel dry/wet FGR combustion conditions remained even and uniform at the same load due to the changes of the specific heat, the diffusion, and the radiation coefficient of the three atomic gases (CO_2 and H_2O). The flue gas temperature in the oxy-fuel wet FGR combustion was much higher than that in the dry FGR situation, and the metal temperature in the oxy-fuel dry FGR condition was lower than that in other two operation modes. Therefore, the wall heat flux distribution and the wall temperature distribution of the oxy-fuel dry FGR combustion were the most uniform and reasonable.

(3) This subject effectively integrated the numerical simulation information of the furnace side with the thermal system model of the boiler side, and the research results provided novel guidance for the design, operation, and diagnosis of large capacity oxy-fuel combustion boilers.

Author Contributions: Project administration, Z.L.; Investigation, B.W.; Writing—original draft, Z.F. and Q.C.; Writing—review and editing, Z.L. All authors have read and agreed to the published version of the manuscript.

Funding: This research was funded by the National Natural Science Foundation of China (no. 51776078).

Conflicts of Interest: The authors declare no conflict of interest.

Nomenclature

c_p	specific isobaric heat capacity $\left[\text{J kg}^{-1}\text{ K}^{-1}\right]$
d	diameter [m]
D	mass flow rate $\left[\text{kg s}^{-1}\right]$
F	flow area $\left[\text{m}^2\right]$
g	acceleration of gravity $\left[\text{m s}^{-2}\right]$
G	mass velocity $\left[\text{kg m}^{-2}\text{ s}^{-1}\right]$
h	specific enthalpy $\left[\text{J kg}^{-1}\right]$
H	height [m]
M	mass [kg]
Nu	Nusselt number
p	pressure [Pa]
Pr	Prandtl number
q	heat flux $\left[\text{W m}^{-2}\right]$
Q	heat flow rate [W]
Re	Reynolds number
S	surface area $\left[\text{m}^2\right]$
T	absolute temperature [K]
u	specific internal energy $\left[\text{J kg}^{-1}\right]$
V	volume $\left[\text{m}^3\right]$
x	steam quality
X_{tt}	Martinelli number
h_{LG}	latent heat of vaporization
k	heat conductivity coefficient
F	Reynolds factor

Greek symbols

α	heat transfer coefficient $\left[\mathrm{W\,m^{-2}\,K^{-1}}\right]$
β	frictional resistance coefficient
Δ	difference in any quantity
ε	emissivity
κ_a	absorption coefficient
λ	thermal conductivity $\left[\mathrm{W\,m^{-1}\,K^{-1}}\right]$
μ	dynamic viscosity $\left[\mathrm{N\,m^{-1}\,s^{-1}}\right]$
ξ	resistance coefficient
ρ	density $\left[\mathrm{kg\,m^{-3}}\right]$
σ	Stefan-Boltzmann constant
ω	velocity $\left[\mathrm{m\,s^{-1}}\right]$

Subscripts

i, j	number of element
l	liquid
G	gas
max	maximum value
in	inner
out	outer
w	water wall
s	saturation
1	inlet
2	outlet

References

1. Li, S.; Xu, Y.; Gao, Q. Measurements and modelling of oxy-fuel coal combustion. *Proc. Combust. Inst.* **2019**, *37*, 2643–2661. [CrossRef]
2. Guo, J.; Liu, Z.; Wang, P.; Huang, X.; Li, J.; Xu, P.; Zheng, C. Numerical investigation on oxy-combustion characteristics of a 200MWe tangentially fired boiler. *Fuel* **2015**, *140*, 660–668. [CrossRef]
3. Nakod, P.; Krishnamoorthy, G.; Sami, M.; Orsino, S. A comparative evaluation of gray and non-gray radiation modeling strategies in oxy-coal combustion simulations. *Appl. Therm. Eng.* **2013**, *54*, 422–432. [CrossRef]
4. Black, S.; Szuhánszki, J.; Pranzitelli, A.; Ma, L.; Stanger, P.; Ingham, D.; Pourkashanian, M. Effects of firing coal and biomass under oxy-fuel conditions in a power plant boiler using CFD modelling. *Fuel* **2013**, *113*, 780–786. [CrossRef]
5. Gani, Z.F.A.; Pandian, P.A. Modelling of coal combustion in a drop tube furnace in air and oxy fuel environment. *Mater. Today Proc.* **2021**, *47*, 4431–4437. [CrossRef]
6. Ahn, J.; Kim, H.-J. Combustion characteristics of 0.5 MW class oxy-fuel FGR (Flue Gas Recirculation) boiler for CO_2 capture. *Energies* **2021**, *14*, 4333. [CrossRef]
7. De Kerret, F.; Béguin, C.; Etienne, S. Two-phase flow pattern identification in a tube bundle based on void fraction and pressure measurements, with emphasis on churn flow. *Int. J. Multiph. Flow* **2017**, *94*, 94–106. [CrossRef]
8. Guzella, M.; Czelusniak, L.; Mapelli, V.; Alvariño, P.; Ribatski, G.; Cabezas-Gómez, L. Simulation of boiling heat transfer at different reduced temperatures with an improved pseudopotential lattice boltzmann method. *Symmetry* **2020**, *12*, 1358. [CrossRef]
9. Weise, S.; Dietrich, B.; Wetzel, T. Flow-pattern based prediction of flow boiling heat transfer in horizontal tubes with circumferentially varying heat flux. *Int. J. Heat Mass Transf.* **2020**, *148*, 119018. [CrossRef]
10. Wu, J.; Li, X.; Liu, H.; Zhao, K.; Liu, S. Calculation method of gas–liquid two-phase boiling heat transfer in helically-coiled tube based on separated phase flow model. *Int. J. Heat Mass Transf.* **2020**, *161*, 120242. [CrossRef]
11. Zheng, J.X.; Chen, T.K.; Chen, X.J.; Yang, D. Dynamic thermal-hydraulic models for single-phase heated tubes. *Adv. Mech.* **1997**, *27*, 538–547.
12. Li, H.; Huang, X.; Zhang, L. A lumped parameter dynamic model of the helical coiled once-through steam generator with movable boundaries. *Nucl. Eng. Des.* **2008**, *238*, 1657–1663. [CrossRef]
13. Yu, T.; Yuan, J. Evaporation system modeling of the utility boiler aiming at real-time estimation of the heat flux into water walls. *IFAC Proc. Vol.* **2013**, *46*, 581–584. [CrossRef]
14. Laubscher, R.; Rousseau, P. CFD study of pulverized coal-fired boiler evaporator and radiant superheaters at varying loads. *Appl. Therm. Eng.* **2019**, *160*, 114057. [CrossRef]
15. Tang, G.; Zhang, M.; Gu, J.; Wu, Y.; Yang, H.; Zhang, Y.; Wei, G.; Lyu, J. Thermal-hydraulic calculation and analysis on evaporator system of a 660 MWe ultra-supercritical CFB boiler. *Appl. Therm. Eng.* **2019**, *151*, 385–393. [CrossRef]
16. Chu, Y.-T.; Lou, C.; Cheng, Q.; Zhou, H.-C. Distributed parameter modeling and simulation for the evaporation system of a controlled circulation boiler based on 3-D combustion monitoring. *Appl. Therm. Eng.* **2008**, *28*, 164–177. [CrossRef]

17. Zheng, S.; Luo, Z.; Zhang, X.; Zhou, H. Distributed parameters modeling for evaporation system in a once-through coal-fired twin-furnace boiler. *Int. J. Therm. Sci.* **2011**, *50*, 2496–2505. [CrossRef]
18. Shu, Z.; Zixue, L.; Yanxiang, D.; Huaichun, Z. Development of a distributed-parameter model for the evaporation system in a supercritical W-shaped boiler. *Appl. Therm. Eng.* **2014**, *62*, 123–132. [CrossRef]
19. Laubscher, R.; van der Merwe, S. Heat transfer modelling of semi-suspension biomass fired industrial watertube boiler at full- and part-load using CFD. *Therm. Sci. Eng. Prog.* **2021**, *25*, 100969. [CrossRef]
20. Laubscher, R.; De Villiers, E. Integrated mathematical modelling of a 105 t/h biomass fired industrial watertube boiler system with varying fuel moisture content. *Energy* **2021**, *228*, 120537. [CrossRef]
21. Guo, J.; Liu, Z.; Huang, X.; Zhang, T.; Luo, W.; Hu, F.; Li, P.; Zheng, C. Experimental and numerical investigations on oxy-coal combustion in a 35 MW large pilot boiler. *Fuel* **2017**, *187*, 315–327. [CrossRef]

Communication

Branched versus Linear Structure: Lowering the CO_2 Desorption Temperature of Polyethylenimine-Functionalized Silica Adsorbents

Jannis Hack, Seraina Frazzetto, Leon Evers, Nobutaka Maeda * and Daniel M. Meier *

Institute of Materials and Process Engineering (IMPE), School of Engineering (SoE), Zurich University of Applied Sciences (ZHAW), CH-8400 Winterthur, Switzerland; hacj@zhaw.ch (J.H.); serainafrazzetto@outlook.com (S.F.); ever@zhaw.ch (L.E.)
* Correspondence: maeo@zhaw.ch (N.M.); meid@zhaw.ch (D.M.M.)

Abstract: Lowering the regeneration temperature for solid CO_2-capture materials is one of the critical tasks for economizing CO_2-capturing processes. Based on reported pKa values and nucleophilicity, we compared two different polyethylenimines (PEIs): branched PEI (BPEI) and linear PEI (LPEI). LPEI outperformed BPEI in terms of adsorption and desorption properties. Because LPEI is a solid below 73–75 °C, even a high loading amount of LPEI can effectively adsorb CO_2 without diffusive barriers. Temperature-programmed desorption (TPD) demonstrated that the desorption peak top dropped to 50.8 °C for LPEI, compared to 78.0 °C for BPEI. We also revisited the classical adsorption model of CO_2 on secondary amines by using in situ modulation excitation IR spectroscopy, and proposed a new adsorption configuration, R1(R2)-NCOOH. Even though LPEI is more expensive than BPEI, considering the long-term operation of a CO_2-capturing system, the low regeneration temperature makes LPEI attractive for industrial applications.

Keywords: CO_2 capture; polyethylenimine; regeneration temperature

Citation: Hack, J.; Frazzetto, S.; Evers, L.; Maeda, N.; Meier, D.M. Branched versus Linear Structure: Lowering the CO_2 Desorption Temperature of Polyethylenimine-Functionalized Silica Adsorbents. *Energies* **2022**, *15*, 1075. https://doi.org/10.3390/en15031075

Academic Editors: Dongdong Feng, Jian Sun and Zijian Zhou

Received: 23 November 2021
Accepted: 27 January 2022
Published: 31 January 2022

1. Introduction

A record-high level of CO_2 concentration in the atmosphere (ca. 415 ppm) poses a global concern about how to suppress or stop anthropogenic CO_2 emissions [1]. CO_2 capture technology is considered to help to reduce the CO_2 concentration in the atmosphere. It falls into two categories: direct CO_2 capture from the air (ca. 415 ppm CO_2) [2] and CO_2 capture from power plants and other industrial sectors (4–100% CO_2) [3,4]. In either case, the regeneration temperature to collect the adsorbed CO_2 is as high as 80–120 °C. With regard to adsorbent degradation and operation cost, lowering the regeneration temperature is strongly desired [5,6]. Traditionally, liquid amine scrubbing has been employed for CO_2 capture [5,7]. However, its downsides, such as loss of volatile amine, corrosion, and high energy consumption in the regeneration step, make industries long for an alternative capturing system [8]. In this regard, solid adsorbents such as mesoporous oxides, zeolites, and metal–organic frameworks (MOFs) have recently gained attention due to their feasibility for desorbing CO_2 from a material surface by simply heating the material. In particular, porous materials functionalized with amines are under the spotlight due to their high adsorption capacity and simple synthetic procedure (wet impregnation) [8–11]. Among amines reported so far, polyethylenimine (PEI) supported on mesoporous support is advantageous over amines with low molecular weight, such as diethylenetriamine and tetraethylenepentamine, because PEI is more thermally stable and tolerant to oxidative conditions [12–14]. Its unique characteristic of enhanced CO_2 adsorption in the presence of moisture is also attractive for industrial applications. Branched PEI (BPEI) contains primary, secondary, and tertiary amines, and their CO_2 adsorption property is in the order of primary > secondary > tertiary [15]. Tertiary amines of BPEI do not adsorb CO_2

under dry conditions. So far, its application has been limited to CO_2 removal in spacecraft by NASA [16]. Large-scale industrial applications of BPEI for CO_2 capture demand significant improvements in the regeneration process in terms of energy consumption and long-term durability. CO_2 desorption from BPEI requires a relatively high temperature (>90 °C) [15]. Lowering the regeneration temperature contributes to a considerable decrease in the energy input and the total cost of capturing systems when long-term operation is considered without exchanging adsorbent materials for several years. In spite of its importance, lowering the regeneration temperature has not been the main focus in this field. One way to enhance CO_2 desorption was reported by using co-functionalization of BPEI with 1,2-epoxybutane [14]. The impregnation of BPEI onto a resin can also lower the desorption temperature down to 70 °C [17]. Previously, we also reported a reduction in the desorption temperature (50 °C) by co-impregnating BPEI with ionic liquids [6]. However, high toxicity and poor biodegradability of ionic liquids make their use in the environment unattractive [18]. PEI has another structure, which is linear PEI (LPEI). The utilization of LPEI for CO_2 capture was also extensively studied [19–27]. The adsorption capacity of LPEI is slightly low compared to BPEI, but LPEI has much better tolerance against oxidative conditions [24]. Similar to BPEI, LPEI also has a better adsorption capacity under humid conditions [25]. These properties make LPEI attractive for practical use. All the reports on LPEI focused on adsorption capacity, and its desorption property has never been a main target. In light of this, we focused on desorption properties of BPEI and LPEI in this study.

Among the three different amines of BPEI, primary amines are the most nucleophilic, with the highest pKa value [28,29]. Therefore, LPEI with high content of secondary amines would allow CO_2 desorption at a lower temperature. We herein report a comparative study of BPEI and LPEI, with a particular focus on CO_2 desorption properties.

2. Materials and Methods

2.1. Materials

Branched polyethylenimine (BPEI, Sigma-Aldrich, St. Louis, MO, USA; average Mw: ~25,000, average Mn: ~10,000), linear polyethlenimine (LPEI, Polysciences Inc., Warrington, PA, USA; average Mw: 250,000), silica (Sigma-Aldrich, St. Louis, MO, USA, fumed powder, average particle size: 0.2–0.3 μm), and ethanol (Acros Organics, Pittsburgh, PA, USA, 99.8%, anhydrous) were used as received without any purification. Helium (PanGas, Winterthur, Switzerland, ≥99.9999%) and 5000 ppm CO_2 in He balance (PanGas, Winterthur, Switzerland, ≥99.995% CO_2, ≥99.9999% He) were also utilized without any further purification.

2.2. Synthesis

PEI/SiO_2 adsorbents were prepared by a conventional wet-impregnation method as previously reported [6]. BPEI or LPEI was dissolved in anhydrous ethanol and stirred with a magnetic stirrer (IKA, Staufen, Germany, RCT standard) at room temperature for 30 min. The solution was then poured into a flask containing silica powders. The resulting suspension was vigorously stirred at room temperature for 1 h. The suspension was then transferred to a rotary evaporator (Büchi, Flawil, Switzerland, Rotavapor R-210) and slowly evacuated at 80 mbar and 32 °C to remove the ethanol solvent for 1 h. To completely dry the sample, it was kept under vacuum at 25–30 mbar at 50 °C for 20 min.

2.3. Adsorption and Desorption

The experimental set-up and procedure were exactly the same as in the previous report [6]. Briefly, CO_2 adsorption was performed in a fixed-bed plug-flow reactor. The CO_2 breakthrough was monitored by FT-IR spectrometer (Bruker, Billerica, MA, USA, ALPHA) equipped with a transmission gas-analysis module and deuterated triglycine sulfate (Bruker, Billerica, MA, USA, DTGS) detector. The adsorbent samples (300 mg) were placed into the tubular reactor and pretreated with a pure helium flow (200 mL/min) at 100 °C for 10 min (ramping rate: 5 °C/min). After pretreatment, the reactor was cooled down to 25 °C, and the gas flow was switched to 5000 ppm CO_2 in He balance. Temperature-programmed desorption (TPD) was carried out by heating the reactor from 25 °C to 100 °C in a He flow at a ramping rate of 5 °C/min. Experiments under a humid condition were operated with a relative humidity (RH) of 73.9% (dew point: 20 °C). The details of the modulation excitation spectroscopy (MES) and diffuse reflectance infrared Fourier transform spectroscopy (ST Japan, Tokyo, Japan, DRIFTS) can be found in the previous report [6].

3. Results and Discussion

Figure 1a displays the molecular structures of BPEI and LPEI. BPEI consists of primary amines at branch ends, and secondary and tertiary amines in the backbone. In contrast, LPEI only contains secondary amines in the repetitive moiety. Both structures have one nitrogen atom in every three atoms in their backbone. Figure 1b shows a comparative study of CO_2 adsorption on 30 wt % BPEI/SiO_2 and 30 wt % LPEI/SiO_2 under dry and humid conditions. As commonly known [17], in the humid condition, CO_2 adsorption capacity increased by a factor of ca. 1.8 for both BPEI and LPEI (see Table 1). The presence of moisture accelerates the formation of bicarbonate species, and thus contributes to the increase in the adsorption capacity [17,25,30–32]. An important characteristic in the adsorption profile is that BPEI slowly adsorbed more CO_2 after reaching the CO_2 breakthrough point, most likely due to gas diffusion barriers in the highly viscous liquid BPEI layers on SiO_2 support (Figure 1b, red lines). On the other hand, LPEI is a solid polymer, and the gas diffusion in the solid phase is faster compared to a liquid layer. After saturating the materials with CO_2, temperature-programmed desorption (TPD) was carried out as shown in Figure 1c. As expected, the TPD peak top for LPEI (50.8 °C) dropped by ca. 30 °C, compared to BPEI (78.0 °C). As rationalized earlier by the nucleophilicity and pKa values [28,29], LPEI with a high concentration of secondary amine tremendously lowered the regeneration temperature. Both BPEI and LPEI adsorbed more CO_2 at 45 °C, as reported by the literature [26]. Previously, we reported lowering the regeneration temperature of BPEI by ionic liquid (IL) down to 50 °C [6]. However, BPEI-IL composite, in return, deteriorates the CO_2 adsorption property. LPEI can overcome this disadvantage by capturing relatively high amounts of CO_2 (Table 1). For example, LPEI showed high capturing performance (76.8 mg/g) at 45 °C. TPD could not be measured at 45 °C for LPEI, since all the CO_2 was released by switching to He. LPEI also desorbed CO_2 from the surface very slowly at 25 °C, which can be seen in Figure 1c. Hence, the interaction of CO_2 and secondary amines of LPEI was considered to be fairly weak. The N atom efficiency in Table 1 was estimated based on how many N atoms were actually used to capture CO_2. The maximum efficiency (41.5%) was achieved with 30 wt % BPEI/SiO_2 at 45 °C under a humid condition. At 25 °C, the N atom efficiency was nearly the same for both BPEI and LPEI under dry and humid conditions.

To evaluate the gas diffusion property, samples bearing 60 wt % PEI were also tested. As in Figure 2a and Table 1, 60 wt % BPEI adsorbed only 90.6 mg/g of CO_2, while 60 wt % LPEI captured 142.1 mg/g of CO_2. The increase in the loading amount had almost no positive effect for BPEI. As shown in Table 1, the N atom efficiency even dropped from 25.3% to 14.8% at 25 °C. This data implied that 30 wt % BPEI covered the entire surface of SiO_2, and further addition formed liquid overlayers, as schematically illustrated in Figure 2c. The high viscosity of liquid BPEI might make CO_2 diffusion inside the liquid layer difficult. Most likely, only the surface layer was accessible for CO_2 capture. This accounted for the reason for the low N atom efficiency for 60 wt % BPEI. However, LPEI with a melting point of 73–75 °C still seemed to have a permeable solid layer on SiO_2, and therefore, all the secondary amines were accessible for CO_2 adsorption. The N atom efficiency for LPEI was not influenced by the loading amount, as seen in Table 1. As shown in Figure 2b, TPD peak tops for both BPEI and LPEI shifted toward higher temperatures due to the diffusive barriers by loading 60 wt %.

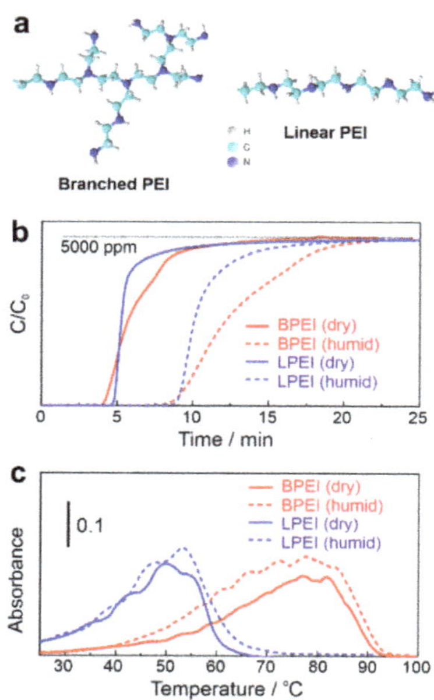

Figure 1. (**a**) BPEI and LPEI molecular structures; (**b**) CO_2 capture (5000 ppm CO_2 in He balance with/without H_2O vapor, 200 mL/min) at 25 °C; (**c**) TPD profile (5 °C/min, 200 mL/min of pure He).

Table 1. Adsorption and desorption performance of BPEI and LPEI supported on SiO_2 under different conditions. Note: 5000 ppm of CO_2 in He balance for adsorption and pure He for desorption. RH = 73.9% (dew point: 20 °C).

Adsorbent	Condition	Adsorption Capacity (mg/g) [#1]	N Atom Efficiency (%) [#4]	Desorption Temperature (°C) [#2]
30 wt % BPEI/SiO_2	25 °C, dry	41.2	13.5	78.0
30 wt % BPEI/SiO_2	25 °C, humid	77.5	25.3	78.6
30 wt % BPEI/SiO_2	45 °C, humid	127.0	41.5	67.1
60 wt % BPEI/SiO_2	25 °C, humid	90.6	14.8	96.7
30 wt % LPEI/SiO_2	25 °C, dry	37.6	12.3	50.8
30 wt % LPEI/SiO_2	25 °C, humid	68.6	22.4	54.4
30 wt % LPEI/SiO_2	45 °C, humid	76.8	25.1	– [#3]
60 wt % LPEI/SiO_2	25 °C, humid	142.1	23.2	73.0

[#1] Amount of CO_2 captured (mg) per adsorbent used (g); [#2] peak centers of TPD profiles; [#3] all the CO_2 molecules were desorbed at 45 °C; [#4] the N atom efficiency was estimated based on the total number of N atoms in the samples. The maximum adsorption capacities were 306.3 mg/g for 30 wt % BPEI and LPEI and 612.7 mg/g for 60 wt % BPEI and LPEI.

To understand adsorption–desorption mechanisms at the molecular level, we applied in situ diffuse reflectance Fourier transform infrared spectroscopy (DRIFTS) combined with modulation excitation spectroscopy (MES) [33–35]. In situ MES-DRIFTS enhances the signal-to-noise (S/N) ratio, and thus allows in situ monitoring of a trace amount of active species involved in chemical reactions [33]. Figure 3a,b display time-domain IR spectra of BPEI and LPEI during repeated adsorption-desorption cycles at 50 °C. The spectral characteristics between BPEI and LPEI differed considerably; the absorbance for LPEI was higher by a factor of 10. This enhanced adsorption–desorption during modulation experiment accounted for the lowered desorption temperature, releasing all the CO_2 at 50 °C within 200 s in the He flow. The IR bands observed for BPEI were well comparable to the ones previously reported [6]: ν(N-H) of carbamate and ammonium ion centered at around 3400 cm^{-1}, ν(C = O) of carbamic acid at 1680 cm^{-1}, ν(COO^-) of carbamate at 1558 cm^{-1}, δ(NH_3^+) of ammonium ion at 1530 cm^{-1}, and overlapped δ(N-H) and ν(C-N) of carbamate at 1502 cm^{-1}. As shown in Figure 3b, LPEI showed a completely different spectral feature. There were no IR bands assignable to carbamate or carbamic acid detected. Interestingly, ν(N-H) at around 3400 cm^{-1} completely disappeared. The band at 1405 cm^{-1} can also be assigned to ν(C-N). CO_2 adsorption with primary and secondary amines is considered to occur as follows [5]:

(Primary amines)

$$CO_2 + \text{R-NH}_2 \rightarrow \text{R-NHCOOH and/or R-NH}_2^+\text{COO}^- \quad (1)$$

$$CO_2 + 2\ \text{R-NH}_2 \rightarrow \text{R-NH}_3^+ + \text{R-NHCOO}^- \quad (2)$$

(Secondary amines)

$$CO_2 + 2\text{R1(R2)-NH} \rightarrow \text{R1(R2)-NCOO}^- + \text{R1(R2)-NH}_2^+ \quad (3)$$

Reaction (1) forms carbamic acid or carbamate ion involving one amine group. Reactions (2) and (3) involve two amine groups. The spectral analysis based on Figure 3 led us to come to an assumption with the following reaction path:

$$CO_2 + \text{R1(R2)-NH} \rightarrow \text{R1(R2)-NCOOH} \quad (4)$$

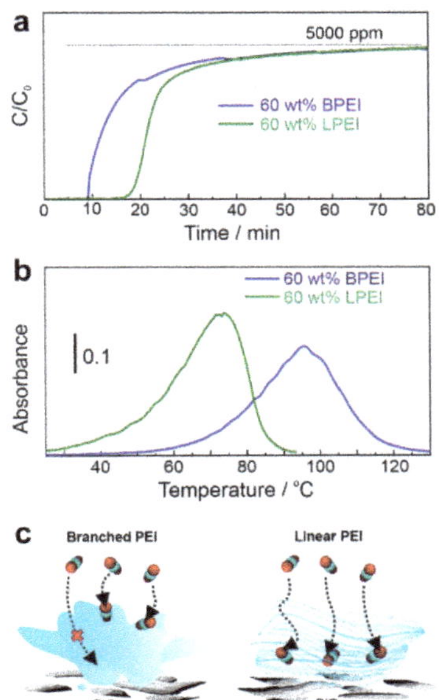

Figure 2. Effect of loading amount of PEI on (**a**) CO_2 capture and (**b**) desorption (conditions are identical to the experiments in Figure 1 with H_2O). (**c**) Schematic illustration of CO_2 diffusion.

This model would fit to our spectral observation. This assumption was further supported by the detection of the broad band of ν(O-H) at 3000–3200 cm^{-1}. To the best of our knowledge, there has been no previous report proposing such an adsorption configuration together with spectroscopic evidence. This configuration provides a rational basis for future molecular design of CO_2 adsorbents for low-temperature regeneration.

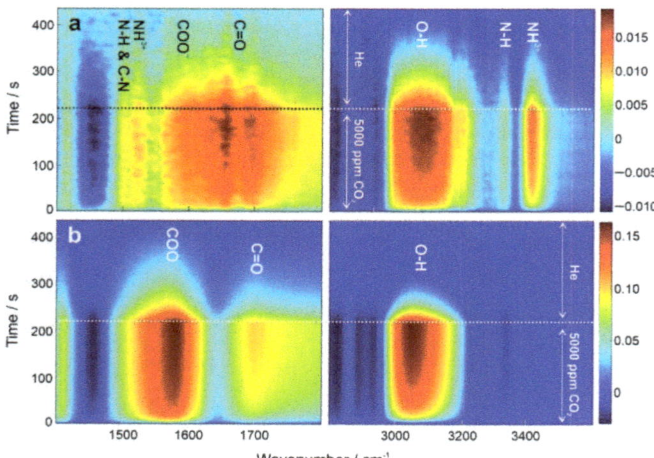

Figure 3. Time-domain IR contour plots of (**a**) 30 wt % BPEI/SiO_2 and (**b**) 30 wt % LPEI/SiO_2 during CO_2 adsorption–desorption at 50 °C (5000 ppm CO_2 in He balance: 200 mL/min, pure He: 200 mL/min). The unit of the color bar is absorbance.

Finally, we tested 30 wt % LPEI/SiO_2 for recyclability. The adsorption–desorption cycles at 45 °C under humid conditions were repeated; 5000 ppm CO_2 in He balance with H_2O vapor at 200 mL/min was fed into the reactor and then switched to pure He for desorption cycles (Supplementary Materials). As seen in Figure 4, within 100 cycles there was no performance deterioration observed. LPEI could release all the CO_2 molecules at 45 °C, and therefore has a high potential for industrial applications.

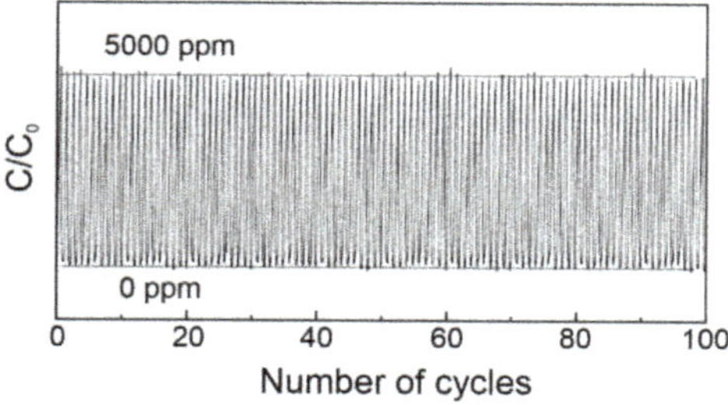

Figure 4. The recyclability test for adsorption and desorption of CO_2 at 45 °C (50 mg, 5000 ppm CO_2 in He balance: 200 mL/min, pure He: 200 mL/min).

4. Conclusions

We compared BPEI and LPEI supported on SiO_2, aiming at low-temperature regeneration. We found that 30 wt % LPEI/SiO_2 could be regenerated below 50 °C. This unique property originated from different adsorption configurations of CO_2 between BPEI and LPEI. MES-DRIFTS analysis proved that CO_2 adsorption on BPEI fell into the classical adsorption model involving carbamic acid, carbamate, and ammonium ions. On the other hand, such a model could not be applied for interpreting the CO_2 adsorption on LPEI.

Based on the in situ IR spectral analysis, we propose the intermediate species to be R1(R2)-NCOOH, which can be decomposed at 25–50 °C to release CO_2. The newly proposed model will pave the way for rational molecular design of amine-based polymers targeting low-temperature regeneration in CO_2 capture.

Supplementary Materials: The following are available online at https://www.mdpi.com/article/10.3390/en15031075/s1, Figure S1. The desorption of CO_2 for 30 wt % LPEI/SiO_2 (300 mg) at 45 °C.

Author Contributions: Conceptualization, N.M.; investigation, J.H., S.F. and L.E.; writing—original draft, J.H.; writing—review and editing, N.M. and D.M.M.; supervision, N.M. and D.M.M.; project administration, N.M. and D.M.M.; funding acquisition, D.M.M. All authors have read and agreed to the published version of the manuscript.

Funding: This research was funded internally by ZHAW.

Data Availability Statement: The data presented in this study are available on request from the corresponding author.

Conflicts of Interest: The authors declare no conflict of interest.

References

1. Cui, Y.; Schubert, B.A.; Jahren, A.H. A 23 m.y. record of low atmospheric CO_2. *Geology* **2020**, *48*, 888–892. [CrossRef]
2. Wilson, S.M.W.; Tezel, F.H. Direct Dry Air Capture of CO_2 Using VTSA with Faujasite Zeolites. *Ind. Eng. Chem. Res.* **2020**, *59*, 8783–8794. [CrossRef]
3. Yoro, K.O.; Daramola, M.O.; Sekoai, P.T.; Armah, E.K.; Wilson, U.N. Advances and emerging techniques for energy recovery during absorptive CO_2 capture: A review of process and non-process integration-based strategies. *Renew. Sustain. Energy Rev.* **2021**, *147*, 111241. [CrossRef]
4. Bains, P.; Psarras, P.; Wilcox, J. CO_2 capture from the industry sector. *Prog. Energy Combust. Sci.* **2017**, *63*, 146–172. [CrossRef]
5. Heldebrant, D.J.; Koech, P.K.; Glezakou, V.-A.; Rousseau, R.; Malhotra, D.; Cantu, D.C. Water-Lean Solvents for Post-Combustion CO_2 Capture: Fundamentals, Uncertainties, Opportunities, and Outlook. *Chem. Rev.* **2017**, *117*, 9594–9624. [CrossRef]
6. Weisshar, F.; Gau, A.; Hack, J.; Maeda, N.; Meier, D.M. Toward Carbon Dioxide Capture from the Atmosphere: Lowering the Regeneration Temperature of Polyethylenimine-Based Adsorbents by Ionic Liquid. *Energy Fuels* **2021**, *35*, 9059–9062. [CrossRef]
7. Rochelle, G.T. Amine Scrubbing for CO_2 Capture. *Science* **2009**, *325*, 1652. [CrossRef]
8. D'Alessandro, D.M.; Smit, B.; Long, J.R. Carbon Dioxide Capture: Prospects for New Materials. *Angew. Chem. Int. Ed.* **2010**, *49*, 6058–6082. [CrossRef]
9. Hedin, N.; Bacsik, Z. Perspectives on the adsorption of CO_2 on amine-modified silica studied by infrared spectroscopy. *Curr. Opin. Green Sustain. Chem.* **2019**, *16*, 13–19. [CrossRef]
10. Kuang, Y.; He, H.; Chen, S.; Wu, J.; Liu, F. Adsorption behavior of CO_2 on amine-functionalized polyacrylonitrile fiber. *Adsorption* **2019**, *25*, 693–701. [CrossRef]
11. Tanthana, J.; Chuang, S.S.C. In Situ Infrared Study of the Role of PEG in Stabilizing Silica-Supported Amines for CO_2 Capture. *ChemSusChem* **2010**, *3*, 957–964. [CrossRef]
12. Liu, Z.L.; Teng, Y.; Zhang, K.; Chen, H.G.; Yang, Y.P. CO_2 adsorption performance of different amine-based siliceous MCM-41 materials. *J. Energy Chem.* **2015**, *24*, 322–330. [CrossRef]
13. Olea, A.; Sanz-Perez, E.S.; Arencibia, A.; Sanz, R.; Calleja, G. Amino-functionalized pore-expanded SBA-15 for CO_2 adsorption. *Adsorption* **2013**, *19*, 589–600. [CrossRef]
14. Choi, W.; Min, K.; Kim, C.; Ko, Y.S.; Jeon, J.W.; Seo, H.; Park, Y.-K.; Choi, M. Epoxide-functionalization of polyethyleneimine for synthesis of stable carbon dioxide adsorbent in temperature swing adsorption. *Nat. Commun.* **2016**, *7*, 12640. [CrossRef]
15. Wang, X.; Song, C. Temperature-programmed desorption of CO_2 from polyethylenimine-loaded SBA-15 as molecular basket sorbents. *Catal. Today* **2012**, *194*, 44–52. [CrossRef]
16. Satyapal, S.; Filburn, T.; Trela, J.; Strange, J. Performance and properties of a solid amine sorbent for carbon dioxide removal in space life support applications. *Energy Fuels* **2001**, *15*, 250–255. [CrossRef]
17. Wang, W.; Liu, F.; Zhang, Q.; Yu, G.; Deng, S. Efficient removal of CO_2 from indoor air using a polyethyleneimine-impregnated resin and its low-temperature regeneration. *Chem. Eng. J.* **2020**, *399*, 125734. [CrossRef]
18. Thi, P.T.P.; Cho, C.W.; Yun, Y.S. Environmental fate and toxicity of ionic liquids: A review. *Water Res.* **2010**, *44*, 352–372. [CrossRef]
19. Buijs, W. Molecular Modeling Study to the Relation between Structure of LPEI, Including Water-Induced Phase Transitions and CO2 Capturing Reactions. *Ind. Eng. Chem. Res.* **2021**, *60*, 11309–11316. [CrossRef]
20. Prud'homme, A.; Nabki, F. Comparison between Linear and Branched Polyethylenimine and Reduced Graphene Oxide Coatings as a Capture Layer for Micro Resonant CO_2 Gas Concentration Sensors. *Sensors* **2020**, *20*, 1824. [CrossRef]

21. Rosu, C.; Pang, S.H.; Sujan, A.R.; Sakwa-Novak, M.A.; Ping, E.W.; Jones, C.W. Effect of Extended Aging and Oxidation on Linear Poly(propylenimine)-Mesoporous Silica Composites for CO_2 Capture from Simulated Air and Flue Gas Streams. *ACS Appl. Mater. Interfaces* **2020**, *12*, 38085–38097. [CrossRef] [PubMed]
22. Sharma, P.; Chakrabarty, S.; Roy, S.; Kumar, R. Molecular View of CO_2 Capture by Polyethylenimine: Role of Structural and Dynamical Heterogeneity. *Langmuir* **2018**, *34*, 5138–5148. [CrossRef] [PubMed]
23. Subagyono, D.J.N.; Marshall, M.; Knowles, G.P.; Chaffee, A.L. CO_2 adsorption by amine modified siliceous mesostructured cellular foam (MCF) in humidified gas. *Microporous Mesoporous Mater.* **2014**, *186*, 84–93. [CrossRef]
24. Yoo, C.-J.; Park, S.J.; Jones, C.W. CO_2 Adsorption and Oxidative Degradation of Silica-Supported Branched and Linear Aminosilanes. *Ind. Eng. Chem. Res.* **2020**, *59*, 7061–7071. [CrossRef]
25. Zhang, H.; Goeppert, A.; Olah, G.A.; Prakash, G.K.S. Remarkable effect of moisture on the CO2 adsorption of nano-silica supported linear and branched polyethylenimine. *J. CO2 Util.* **2017**, *19*, 91–99. [CrossRef]
26. Zhang, H.; Goeppert, A.; Prakash, G.K.S.; Olah, G. Applicability of linear polyethylenimine supported on nano-silica for the adsorption of CO_2 from various sources including dry air. *RSC Adv.* **2015**, *5*, 52550–52562. [CrossRef]
27. Zhou, Z.; Balijepalli, S.K.; Nguyen-Sorenson, A.H.T.; Anderson, C.M.; Park, J.L.; Stowers, K.J. Steam-Stable Covalently Bonded Polyethylenimine Modified Multiwall Carbon Nanotubes for Carbon Dioxide Capture. *Energy Fuels* **2018**, *32*, 11701–11709. [CrossRef]
28. Suh, J.; Paik, H.J.; Hwang, B.K. Ionization of Poly(ethylenimine) and Poly(allylamine) at Various pH's. *Bioorg. Chem.* **1994**, *22*, 318–327. [CrossRef]
29. Sun, C.; Tang, T.; Uludağ, H.; Cuervo, J.E. Molecular Dynamics Simulations of DNA/PEI Complexes: Effect of PEI Branching and Protonation State. *Biophys. J.* **2011**, *100*, 2754–2763. [CrossRef]
30. Li, K.J.; Kress, J.D.; Mebane, D.S. The Mechanism of CO_2 Adsorption under Dry and Humid Conditions in Mesoporous Silica-Supported Amine Sorbents. *J. Phys. Chem. C* **2016**, *120*, 23683–23691. [CrossRef]
31. Sayari, A.; Belmabkhout, Y. Stabilization of Amine-Containing CO_2 Adsorbents: Dramatic Effect of Water Vapor. *J. Am. Chem. Soc.* **2010**, *132*, 6312–6314. [CrossRef]
32. Veneman, R.; Frigka, N.; Zhao, W.Y.; Li, Z.S.; Kersten, S.; Brilman, W. Adsorption of H_2O and CO_2 on supported amine sorbents. *Int. J. Greenh. Gas Control.* **2015**, *41*, 268–275. [CrossRef]
33. Maeda, N.; Meemken, F.; Hungerbuhler, K.; Baiker, A. Spectroscopic Detection of Active Species on Catalytic Surfaces: Steady-State versus Transient Method. *Chimia* **2012**, *66*, 664–667. [CrossRef]
34. Muller, P.; Hermans, L. Applications of Modulation Excitation Spectroscopy in Heterogeneous Catalysis. *Ind. Eng. Chem. Res.* **2017**, *56*, 1123–1136. [CrossRef]
35. Urakawa, A.; Burgi, T.; Baiker, A. Sensitivity enhancement and dynamic behavior analysis by modulation excitation spectroscopy: Principle and application in heterogeneous catalysis. *Chem. Eng. J.* **2008**, *63*, 4902–4909. [CrossRef]

MDPI
St. Alban-Anlage 66
4052 Basel
Switzerland
Tel. +41 61 683 77 34
Fax +41 61 302 89 18
www.mdpi.com

Energies Editorial Office
E-mail: energies@mdpi.com
www.mdpi.com/journal/energies